PROF. E. McSQUARED'S

GRADE EXPANDED INTERGALACTIC VERSION!

A CALCULUS PRIMER

by Howard Swann & John Johnson

Janson Publications, Inc.
222 Richmond Street, Suite 105, Providence, RI 02903

BEWARE!
THIS IS A GENUINE **CALCULUS** BOOK,
specially designed for anyone who recalls a little high school algebra and is curious about seeing the peculiar way that mathematicians have wired calculus together. An added bonus — anyone who makes it through Professor McSquared's unusual course here will have a fearless start on the usual calculus text!

Our original idea was this: if we could find characters for each mathematical concept in differential calculus and set them all to work, the result would be far more lively and involving than the usual textbook trip. What happened along the way was that the characters acquired more life than we had expected and sometimes seem to charge off in their own directions. So, if they lead you astray, go back and re-read what you have already done and try the exercises — we have left room to work them out in the book, and the solutions are given at the end of the book.

This is a new expanded version of McSquared's original book. At the end of each chapter you will find new extra exercises and excursions. Although they are not needed to get a sense of what calculus is all about, they will give you more practice and lead you through more limit and derivative theorems.

We hope you enjoy the book.
Prof. E. McSquared
H. Swann
J. Johnson

Janson Publications, Inc.
222 Richmond Street, Suite 105
Providence, Rhode Island 02903

Printed in the United States of America

95 94 93 92 91 90 89 8 7 6 5 4 3 2 1

ISBN 0-939765-12-8

Bertrand Russell said:
THE SOLUTION OF THE DIFFICULTIES WHICH FORMERLY SURROUNDED THE MATHEMATICALLY INFINITE IS PROBABLY THE GREATEST ACHIEVEMENT OF WHICH OUR AGE HAS TO BOAST.

To help understand Dr. Russell, take our course!
SEE ABSTRACT THEORIES COME TO LIFE BEFORE YOUR VERY EYES!
INNERMOST SECRETS OF STRANGE MATHEMATICAL RITUALS LAID BARE AT LAST!
HERE IS THE GENUINE INSIDE DOPE
FOR YOUR MATHEMATICAL EDIFICATION AND INTELLECTUAL ENLIGHTENMENT.
ACCEPT NO SUBSTITUTES
INSIDE DOPE
g()
Grover

TABLE OF CONTENTS
IN WHICH THE CALCULUS PLOT IS REVEALED
f()
ALFRED

Calculus is concerned with studying VERY CAREFULLY relationships of the sort that can be put on a graph.

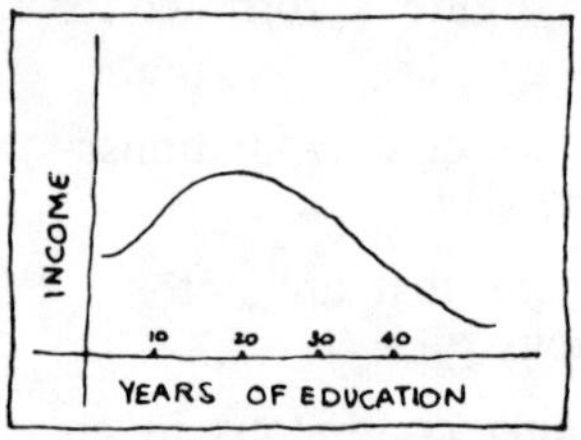

In the course of studying these relationships, mathematicians tried to answer the following three questions:

The FIRST question is...

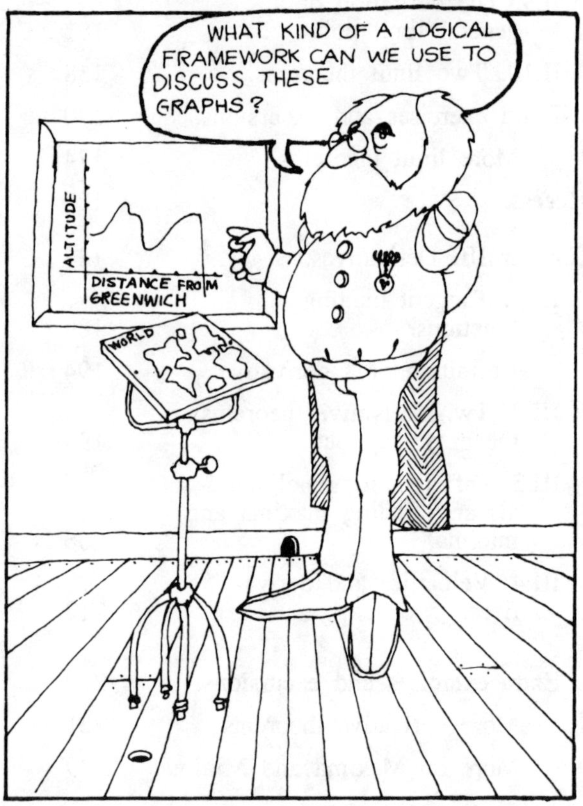

Pictures are not precise enough and computers can't compute with them.

Assuming a logical answer to the first question, the SECOND question is . . .

Mathematicians floundered around for years over this one and finally came up with a clever if somewhat slippery answer that also gave them a way to deal with the THIRD question . . .

The answers to these three questions provided the basis for what is called "Differential Calculus," and these answers were just what mathematicians needed to talk about velocities, gravity, etc., and thus get the whole scientific trip off the ground.

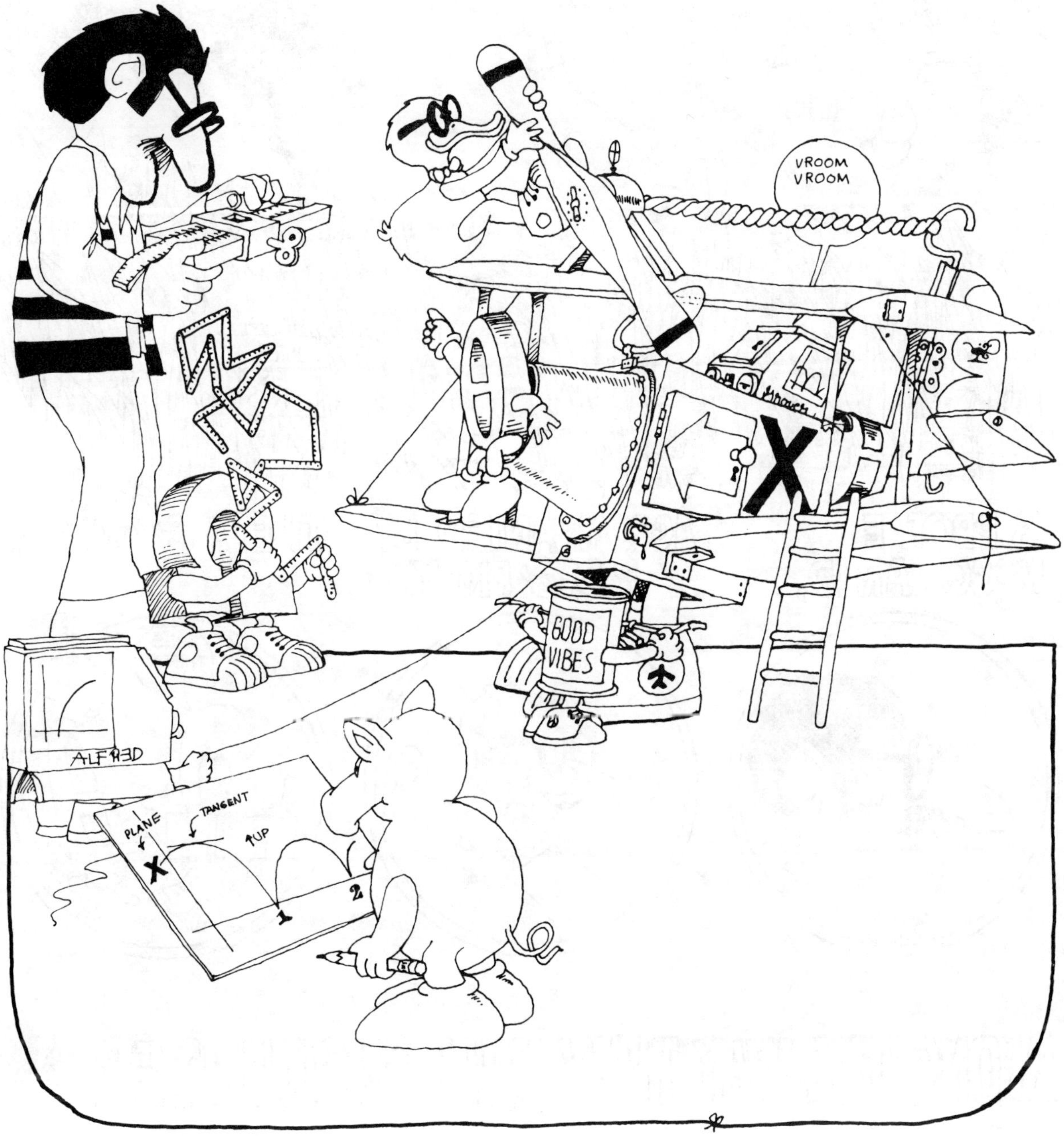

LEADING THIS TRIP THROUGH THE STRANGE WORLD OF...

IS PROF. E. McSQUARED, ACCOMPANIED BY R.D. & PIGGY, THETA & OMEGA.

ALONG THE WAY THEY WILL ENCOUNTER

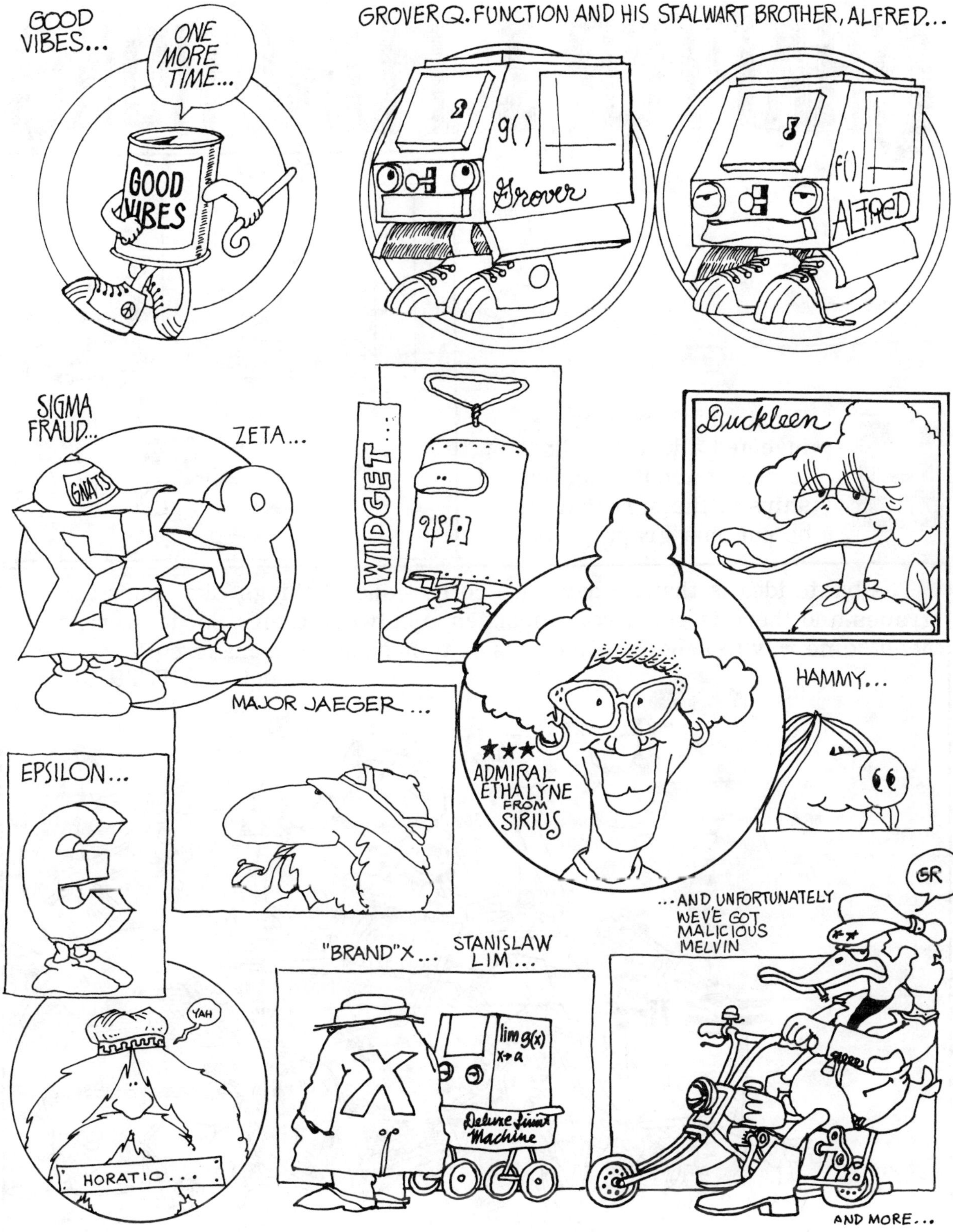
GOOD VIBES...
ONE MORE TIME...
GOOD VIBES
GROVER Q. FUNCTION AND HIS STALWART BROTHER, ALFRED...
g()
Grover
f()
ALFRED
SIGMA FRAUD...
GNATS
ZETA...
WIDGET...
Duckleen
MAJOR JAEGER...
HAMMY...
ADMIRAL ETHALYNE FROM SIRIUS
EPSILON...
"BRAND" X...
STANISLAW LIM...
lim g(x)
x→a
Deluxe Limit Machine
...AND UNFORTUNATELY WE'VE GOT MALICIOUS MELVIN
GR
YAH
HORATIO...
AND MORE...

CHAPTER I: FUNCTIONS

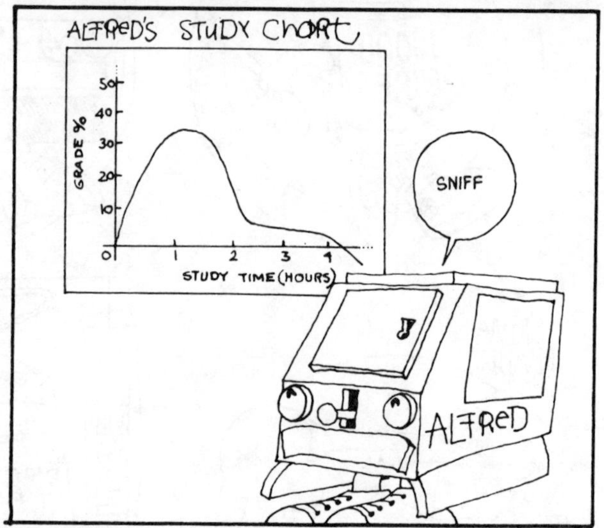

I·1

First, mathematicians invented a way to talk logically about relationships of the sort that can be put on a graph.

The basic idea is that we have two collections of things like hours and grades and there is some correspondence between them. So, to start, we need some way to handle collections of things. We can make collections out of ANYTHING.

1·2 SET THEORY

To keep track of collections of things, we put the things in bags called SETS. The mathematical symbol jargon for "A is the set containing cube, Theta, six, and Good Vibes" is

$$A = \left\{ \square, \theta, 6, \boxed{GV} \right\}.$$

Think of the squiggly brackets $\{$ and $\}$ as the sides of the bag and think of A as a label.

If we want to mention that some particular thing is in a particular bag, like "Theta is in set A," the symbol-jargon for this statement is

$$\theta \in A$$

where the symbol $\in$ means "is a member of the set."

Mathematicians do all sorts of operations with sets, like trying to figure out what things, if any, are common to two different sets . . .

. . . or dumping two sets together to form a new set.

The study of such maneuvers, which can get very complicated, is called SET THEORY. Here are some of the ground rules for the set theory game.

GROUND RULE 1 $\{2, \theta, \Omega\} = \{\theta, \Omega, 2\} = \{\Omega, 2, \theta\}$, ETC., SINCE THE BAGS DON'T CARE HOW THE THINGS ARE ARRANGED INSIDE THEM.

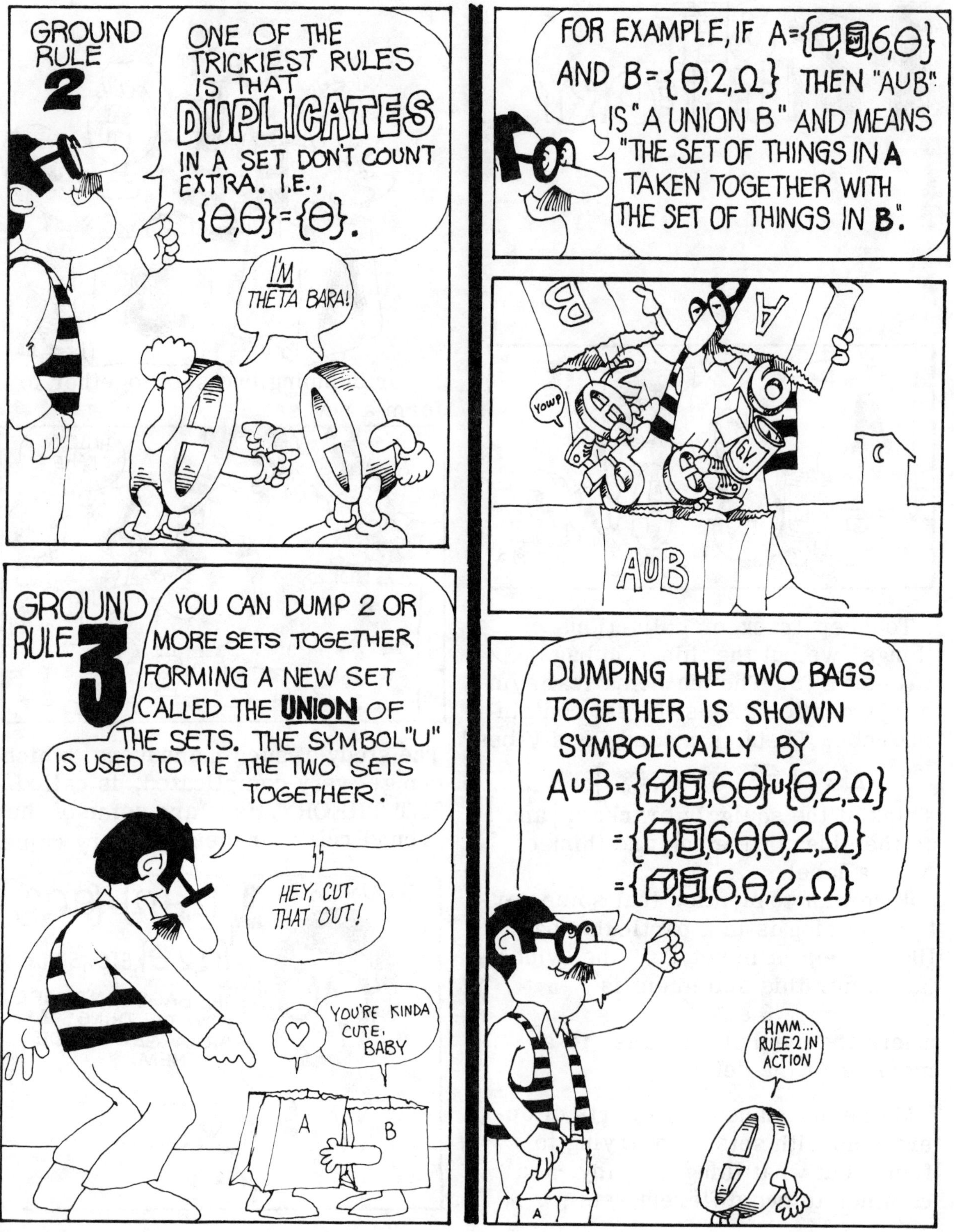
GROUND RULE 2
ONE OF THE TRICKIEST RULES IS THAT DUPLICATES IN A SET DON'T COUNT EXTRA. I.E., $\{\Theta,\Theta\}=\{\Theta\}$.
I'M THETA BARA!
GROUND RULE 3
YOU CAN DUMP 2 OR MORE SETS TOGETHER FORMING A NEW SET CALLED THE UNION OF THE SETS. THE SYMBOL "U" IS USED TO TIE THE TWO SETS TOGETHER.
HEY, CUT THAT OUT!
YOU'RE KINDA CUTE, BABY
A
B
FOR EXAMPLE, IF A={□,▯,6,Θ} AND B={Θ,2,Ω} THEN "AUB" IS "A UNION B" AND MEANS "THE SET OF THINGS IN A TAKEN TOGETHER WITH THE SET OF THINGS IN B".
B
A
YOWP
AUB
DUMPING THE TWO BAGS TOGETHER IS SHOWN SYMBOLICALLY BY
A∪B={□,▯,6,Θ}∪{Θ,2,Ω}
={□,▯,6,Θ,Θ,2,Ω}
={□,▯,6,Θ,2,Ω}
HMM... RULE 2 IN ACTION
A

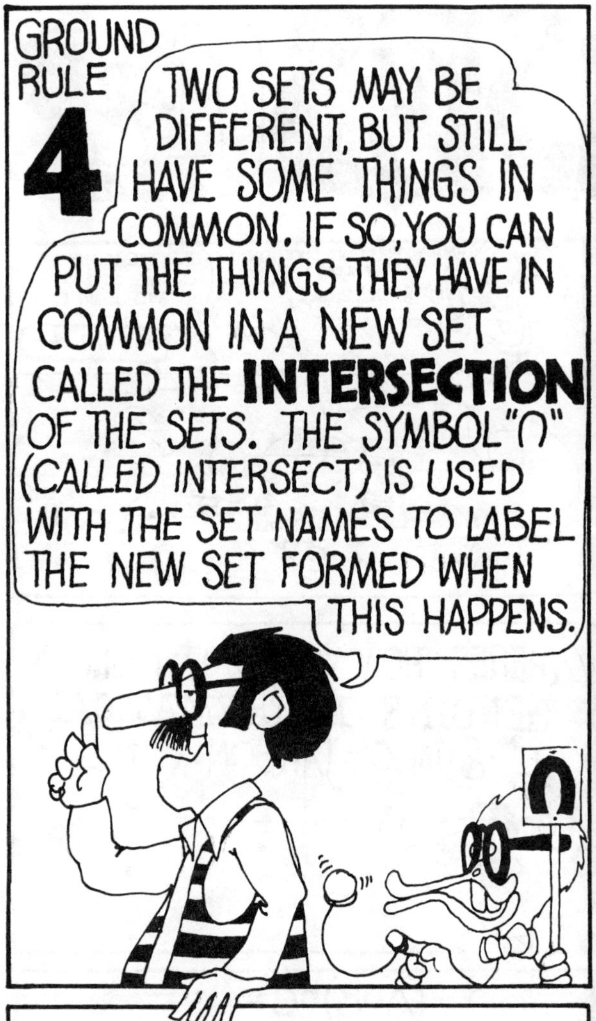

For example, if $A = \{\text{cube}, \text{can}, 6, \Theta\}$ and $B = \{\Theta, 2, \Omega\}$, "A intersect B" is "the set of those things in A that are also in B," and this is the set containing only Theta. In mathematical symbols we have

$$A \cap B = \{\text{cube}, \text{can}, 6, \theta\} \cap \{\theta, 2, \Omega\} = \{\theta\}.$$

It may turn out that two sets have NOTHING in common. In this case we use a handy set that has nothing in it at all. This peculiar set is called the "empty set" and has the symbol $\emptyset$. So, $\emptyset = \{\ \}$.

EXERCISES

I.2.1 Suppose $A = \{$[cube], [can], $6, \theta\}$,
$B = \{\theta, 2, \Omega\}$, $C = \{3, 6, \Omega,$ [can]$\}$
and $D = \{$[can], $3,$ [cube]$\}$.

Figure out

(a) $A \cap C = \{?\}$

(b) $B \cup C = \{?\}$

(c) $B \cap D = \{?\}$

(d) $C \cap \emptyset = \{?\}$

(e) $C \cup \emptyset = \{?\}$

YOU CAN KEEP USING UNION (∪) AND INTERSECTION (∩) OVER AND OVER AGAIN TO BREW UP FIENDISHLY COMPLICATED SETS, USING PARENTHESES TO KEEP TRACK OF WHAT TO DO FIRST. FOR EXAMPLE, R.D. AND PIGGY WILL NOW USE A VENN DIAGRAM TO FIGURE OUT (A∪B)∩C.

VENN WILL IT ALL END?

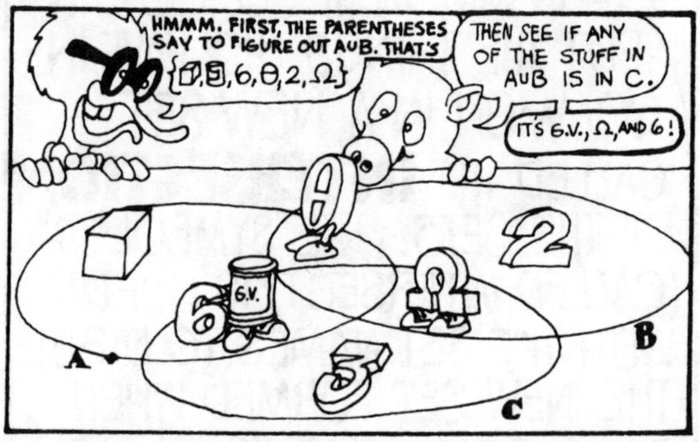

HMMM. FIRST, THE PARENTHESES SAY TO FIGURE OUT A∪B. THAT'S {[cube],[can],6,Θ,2,Ω}
THEN SEE IF ANY OF THE STUFF IN A∪B IS IN C.
IT'S G.V., Ω, AND 6!
G.V.
A
B
C

HERE'S HOW YOU DO IT USING GENUINE MATHEMATICAL SYMBOL-JARGON

$(A \cup B) \cap C =$

$(\{$[cube]$, 6,$ [can]$, \Theta\} \cup \{\Theta, 2, \Omega\}) \cap \{6,$ [can]$, 3, \Omega\}$

$= \{$[cube]$, 6,$ [can]$, \Theta, 2, \Omega\} \cap \{6,$ [can]$, 3, \Omega\}$

$= \{6,$ [can]$, \Omega\}$!

You try some. R. D.'s Venn diagram may be helpful.

I.2.2 $A \cup (B \cap C) =$

This result should be different from $(A \cup B) \cap C$ – you have to be careful about moving parentheses around.

I.2.3 $(A \cap B) \cap C =$

I.2.4 $(A \cap C) \cup (B \cap C) =$

PSST... THIS SHOULD BE THE SAME AS (AUB)∩C

I.2.5 $(A \cup B) \cap (A \cup C) =$

PSST. THIS SHOULD BE THE SAME ANSWER AS 1·2·2

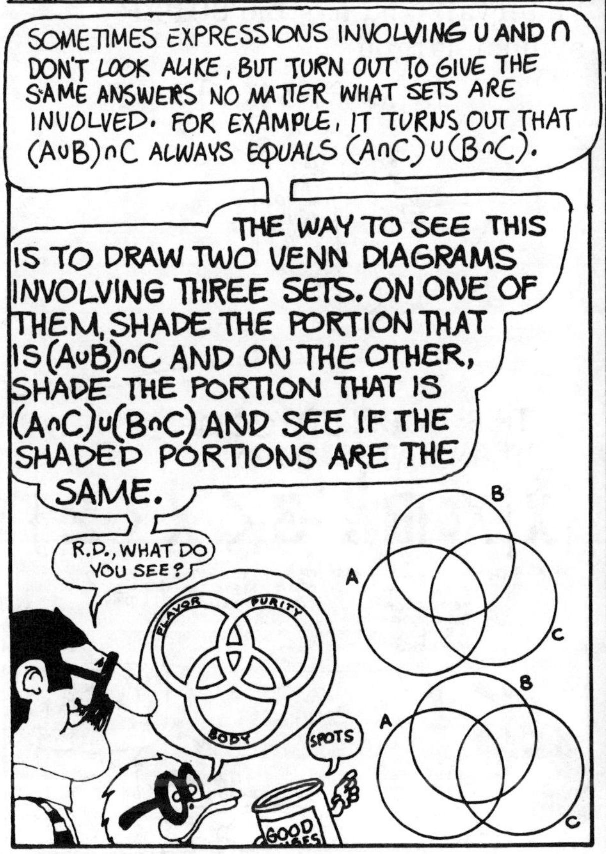

I.2.6 **EXTRA PROBLEMS**

for set theory freaks.

Draw cross-hatched Venn diagrams to check all these results:

(a). $(A \cup B) \cup C = A \cup (B \cup C)$

(more on next page)

(b) $(A \cap B) \cap C = A \cap (B \cap C)$.

(c) $A \cup B = B \cup A$.

(d) $A \cap B = B \cap A$.

(e) $(A \cup B) \cap C = (A \cap C) \cup (B \cap C)$.

(f) $A \cup (B \cap C) = (A \cup B) \cap (A \cup C)$.

1·3 SETS OF NUMBERS,

WHILE STUDYING CALCULUS, THE SETS WE WILL BE USING WILL BE SETS OF REAL NUMBERS.

WHY NOT BOTTLE CAPS? BASEBALL CARDS? MARBLES? COCKROACHES OR MATCHBOOKS, MAYBE

SHUDDUP

THE SET CONTAINING THE ENTIRE NUMBER LINE IS LABELED ℝ (FOR REAL NUMBERS).

YEZZIR FOLKS— GETCHER GEN·U·INE SPECIAL REAL NUMBERS NOW!

NEW ℝ

+∞

−∞

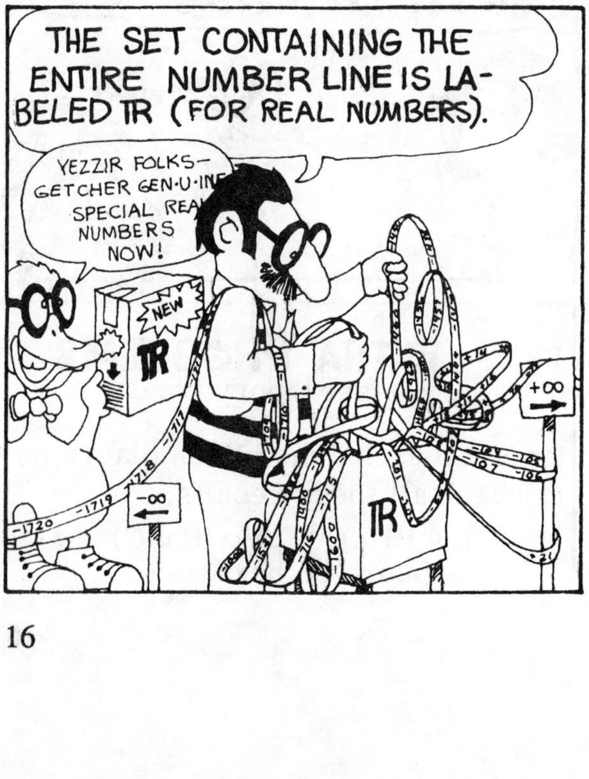

If we take the chunk of numbers from −3 to +2 . . .

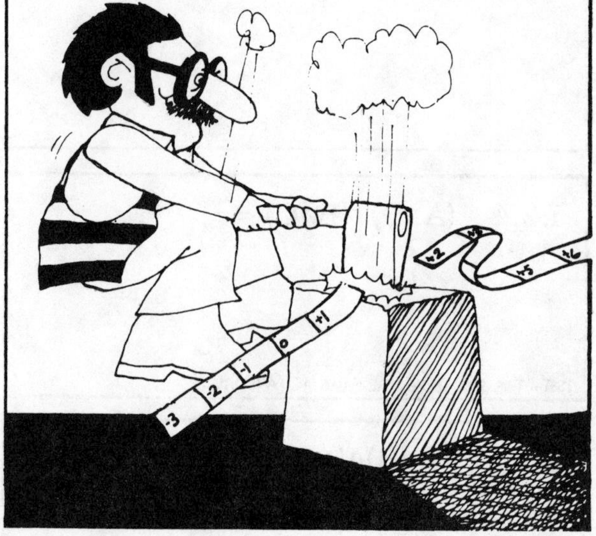

. . . and put it in a bag, the bag containing the chunk is called an "interval" and has the SUPER symbol-jargon

$$\{X \mid X \in \mathbb{R} ; -3 < X < 2\} \,!$$

THIS GOBBLEDYGOOK TRANSLATES INTO

$$\{X \mid X \in \mathbb{R} ; -3 < X < 2\}$$

THE SET OF / X THE X / SUCH THAT / X IS A MEMBER OF THE SET OF REAL NUMBERS / AND / NEGATIVE 3 / IS LESS THAN / THE NUMBER X / IS LESS THAN / 2

仏教の聖典の中では華厳経が一番すぐれ

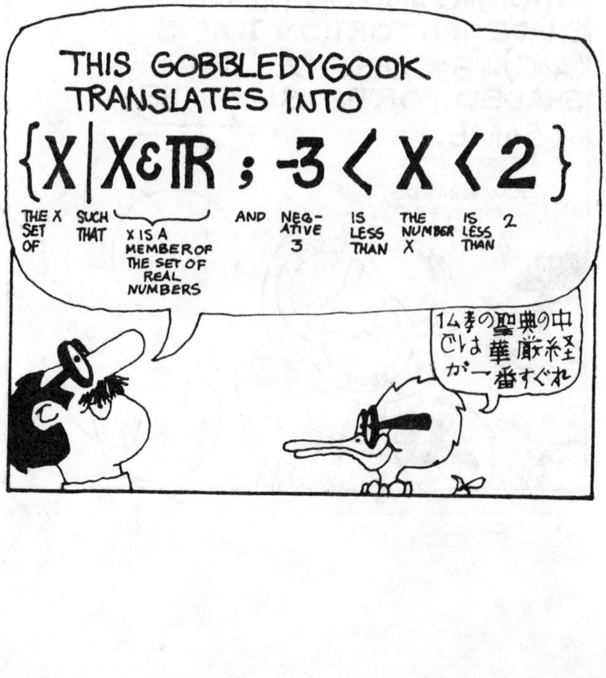

HA! WE JUST USED X FOR THE FIRST TIME... HE'S A PRETTY ELUSIVE CHARACTER!
Ø
X

THE IDENTITY OF X WILL GET PINNED DOWN SOMEWHAT BY WHATEVER APPEARS HERE { | } IN THE PARTICULAR SET WE ARE TALKING ABOUT.
X

In $\{X \mid X \in \mathbb{R};\ -3 < X < 2\}$,

" x " represents ANY number in the interval between -3 and 2 , NOT including -3 and 2 .

X
-3
-2
-1
0
+1
+2

If we want the set that contains the chunk of the number line from -3 to 2 and INCLUDES the numbers -3 and 2 , we use the symbol " ≤ " (less than or equal to) and concoct

$$\{X \mid X \in \mathbb{R};\ -3 \leq X \leq 2\}.$$

Both of these sets have shorter symbols:

(-3, 2) represents $\{X \mid X \in \mathbb{R};\ -3 < X < 2\}$

and [-3, 2] represents $\{X \mid X \in \mathbb{R};\ -3 \leq X \leq 2\}$.

Here, square brackets "[" and "]" are used to mean the same thing as " ≤ " and regular parentheses "(" and ")" are used to mean " <." (Later we will also use symbols like "(-3, 2)" to name "coordinates of points on a grid," which is an entirely different case. You have to interpret the meaning of the symbol from the context. It is usually pretty obvious.)

A handy way to figure out intersections and unions of such intervals of numbers is to chart them on the number line, using square brackets and parentheses to remember where the symbols "≤" and "<" are used.

If $A = (-3,2) = \{X \mid X \in \mathbb{R};\ -3 < X < 2\}$

and $B = [-1,3] = \{X \mid X \in \mathbb{R};\ -1 \leq X \leq 3\}$,

R. D. and Prof. McSquared will figure out $A \cap B$ and $A \cup B$. It's O. K. to use both square brackets and parentheses in the same expression if necessary. For example, (-2, 4] stands for $\{x \mid x \in \mathbb{R};\ -2 < x \leq 4\}$.

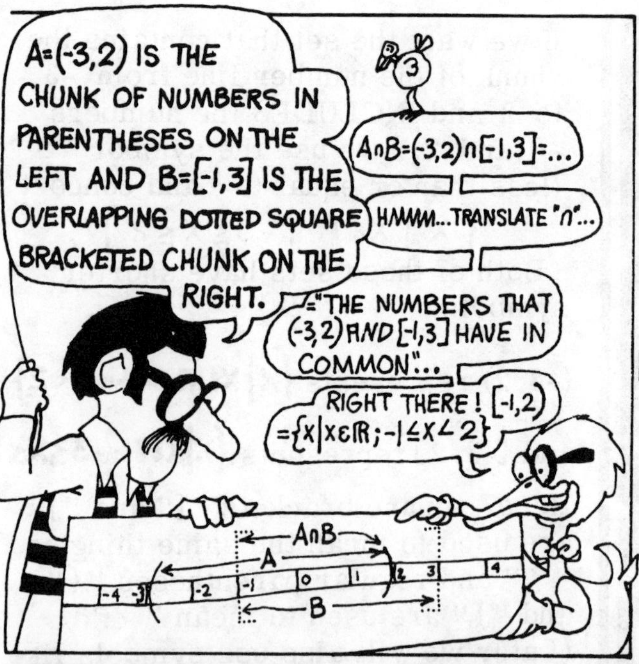

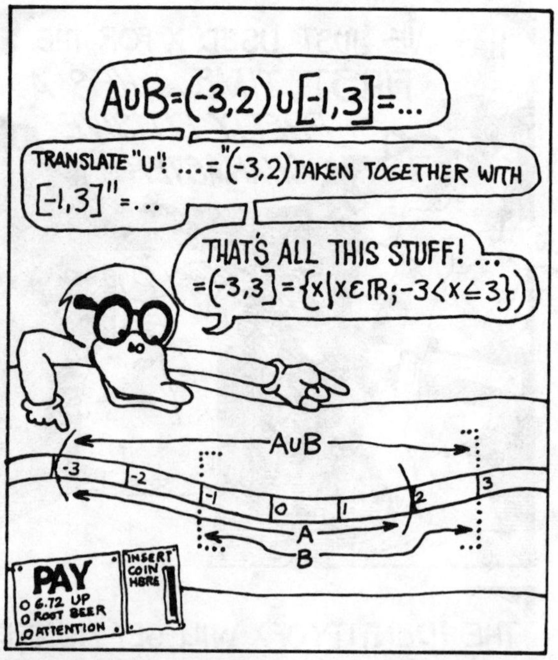

EXERCISES

Suppose A = (-3, 2), B = [-1, 3] and C = (-2, -1] . Indicate A, B and C on the number line and figure out the following — write the answers using both the squiggly bracket { | } set notation and the shorter notation using combinations of "[", "] ", "(", and ") ."

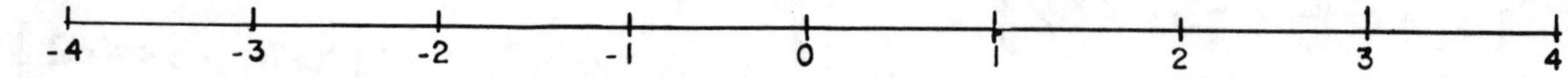

I.3.1 A ∩ C =

I.3.2 A ∪ C =

I.3.3 B ∩ C =

I.3.4 B ∪ C =

I.3.5 (B ∪ C) ∩ A =

I.3.6 (B ∩ C) ∩ A =

I.3.7 A ∪ (B ∩ C) =

I.3.8 (A ∩ B) ∪ C =

Later, we will spend a lot of time with "inequalities" like $0 \leq x < 3$ that come up in descriptions of intervals like $\{x \mid x \in \mathbb{R};\ 0 \leq x < 3\}$.
In particular, we will want to do things like multiplying every number in a set described with inequalities by some constant number and see what we get. For example, what happens if we multiply every number x in

$$\{x \mid x \in \mathbb{R};\ 0 \leq x < 3\} = [0,3)$$

by 2? Each number x originally satisfying inequality $0 \leq x < 3$ will be doubled, covering a new chunk of the number line — it's just like stretching the original interval $[0,3)$ to twice its length, keeping 0 where it is, and seeing where we end up.

Then, although $0 = 2 \cdot 0$ will stay where it is, 3 will stretch up to $2 \cdot 3 = 6$ and the doubled x's will stretch along with 3. (When a dot appears between two numbers, like "$2 \cdot 3$," it means the product " 2 times 3.")

So, if x satisfies $0 \leq x < 3$, then " twice these x's ," which is what " 2x " means, satisfies

(multiply by 2) $\quad 2 \cdot 0 \leq 2x < 2 \cdot 3$
or $\quad 0 \leq 2x < 6.$

While the x's still lie in the interval $0 \leq x < 3$, these (2x)'s will cover the interval $[0,6)$ from 0 up to 6 . We can use symbol-jargon and say

$$\{2x \mid x \in \mathbb{R};\ 0 \leq x < 3\} = [0,6).$$

Notice where "2x" appears!

If the numbers x start at 1 instead of 0 , when we multiply everything in the interval by 2 , the starting point 1 will stretch up to 2 and we get a new chunk of the real number line running from 2 to 6 .
In inequalities, if x satisfies

$$1 \leq x < 3$$

then "2x" satisfies

$$2 \cdot 1 \leq 2x < 2 \cdot 3$$

or $\quad 2 \leq 2x < 6$.

We can also say this by using sets:

$$\{2x \mid x \in \mathbb{R};\ 1 \leq x < 3\} = [2,6).$$

Again, "2x" is inside the squiggly brackets to show that all x's in $[1,3)$ have been doubled.

If the interval we are talking about contains negative numbers as well, like the x's satisfying the inequality

$$-1 < x < 3,$$

then, when we multiply by 2, 3 stretches to 6 and 0 stays at the original 0, but -1 stretches back to $-2 = (2)\cdot(-1)$, i.e., the set of numbers stretches in the negative direction also. Using inequalities again: If x satisfies $-1 < x < 3$ then "2x" satisfies the inequality

$$(2)\cdot(-1) < 2x < 2\cdot 3$$

or $\{2X \mid X \varepsilon \mathbb{R}; -1 < X < 3\} = (-2, 6)$.

Finally, if we multiply everything in an interval by a positive number less than 1, say ⅓, it "shrinks" the interval to one-third its original size; For example, if x satisfies the inequality $-1 < x < 3$, then "⅓x" satisfies the inequality

$$(\tfrac{1}{3})\cdot(-1) < \tfrac{1}{3}x < \tfrac{1}{3}3$$

or $\{\tfrac{1}{3}X \mid X \varepsilon \mathbb{R}; -1 < X < 3\} = (-\tfrac{1}{3}, 1)$.

If the constant number doing the multiplying is negative, things get more complicated. Multiplying a number by -1 "flips" the number around 0 to the opposite side of the number line:

$$2 \text{ goes to } (-1)\cdot(2) = -2$$

while -3 goes to $(-1)\cdot(-3) = 3$.

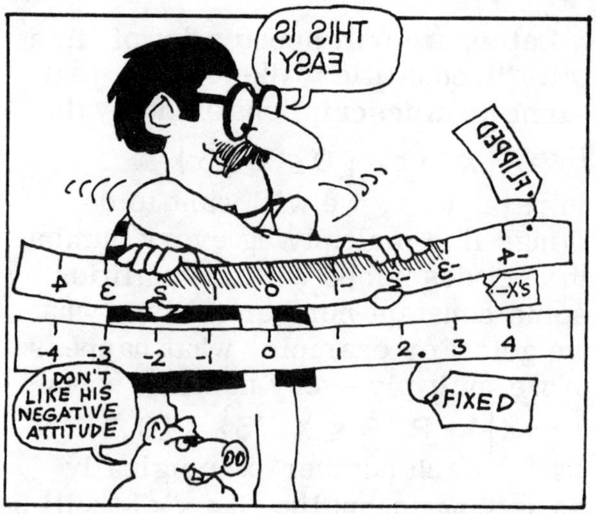

Now this does funny things to inequalities. Take a simple one like

$$-3 < 2,$$

which could be read "-3 is 'to the left of' 2 on the number line." Yet when we multiply through by -1, we get $3 = (-1)\cdot(-3) > (-1)\cdot(2) = -2$, where " > " means "is greater than." So multiplying an inequality by -1 (or any negative number) will FLIP the inequality signs from " < " to " > " or from " ≤ " to " ≥ " (where ≥ means "greater than or equal to ") provided the original numbers -3 and 2 (times factor -1) are left in their original locations in the inequality.

Take a whole chunk of the number line like [-3, 2) and multiply everything by -1. If we take a look at those x's, defined by the inequality

$$-3 \le x < 2$$

THESE WILL FLIP

and multiply these x's by -1, we get

$$(-1)(-3) \ge (-1)(x) > (-1)(2)$$

THESE ARE FLIPPED VERSIONS OF THE ORIGINAL INEQUALITY SIGNS

or $3 \ge -x > -2$.

Thus -x (remember, we're now talking about minused values) satisfies

$$3 \ge -x > -2.$$

We summarize the whole maneuver using set theory:

$$\{-X \mid X \in \mathbb{R}; \ -3 \leq X < 2\} = (-2, 3].$$

Finally, if we multiply everything in an interval by a negative number like -2 , it will both flip the interval and stretch it by two to cover a new chunk of the number line.

For example, if x's satisfy

$$-3 \leq x < 2$$

then, for those x's , (multiply by -2)

$$6 = (-2)\cdot(-3) \geq -2x > (-2)\cdot(2) = -4.$$

Now this inequality works exactly the same as

$$-4 < -2x \leq 6 \quad !$$

So $\{-2X \mid X \in \mathbb{R}; \ -3 \leq X < 2\} = (-4, 6].$
This one is a little harder to picture...

EXERCISES

I.3.9 If x's satisfy

$$-2 < x \leq 1 \ ,$$

what inequality will the set of 2x's satisfy?

What inequality will the set of ($\frac{1}{2}$x)'s satisfy?

What inequality will the set of (-3x)'s satisfy?

I.3.10 If x's satisfy

$$-1 \geq x > -\tfrac{3}{2} \ ,$$

what inequality will the set of 2x's satisfy?

What inequality will the set of ($\frac{1}{2}$x)'s satisfy?

What inequality will the set of (-3x)'s satisfy?

Sometimes a constant gets added to every number in a set described by inequalities. Adding constants is a lot easier to handle than multiplying everything by a constant. For example, if we add 3 to every number in set

$$\{X \mid X \in \mathbb{R};\ -2 < X < 1\} = (-2, 1)$$

it's just like carrying (-2, 1) three units to the RIGHT.

So, if x satisfies

$$-2 < x < 1 \ ,$$

then x + 3 (3 added to these x's) satisfies (add 3 to everything)

$$-2 + 3 < x + 3 < 1 + 3$$

or $1 < x + 3 < 4$.

So $\{X + 3 \mid X \in \mathbb{R}; -2 < X < 1\} = (1,4)$.

If we add a negative number, like -4, to every number in a set, it just hauls the set 4 units to the LEFT.

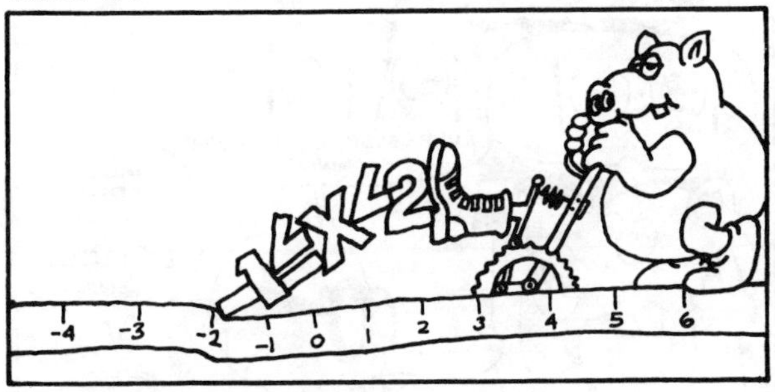

For example, if x satisfies

$$-1 \le x < 2 \ ,$$

then x - 4 (-4 added to these x's) satisfies

$$-1-4 \le x-4 < 2 - 4$$

or $-5 \le x - 4 < -2$

So $\{X-4 \mid X \in \mathbb{R}; -1 \le X < 2\} = [-5,-2)$.

Here are a couple more exercises for practice.

I.3.11 Add -4 to every number in $\{X \mid X \in \mathbb{R}; -5 \le X < 4\}$.

What interval do you get?

I.3.12 Add $\frac{3}{2}$ to every number in $\{X \mid X \in \mathbb{R}; 2 \ge X > -1\}$.

What interval do you get?

IN OUR CONTINUING SEARCH FOR WAYS TO WORK WITH INEQUALITIES, WE DIP INTO OUR INEXHAUSTIBLE SYMBOL SECTION AND COME UP WITH "| |," CALLED
ABSOLUTE VALUE.

The "absolute value" of a number is just its distance to point 0 on the number line, with "distance" always counted positive (or "0" if the number is actually 0). So, to use the symbol,

$|2| = 2$; $|-3| = 3$; $|3| = 3$;etc.

I GET 3.
ROAD CLOTHED
-3
-2
-1
0
1
2
3

Notice that, since $|2| = 2 = |-2|$,

$|-105| = 105 = |-(-105)|$, etc. ,

we have

$|\text{any number}| = |\text{minus that number}|$

no matter what sign "any number" has.

We can put in extra absolute value symbols "|" to separate the absolute value of a product of two numbers into the product of their absolute values.

For example,

$$|(-3)\cdot(2)| = |(-3)|\cdot|(2)|,$$

since both equal 6 : check it out -

$$|\underbrace{(-3)\cdot(2)}| =$$

$$|\ -6\ | = 6$$

and also $|(-3)|\cdot|(2)| =$

$$3 \cdot 2 = 6 .$$

DUM DEE DEE DUM
|(-3)·(2)|
|(-3)| |(2)|
URF-STOP MESSING WITH MY MIND!
THEY ARE BOTH EQUAL YESSIREE!

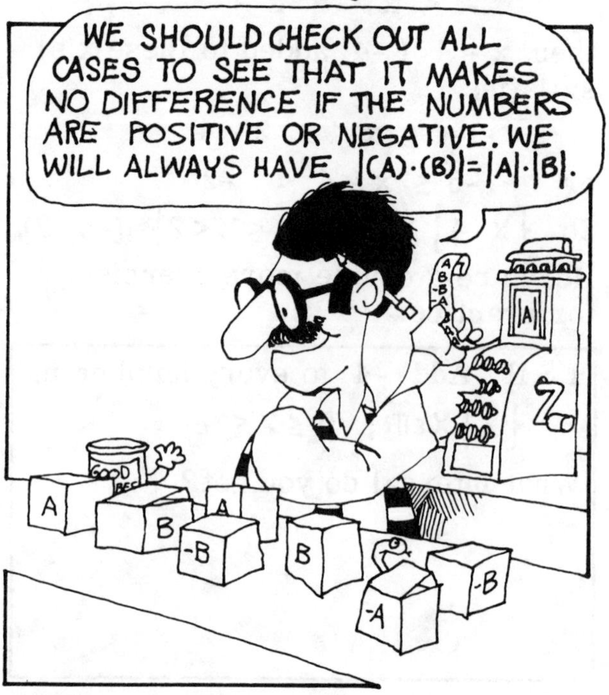
WE SHOULD CHECK OUT ALL CASES TO SEE THAT IT MAKES NO DIFFERENCE IF THE NUMBERS ARE POSITIVE OR NEGATIVE. WE WILL ALWAYS HAVE |(A)·(B)|=|A|·|B|.
A
B
-B
A
B
-A
-B
|A|

SUPPOSE A=-4 AND B=-3... THEN,
2 NEGATIVE CASES
|(A)·(B)| =
|(-4)·(-3)| =
| +12 | = 12
AND
|(A)|·|(B)| =
|(-4)|·|(-3)| =
4 · 3 =12
SINCE BOTH COMPUTATIONS EQUAL 12,
|(-4)·(-3)| = 12 = |(-4)|·|(-3)|
OR |(-4)·(-3)| = |(-4)|·|(-3)|
O.K. TO PUT THESE IN

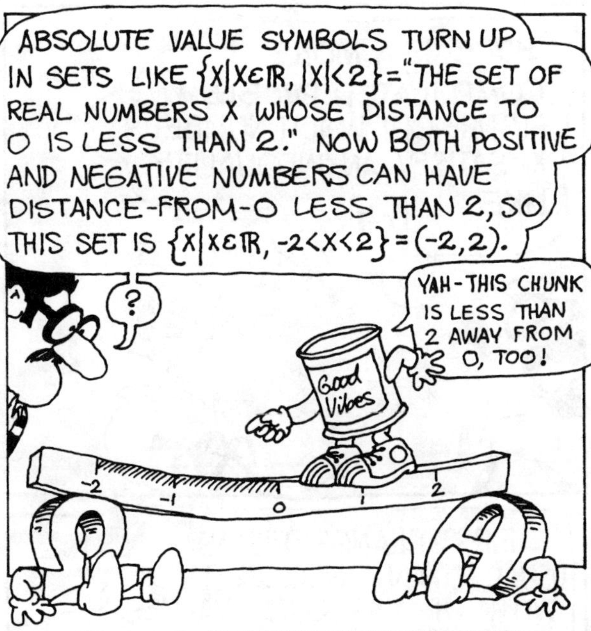
ABSOLUTE VALUE SYMBOLS TURN UP IN SETS LIKE {X|X∈ℝ, |X|<2} = "THE SET OF REAL NUMBERS X WHOSE DISTANCE TO 0 IS LESS THAN 2." NOW BOTH POSITIVE AND NEGATIVE NUMBERS CAN HAVE DISTANCE-FROM-0 LESS THAN 2, SO THIS SET IS {X|X∈ℝ, -2<X<2} = (-2,2).
?
YAH-THIS CHUNK IS LESS THAN 2 AWAY FROM 0, TOO!
Good Vibes
-2
-1
0
1
2

THAT IS, IF X SATISFIES |X|<2 , THEN -2<X<2 .
WE CAN RUN THIS ARGUMENT BACKWARD —ANY TIME WE HAVE X SATISFYING AN INEQUALITY LIKE -3<X<3, WE CAN CONCLUDE THAT |X|<3 (THE DISTANCE FROM X TO 0 IS LESS THAN 3).
BOUNCING BACK AND FORTH BETWEEN THESE TWO EXPRESSIONS IS VERY VALUABLE LATER; ONE CAN ALWAYS BE CONVERTED TO THE OTHER.

However, we CAN'T put in extra absolute value symbols on absolute values of SUMS of numbers without sometimes messing up equations.

$|2| + |3| = 5 = |2 + 3|$ is O.K.

as is

$|-2| + |-3| = 2 + 3 = 5 = |(-2) + (-3)|$,

but if one of the two numbers is negative and the other positive, it doesn't work;

$$|(-2)| + |(3)| = 5$$

is NOT equal to

$$|(-2) + (3)| = |1| = 1.$$

(For any numbers a and b ,

$$|a + b| \le |a| + |b|$$

is true, however .)

HERE IS A MORE COMPLICATED PROBLEM. FIGURE OUT WHICH NUMBERS X SATISFY AN INEQUALITY LIKE |X-1|<2.
FIRST, CHANGE THIS AS USUAL TO AN INEQUALITY WITHOUT ABSOLUTE VALUES: THESE X'S WILL SATISFY -2<X-1<2.
NOW USE ALGEBRA ON THIS INEQUALITY TO GET X ALL ALONE. ADD 1 TO GET RID OF "-1" IN "X-1": -2+1<X-1+1<2+1 OR -1<X<3... OR, MOST IMPORTANT,
A.HEM
Z.
THIS TURNS OUT TO BE THE SET OF NUMBERS X THAT ARE WITHIN DISTANCE 2 OF NUMBER 1.
(CHORTLE)
YUP...
H₂O
-1 0 1 2 3

SO ANYTIME X TURNS UP IN SOMETHING LIKE |X-A|<B, WE CAN CHANGE THIS TO -B<X-A<B OR, (ADD A) A-B<X<A+B. THIS SAYS THAT IF X'S SATISFY |X-A|<B, THEN THOSE X'S LIE WITHIN DISTANCE B OF NUMBER A.
(R.D.)U(H₂O)
WOWK!
X'S IN HERE
B
B
A-B
A
A+B

HERE ARE A FEW EXERCISES— THE FIRST TWO ARE WORKED OUT...BUT WATCH OUT–THE SECOND ONE IS PRETTY TRICKY. *the management.*

EXERCISES

I.3.A What number interval does the inequality

$$|x+2| < 3$$

describe ?

Answer: Since

$$|x+2| = |x-(-2)| \;,$$

these x's should be within distance 3 of number -2 . Check it:

If $|x+2| < 3$

then $-3 < x+2 < 3$ or

(add -2): $-5 = -3-2 < x < 3-2 = 1$.

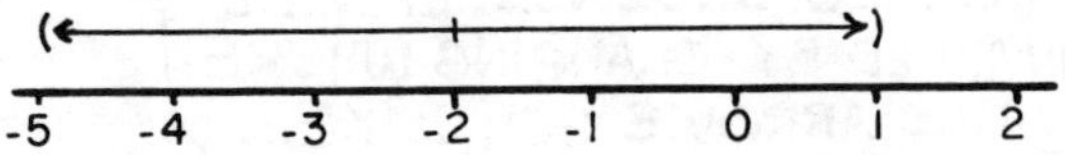

I.3.B What number interval does the inequality

$$|2-3x| < 1$$

describe?

Answer: First, if

$$|2-3x| < 1,$$

then $-1 < 2-3x < 1$.

Now use algebra on this inequality to get x all alone in the middle:

First get rid of the 2 in "2 - 3x" by adding (-2):

$$-2-1 < -2+2-3x < -2+1$$

or $-3 < -3x < -1$.

Now get rid of the -3 in "-3x" by multiplying by "$-\frac{1}{3}$" —WATCH OUT— this will flip the inequality signs!

I.3.B Continued:

$$(-\tfrac{1}{3})\cdot(-3) > (-\tfrac{1}{3})\cdot(-3x) > (-\tfrac{1}{3})\cdot(-1)$$

or $1 > x > \frac{1}{3}$

which is exactly the same as saying

$$\tfrac{1}{3} < x < 1 \;,$$

so we get interval $(\frac{1}{3}, 1)$.

I.3.13 What number interval does the inequality

$$|1-x| < 2$$

describe?

I.3.14 What number interval does the inequality

$$|2x+3| < 4$$

describe?

I.3.15 What number interval does the inequality

$$|-3-\tfrac{1}{2}x| < 5$$

describe?

I.3.16 What number interval does the inequality

$$|\tfrac{2}{3}x+4| < \tfrac{1}{2}$$

describe?

I·4 FUNCTIONS

Now, back to the problem of talking logically about relationships of the sort that can be put on a graph.

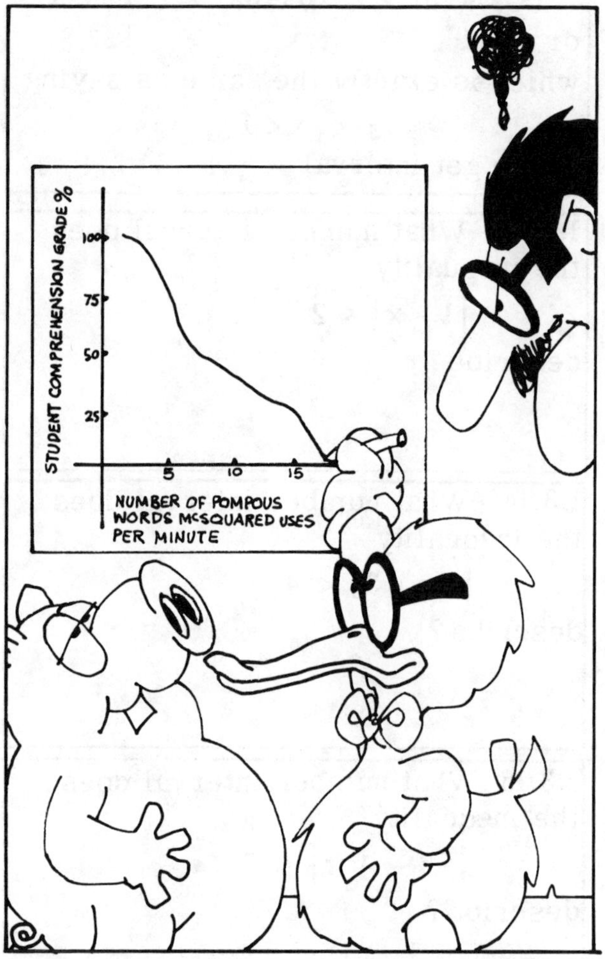

The basic idea is that we have two collections of things (like "numbers of pompous words" and "grades") and there is some correspondence between them. Set theory gives us a way to handle collections of things, but we still need some way to talk about correspondences between two sets.

Start with a simple example: suppose the sets are

A = { 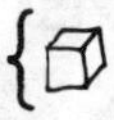 ,  , 6, θ } and B = { θ, 2, Ω }.

We then INVENT a correspondence between A and B and draw arrows to show what corresponds to what.

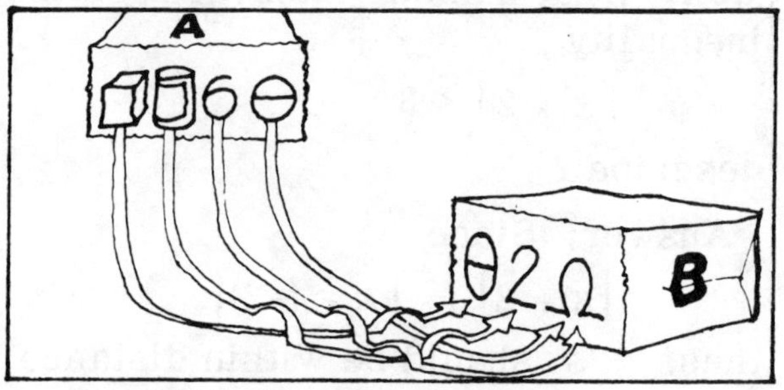

WE CAN INVENT LOTS OF OTHER CORRESPONDENCES BETWEEN SETS A&B THAT ARE JUST AS GOOD-FOR EXAMPLE, A DIFFERENT CORRESPONDENCE CAN BE INVENTED BY CHANGING WHERE THE ARROWS GO...LIKE...
A
B

Mathematicians decided on some

RULES

for their favorite kinds of correspondences, and any particular correspondence that follows these rules is called a

FUNCTION.

RULES FOR FUNCTIONS

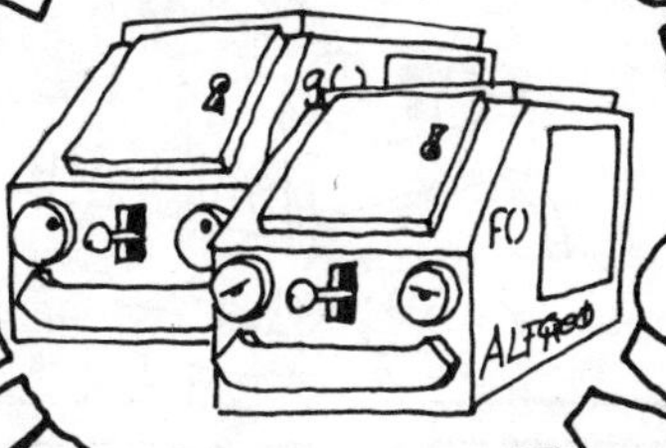

YOUR FRIENDLY NEIGHBORHOOD FUNCTION CONSISTS OF TWO SETS AND A BUNCH OF ARROWS THAT OBEY

RULE 1 THE ARROWS ALWAYS START FROM THE SAME SET, CALLED THE **DOMAIN** AND GO <u>TO</u> THE OTHER SET, CALLED THE **RANGE**.

RULE 2 EVERYTHING IN THE DOMAIN-SET MUST HAVE EXACTLY ONE ARROW FROM IT. EVERYTHING IN THE RANGE-SET MUST HAVE <u>AT LEAST</u> ONE ARROW TO IT.
(IT'S O.K. TO HAVE 2 ARROWS TO 1 THING.)

So two or more arrows can hit the same thing in the range-set, but only one arrow can come from any particular thing in the domain-set.

Using arrows in the RULES unfortunately has its drawbacks — as functions become more elaborate, the arrows can get pretty difficult to follow . . . .

SO, TO TAKE CHARGE OF WHERE THE ARROWS GO, WE NOW INTRODUCE

ALFRED & GROVER

IN THEIR TRUE ROLES AS

Incredible Function-Machines

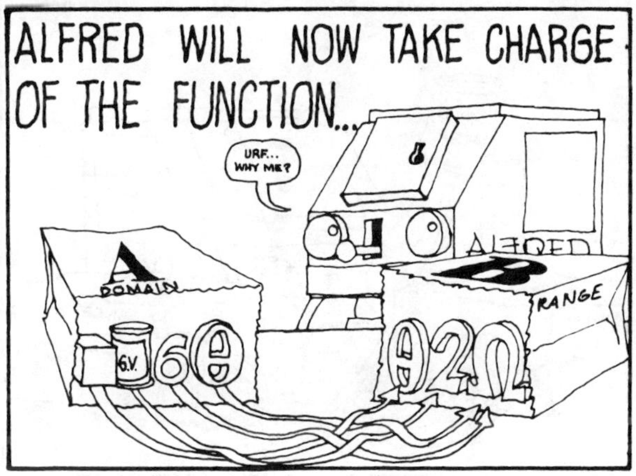

HIS TRIP IS TO **TRANSFORM** ANYTHING FROM THE DOMAIN SET A INTO SOMETHING IN THE RANGE SET B **EXACTLY** THE WAY THE ARROWS *INDICATE.*

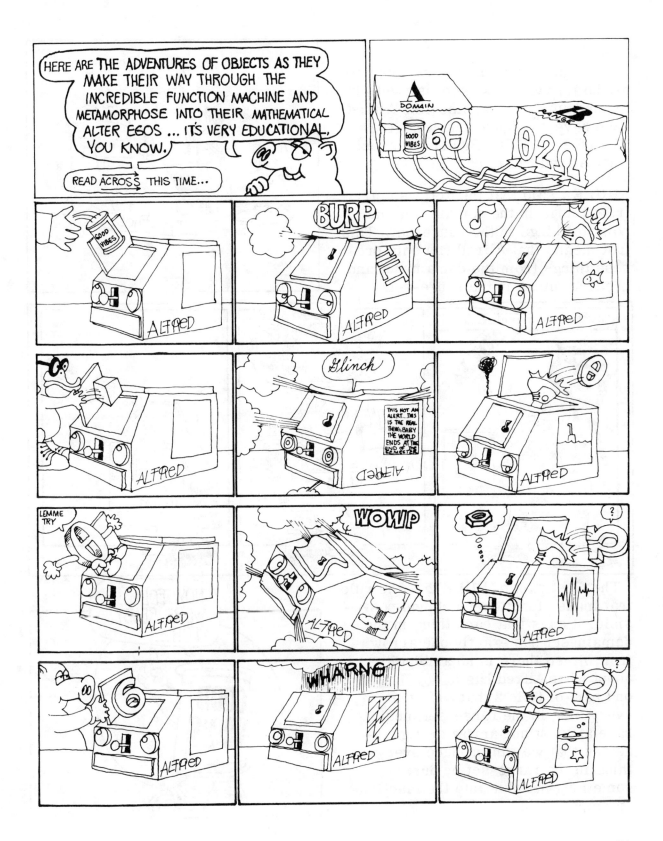

HERE ARE THE ADVENTURES OF OBJECTS AS THEY MAKE THEIR WAY THROUGH THE INCREDIBLE FUNCTION MACHINE AND METAMORPHOSE INTO THEIR MATHEMATICAL ALTER EGOS ... IT'S VERY EDUCATIONAL, YOU KNOW.
READ ACROSS THIS TIME...
A
DOMAIN
GOOD VIBES
RANGE
B
GOOD VIBES
ALFRED
BURP
TILT
ALFRED
ALFRED
ALFRED
Glinch
THIS NOT AN ALERT... THIS IS THE REAL THING, BABY THE WORLD ENDS AT THE END OF THE SEMESTER
ALFRED
ALFRED
LEMME TRY
ALFRED
WOWP
ALFRED
?
ALFRED
ALFRED
WHARNG
ALFRED
?
ALFRED

Thus, instead of having to cope with all the arrows, all we need is a well-informed function-machine (like Alfred).

We have to figure out what the RULES FOR FUNCTIONS say about what Alfred has to do.

RULE 1 said that the arrows start from things in the domain-set and point to things in the range-set. This just means that Alfred can only process things from the domain-set and produce things in the range-set.

The first part of RULE 2 is the rule that makes function-machines possible: it says that each thing in the domain-set has exactly one arrow from it, so there is no ambiguity about what Alfred has to do. The second part of Rule 2 that says that everything in the range-set must have at least one arrow to it guarantees that we can produce everything in the range-set by dumping something suitable into the function-machine.

NEW OFFICIAL DELUXE RULES FOR FUNCTIONS

FEATURING

FUNCTION MACHINES

A function consists of two sets, called the domain-set and the range-set, and a function-machine that follows these rules:

THIS SPACE STILL RENTED BY THE 7TH HOUR MATH CLASS, SABETHA HIGH SCHOOL, SABETHA, KANSAS

RULE 1. The function-machine can only process things in the domain-set and only produce things in the range-set.

RULE 2. The function-machine is unambiguous: the same input always produces the same output. Any particular thing in the range-set can be produced by finding something suitable in the domain-set and running it through the function-machine.

THE SYMBOL FOR ALFRED, IN HIS TRUE ROLE AS A FUNCTION-MACHINE IS f( ), WHICH TRANSLATES "f OF BLANK", OR, JUST PLAIN "f" FOR SHORT.

Alfred can run different functions at different times, but he always keeps the same symbol f( ) . When we change the function he is running, we have to say so.

When Alfred is running a specific function, like

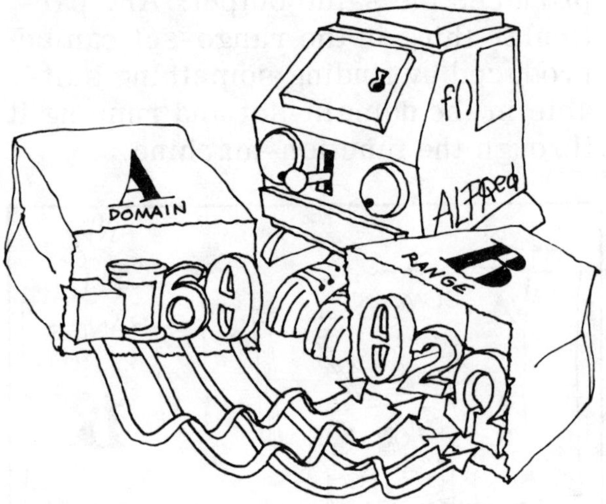

the symbol f( ) is used to label the output corresponding to each specific input like this:

and this labeling is what it means when we write

$$f(6) = \Omega$$

(which is translated " f of six equals Omega "). This is just a way to record the fact that 6 is transformed into Omega by the function f( ). So, with the same function,

$f(\theta) = \Omega$,

f() = θ

and f() = 2 .

Here is Grover in charge of another function g( ) .

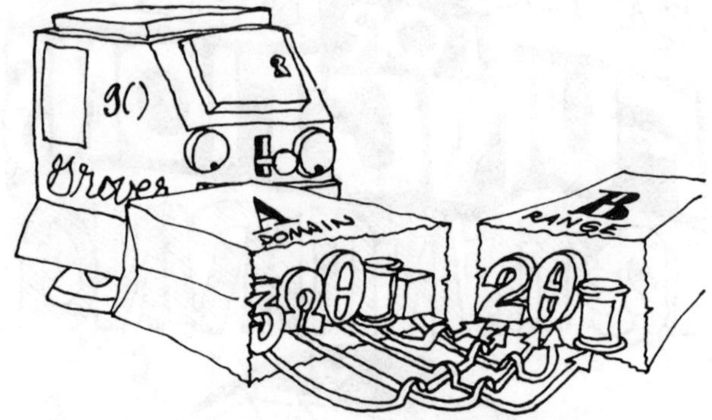

See if you can fill in for all "?."

g() = ? g(3) = ?

g(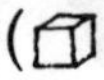) = ? g(?) = 

g(?) = θ .

Can you find another answer to

g(?) = θ ?

The neat thing about using bags and arrows is that if a function is shown using them, we know exactly what corresponds to what in the function. When a function-machine like Alfred takes over, HE knows what the function does, but we DON'T know because we don't have the arrows any more.

When the domain-set and the range-set of a function are chunks of the real number line, there is a clever and arrow-free way of showing what corresponds to what by constructing what is called the "graph" of the function. Prof. McSquared will show how to convert one of these "bags and arrows " functions to a "graph."

HERES THE FUNCTION. WE ONLY SHOW SOME OF THE ARROWS.
3
2
1
0
-1
-2
-3
D
DOMAIN
R
RANGE
1
PROF McSQUARED NOW TILTS THE DOMAIN AND SHOVES IT AROUND...
εD
D
DOMAIN
εR
R
RANGE
2
CENSORED
FOR RENT
εD
εR
3
...UNTIL THE TWO CHUNKS OF THE NUMBER LINE ARE PERPENDICULAR AND THE TWO ZEROS COINCIDE.
-3 -2 -1 0 1 2 3
εD
εR
4

Fix up each arrow so it starts off straight up or down and then makes a right-angle turn directly over to the range.

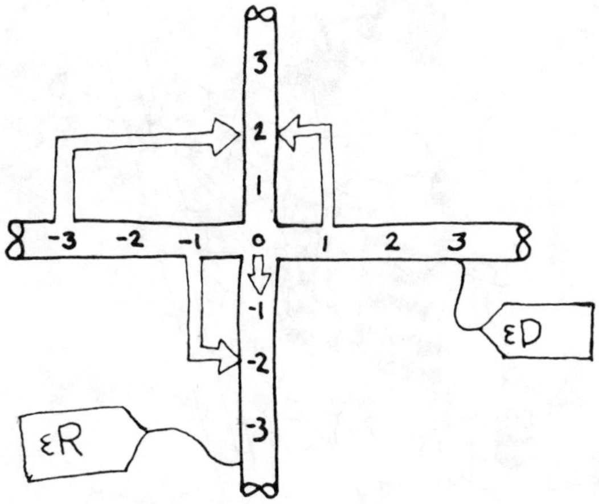

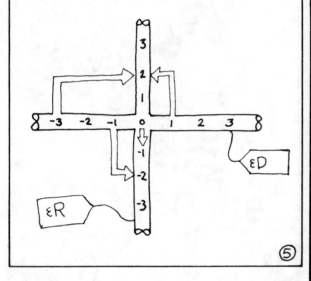

⑤

Now we can preserve ALL the information about the function by just keeping the DOTS where the function-arrows turn.

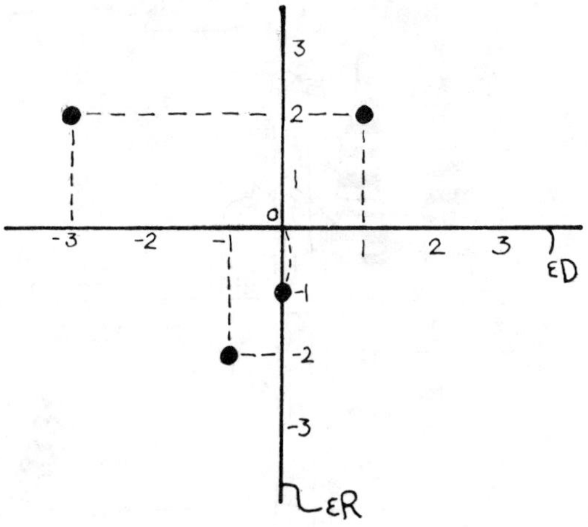

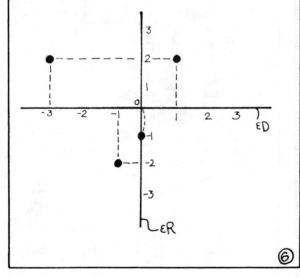

⑥

So the dots tell us all about the function. There are usually many more arrows than we have shown and thus more dots. In fact, usually there are so many that the dots make a solid curve. Showing how such a function works using just arrows from the domain-set to the range-set would really be a problem.

⑦

Remember that the DOMAIN of any function is always part of the horizontal ⟷ line, called the "x-axis" because any arbitrarily chosen thing in the domain-set is usually called "x." The RANGE is always part of the vertical (↕) line, called the "y - axis" because an arbitrary thing in the range-set is usually called "y."

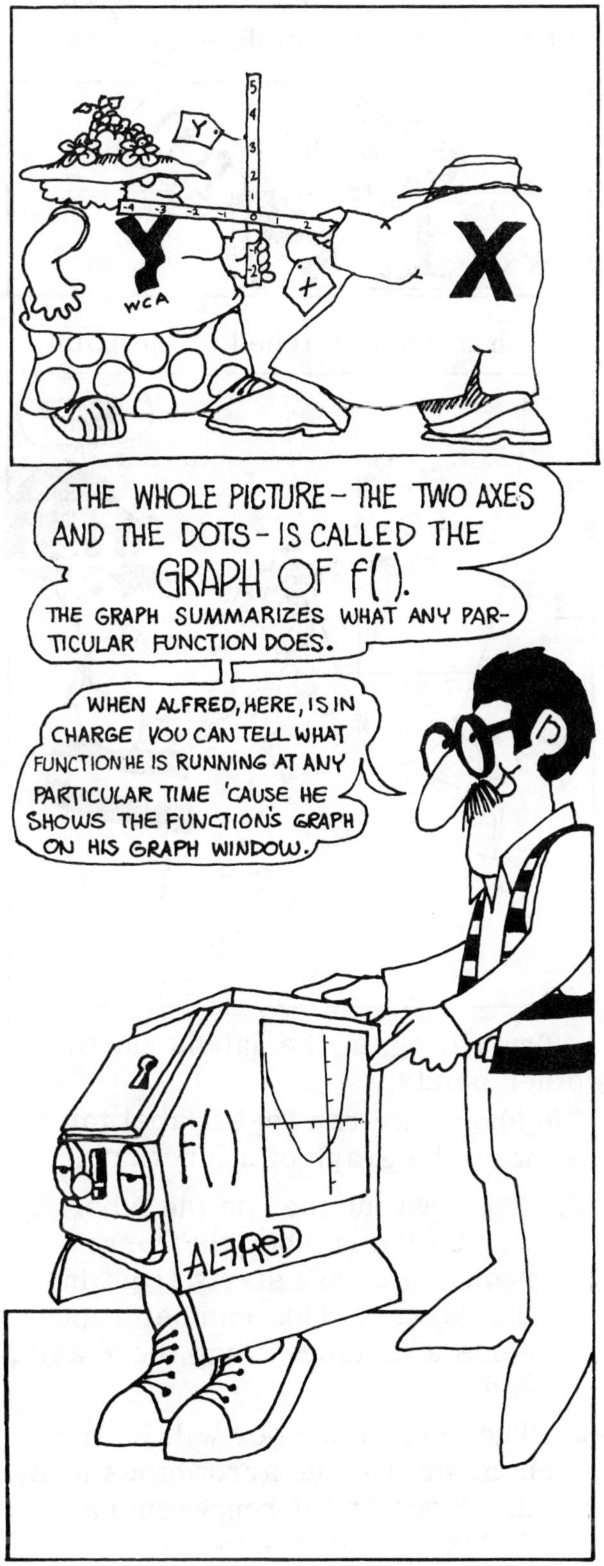
Y
X
WCA
THE WHOLE PICTURE – THE TWO AXES AND THE DOTS – IS CALLED THE GRAPH OF f( ).
THE GRAPH SUMMARIZES WHAT ANY PARTICULAR FUNCTION DOES.
WHEN ALFRED, HERE, IS IN CHARGE YOU CAN TELL WHAT FUNCTION HE IS RUNNING AT ANY PARTICULAR TIME 'CAUSE HE SHOWS THE FUNCTION'S GRAPH ON HIS GRAPH WINDOW.
f( )
ALFRED

So, if we are given the graph of some other function g( ) (for variety), then, in order to "run g( )" we start with anything in the domain of g( ), say -2, and go up (or down) to the appropriate dot on the curve and then over to the y-axis to find out what number is g(-2) in the range of g( ).

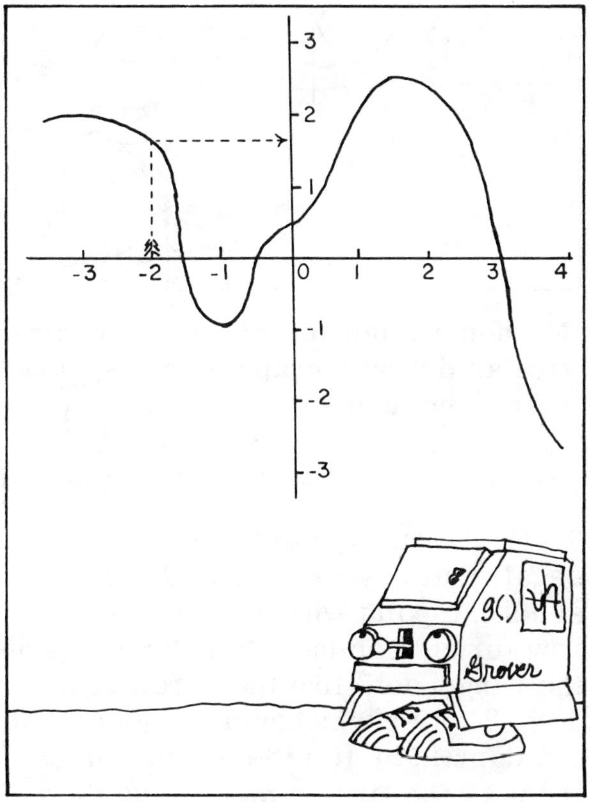
3
2
1
-1
-2
-3
-3
-2
-1
0
1
2
3
4
g( )
Grover

It turns out that

g(-2) is about $\frac{5}{3}$ here.

Try a few?

g(-1) = ?	g(0) = ?
g(½) = ?	g(3) = ?
g(?) = -2	g(?) = 0
g(?) = 2	g(??) = 2
g(???) = 2	

Now, back to the graph of f( ) that McSquared constructed:

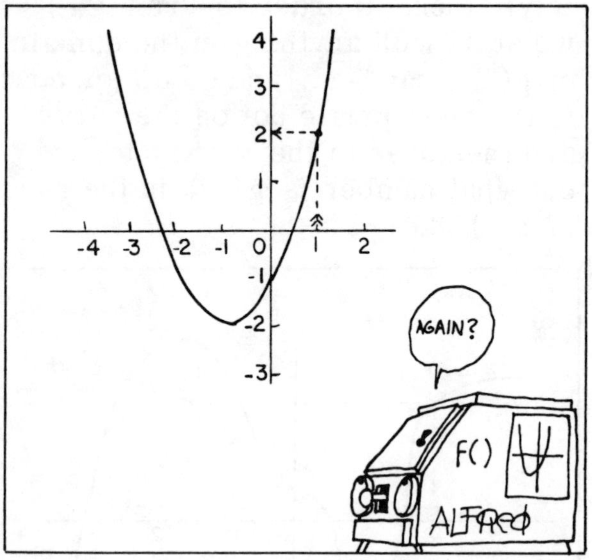

Mathematicians often label any particular dot on a graph with a special pair of numbers

To figure out how to fill in the label, recall that any particular dot represents the point where a function-arrow turns, coming from something in the range-set, like the arrow from 1 to 2 . Use whichever number from the domain of f( ) the arrow comes from as the first number, and then run this number through the function-machine

to find out the number in the range of f( ) on the y—axis that the arrow goes to, and put in this second number to complete the label,

which is then attached to the point.

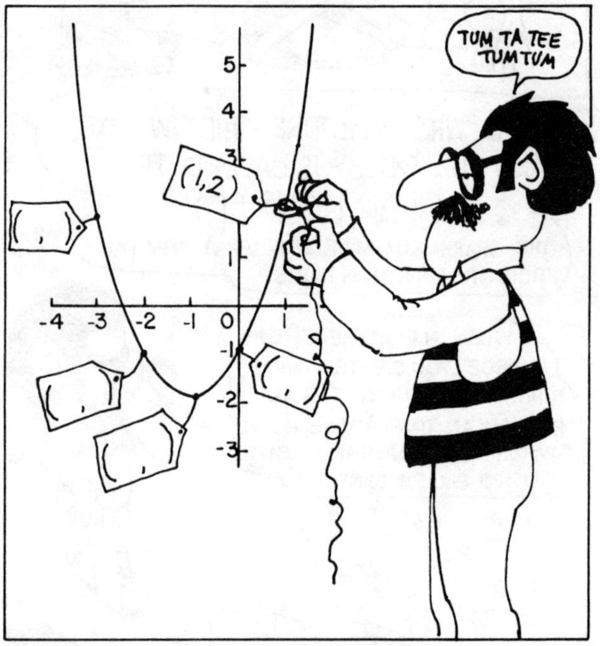

Try filling in the labels on the other points.

In short, we can find a label for any point on the graph of a function:

1. The first number on the label will be the number the corresponding arrow starts from in the domain. This number represents a distance along the x-axis, and
2. The second number will be the number that the arrow goes to in the range and it represents a distance along the y-axis.

Of course, even if a point on the x-axis, y-axis grid is not on the graph of our function, it can still be labeled.

Every point on the grid must sit directly above (or below) some number "a" on the x-axis and directly across from some number "b" on the y-axis.

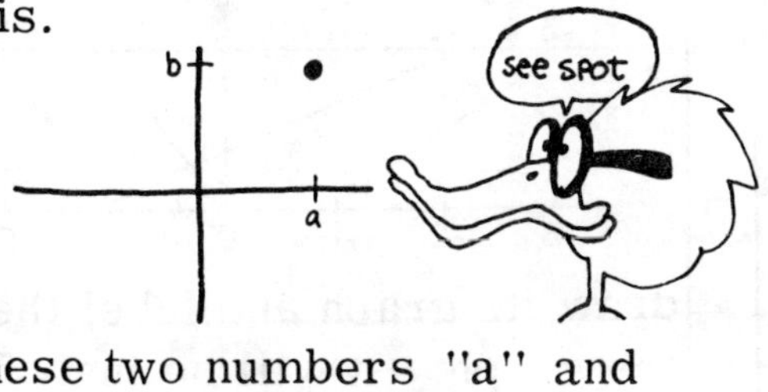

With these two numbers "a" and "b" we can make up a label

that is the official name of the point.

> "a" is called the "x - coordinate of the point" and
>
> "b" is called the "y - coordinate of the point."

A few sample points are labeled -

Have a try at labeling some more points.

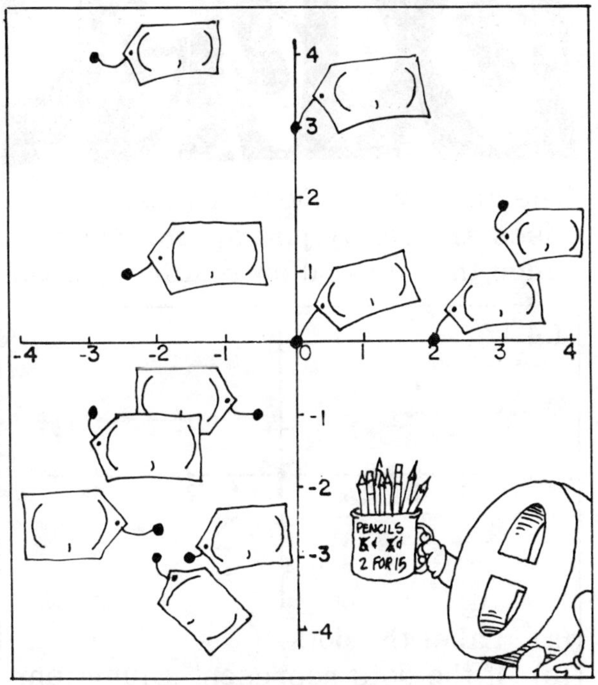

Actually, label tags are not Official Symbols—instead, mathematicians just put the number pairs as close to the point as they can, like this:

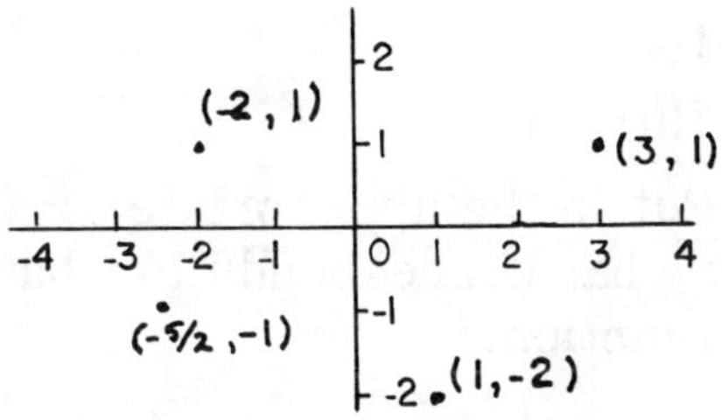

A symbol on the label, like (-2, 1), looks exactly like the notation for a chunk of the number line, like

$$(-2, 1) = \{x \mid x \in \mathbb{R};\ -2 < x < 1\}$$

but the label has nothing to do with such chunks. You have to figure out which way the symbol is being used from the context. It is usually pretty easy.

EXERCISES

The answers to most of these exercises are really just approximate, since they depend on reading graphs.

I.4.1

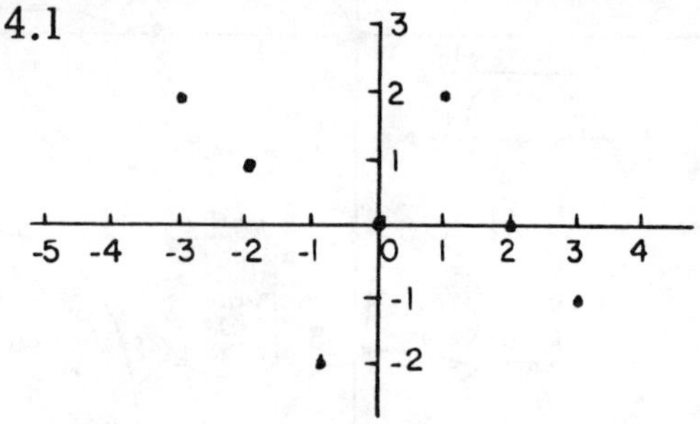

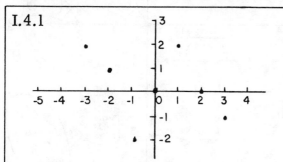

(a) Label the dots.

(b) If the dots represent a function, what is

f(3) = ?

f(-3) = ?

f(1) = ?

f(2) = ?

f(0) = ?

(c) Put in the appropriate arrows to show what corresponds to what in this function.

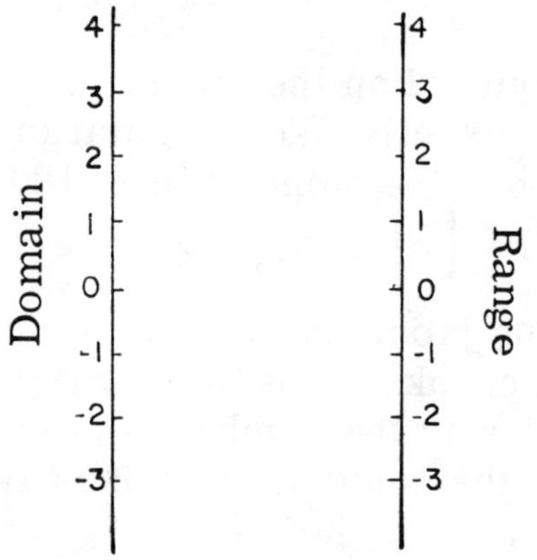

I.4.2 If f( ) is this correspondence;

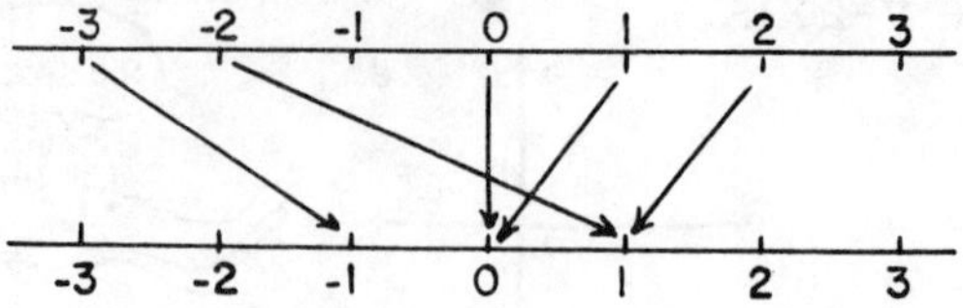

draw its graph and label the dots on the graph. (It will be just 5 dots.)

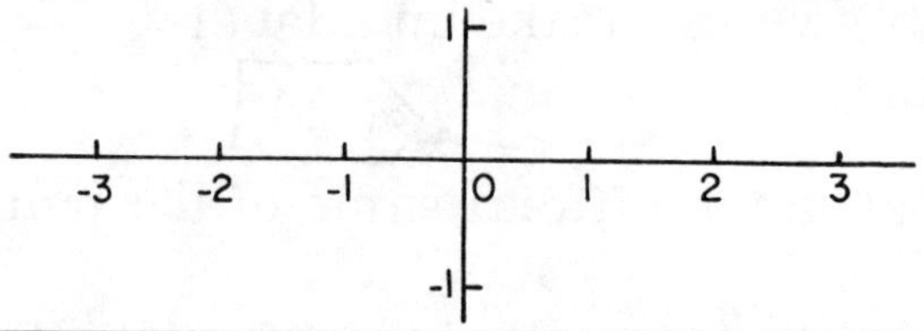

I.4.3 If the domain of f( ) is

$\{-1, 0, \frac{1}{2}, 2, 3\}$

and $f(-1) = -1$, $f(0) = 0$, $f(\frac{1}{2}) = 1$, $f(2) = -2$ and $f(3) = -2$,

(a) Put in the appropriate arrows to show what corresponds to what in this function.

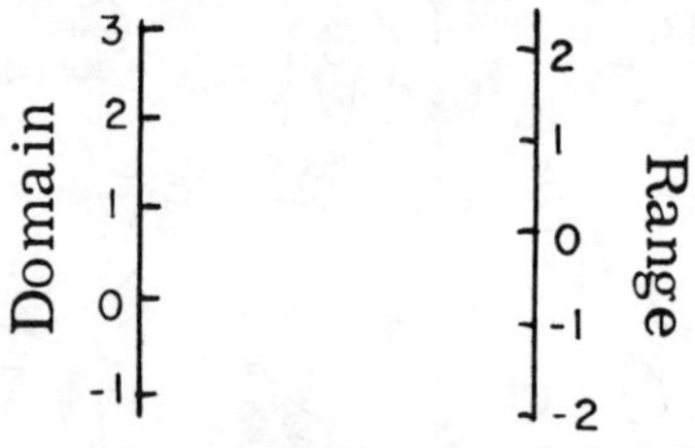

(b) Draw the graph of f( ). (It will just be five dots.)

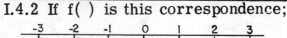

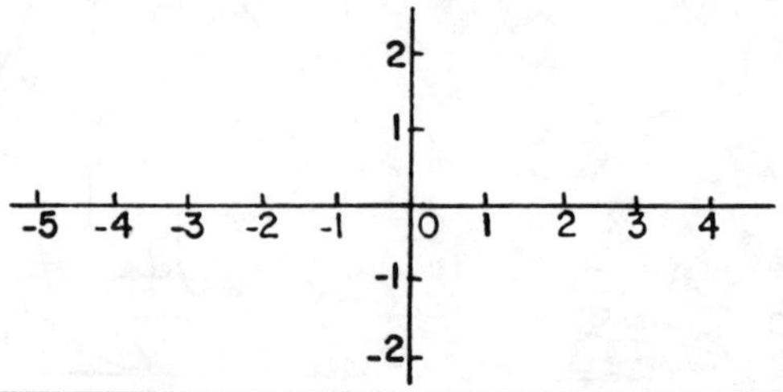

I.4.4 If this is the graph of f( ) ,

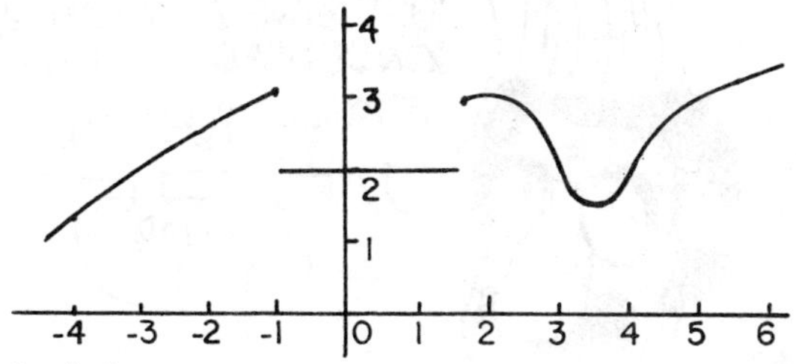

what is

(a) f(2) = ?

(b) f(-2) = ?

(c) If x is any number in

$(0, 1) = \{x \mid x \in \mathbb{R}\ ;\ 0 < x < 1\}$,

what is f(x) = ?

(d) f(?) = 3 .

(e) f(?) = $5/2$.

(f) f(?) = 2 .

The bags (sets) and arrows and function - machines give a fair amount of logical equipment to talk about relationships of the sort that can be put on a graph.

But, so far, we only have a function's graph to tell what output corresponds to any particular input. Graphs just do not give precise enough results and computers can't compute with them.

Fortunately, for many useful functions, a precise recipe using ALGEBRA can be given for finding the exact output that corresponds to any particular input. For example, suppose Alfred is running a function with the following graph:

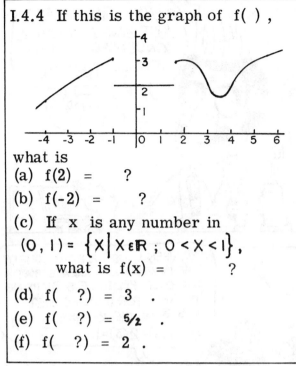

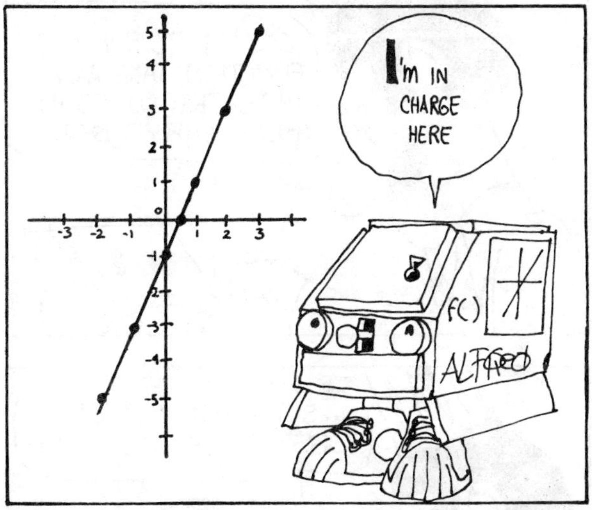

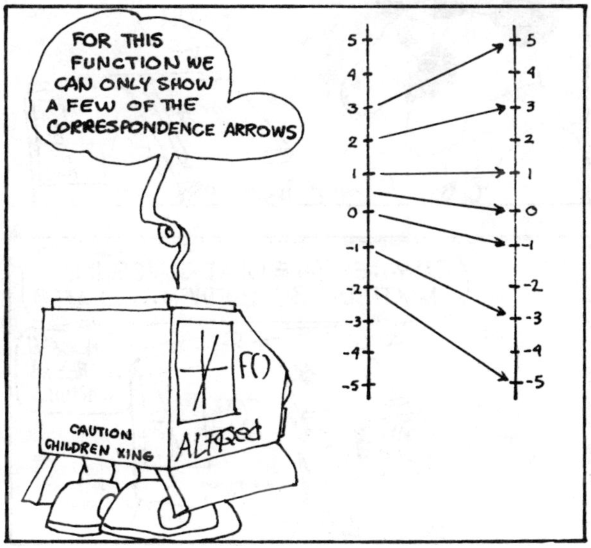

NOW, HERE IS A NEAT RECIPE FOR FINDING OUT WHAT CORRESPONDS TO WHAT IN THIS FUNCTION.
OK, ALF-TO RUN THIS FUNCTION, TAKE ANY THING THAT COMES IN AND MULTIPLY IT BY 2...
F()
... SUBTRACT 1, THEN TOSS OUT WHAT YOU GET- GOT IT?
YUP
F()
EASY... ER, TAKE WHAT COMES IN, MULTIPLY BY 2, THEN... WHAAT?
PHOOEY! WHERE'S MY FORMULA?
F()
ALFRED
OH, YAH... THE RECIPE I JUST GAVE ALFY CAN BE CONDENSED INTO THE FORMULA f()=2·()-1
GOOD LAD, ALF!
NOW, WHEN I TOSS ALF THE NUMBER 3, HE IMMEDIATELY KNOWS THAT HE SHOULD PRODUCE 5 BY USING THIS NEAT FORMULA. HE JUST PUTS 3 IN THE BLANKS: f(3)=2(3)-1.
3
THEN HE TREATS 2·(3)-1 USING ARITHMETIC: f(3)=2·(3)-1=6-1=5. THIS IS EXACTLY WHAT THE RECIPE SAYS TO DO WITH ANY NUMBER-IN THIS CASE "3". IT SAYS "MULTIPLY THE INPUT '3' BY 2 AND SUBTRACT 1." 5 IS THE CORRECT OUTPUT-CHECK THE GRAPH TO SEE!
WELL DONE!
I ACTUALLY PREFER MEDIUM RARE
WHADDA I DO WITH THE 5?
F(

Check a few more: using the graph of the function, we can figure out that

$$3 = f(2)$$
$$1 = f(1)$$
$$-3 = f(-1)$$
$$0 = f(\tfrac{1}{2}) \; .$$

If we use our formula, instead of the graph, we find that:

$$f(2) = 2 \cdot (2) - 1 = 4 - 1 = 3$$
$$f(1) = 2 \cdot (1) - 1 = 2 - 1 = 1$$
$$f(-1) = 2 \cdot (-1) - 1 = -2 - 1 = -3$$
$$f(\tfrac{1}{2}) = 2 \cdot (\tfrac{1}{2}) - 1 = 1 - 1 = 0$$

and this agrees with the results from the graph right down the line!

Now this is exceptionally neat! We will see in a minute that any function whose graph turns out to be a straight line can be described by a handy formula like this one for

$$f(\;) = 2 \cdot (\;) - 1 \; .$$

When we can find such a convenient formula as this for a function, we often use it instead of "f( )" as the symbol for the function. Such formula-symbols are often written with "x" in the blanks, like

$$f(x) = 2 \cdot (x) - 1 = 2x - 1$$

or even like

$$y = f(x) = 2x - 1$$

since such formulas give the function's rule for finding the number "y" on the "y-axis" corresponding to any particular value of "x" on the "x-axis."

UNFORTUNATELY, NOT ALL FUNCTIONS HAVE NICE FORMULAS LIKE f()=2()-1. TRIGONOMETRY GIVES AN EXAMPLE OF THIS UNFORTUNATE FACT— FUNCTION "SIN()" OR, CONVENTIONALLY, "SIN (x)" DOESN'T HAVE A SIMPLE FORMULA—WE HAVE TO LOOK UP VALUES OF SIN() IN TABLES.
MY AZIMUTH IS BOTHERING ME
GOOD SHIP LOLLIPOP

Fortunately, however, we will be using mostly functions whose symbols give an exact formula for how to compute the right output for any specific input, like

$f(\;) = 2(\;) - 1$ (same as $y = f(x) = 2x - 1$).

Here are some simple but important members of the family of functions that all turn out to have straight lines as their graphs.

1. The "CONSTANT" FUNCTION: (constant "1" in this case)

Formula-symbol:

$f(\;) = 1$ (same as $y = f(x) = 1$).

Some of the correspondence-arrows:

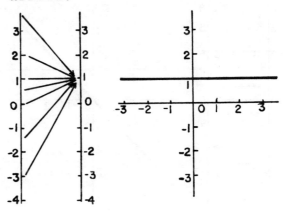

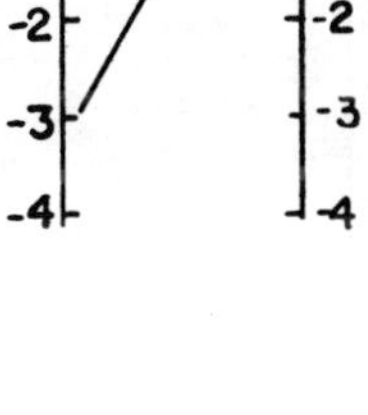
3
2
1
0
-1
-2
-3
-4
3
2
1
0
-1
-2
-3
-4

Graph:

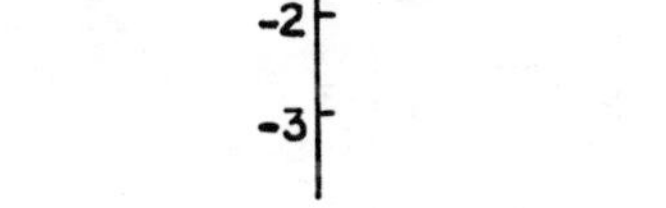
3
2
1
-3 -2 -1 0 1 2 3
-1
-2
-3

Using the formula a bit:

$f(2) = 1$; $f(1) = 1$; $f(\frac{3}{5}) = 1$;

$f(0) = 1$; $f(-3) = 1$

since the output of the function-machine is ALWAYS ONE.

2. The "IDENTITY" FUNCTION.

Formula-symbol:

$f(\ ) = (\ )$ (same as $y = f(x) = x$).

Some of the correspondence-arrows:

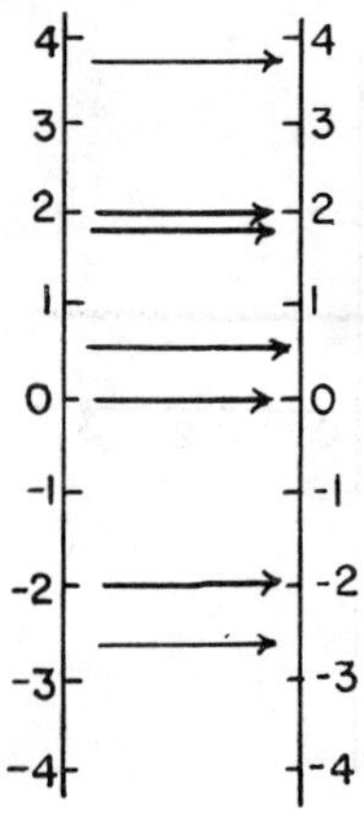

Graph:

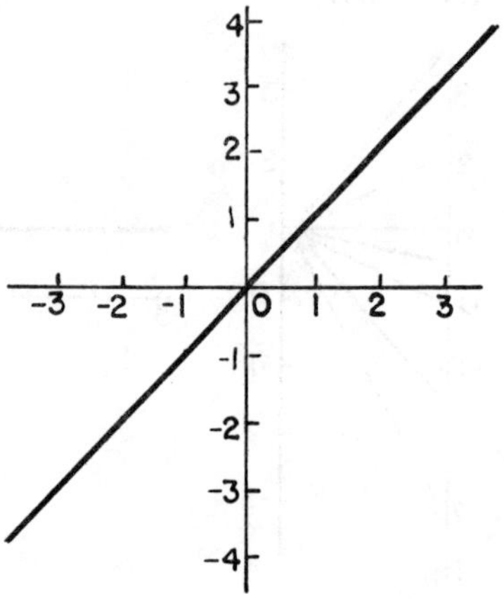

Using the formula a bit:

$f(2) = (2) = 2$; $f(-52) = (-52) = -52$;

$f(0) = (0) = 0$; $f(\frac{9}{5}) = (\frac{9}{5}) = \frac{9}{5}$,

since the function puts out exactly the same thing that comes in.

3. The "MULTIPLYING-BY-A-CONSTANT" FUNCTION (½ in this case)

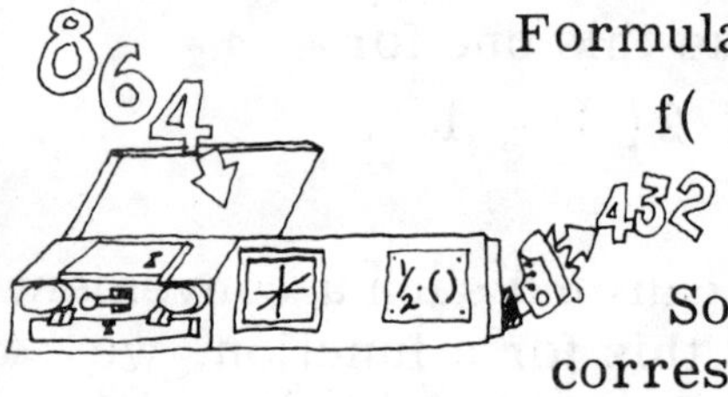

Formula - symbol:

$f(\ ) = \frac{1}{2}\cdot(\ )$

($y = \frac{1}{2}x$).

Some of the correspondence-arrows:

Graph:

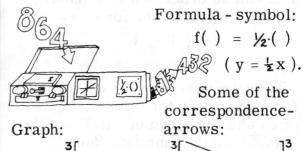

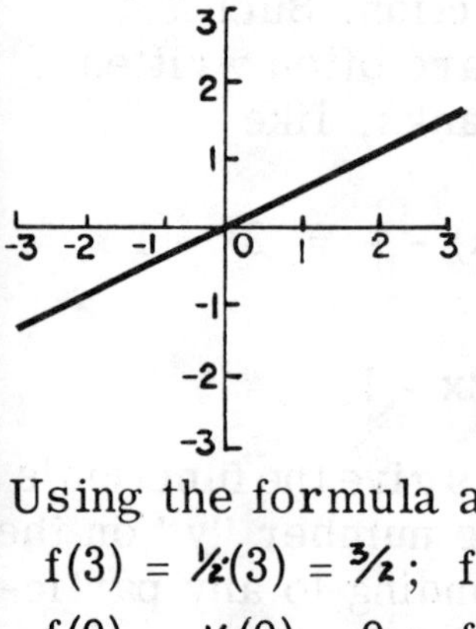

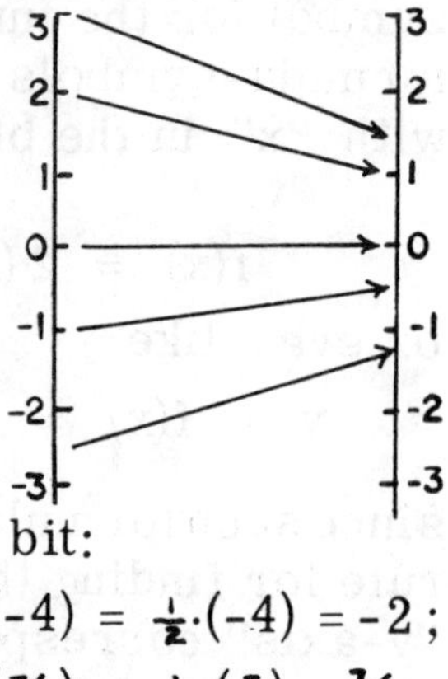

Using the formula a bit:

$f(3) = \frac{1}{2}(3) = \frac{3}{2}$; $f(-4) = \frac{1}{2}\cdot(-4) = -2$;

$f(0) = \frac{1}{2}(0) = 0$; $f(\frac{7}{3}) = \frac{1}{2}\cdot(\frac{7}{3}) = \frac{7}{6}$.

EXERCISES

I.4.5

Formula - symbol:

f( ) = 2·( ) .

Show some of the correspondence-arrows?

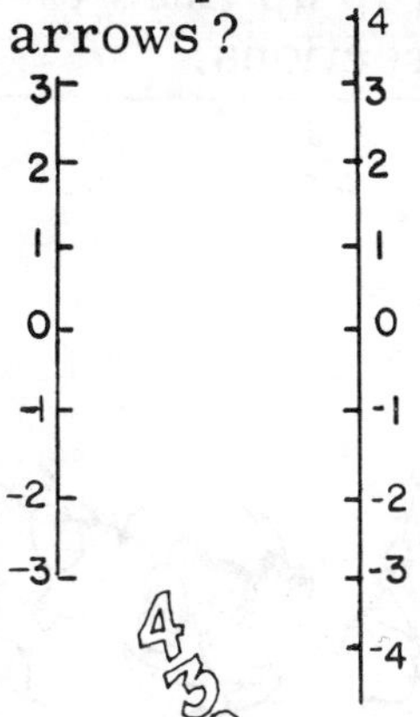

Graph?

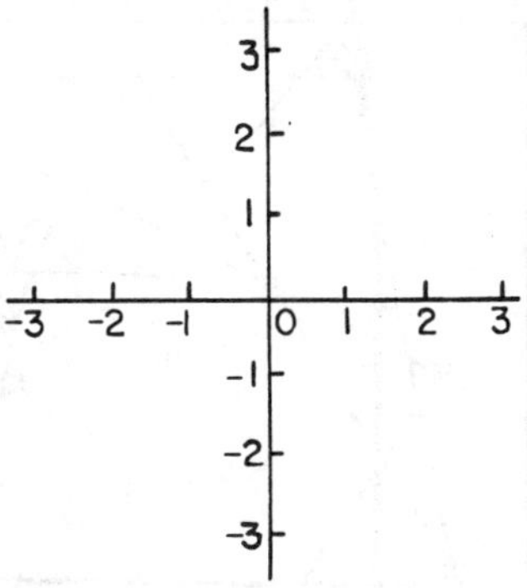

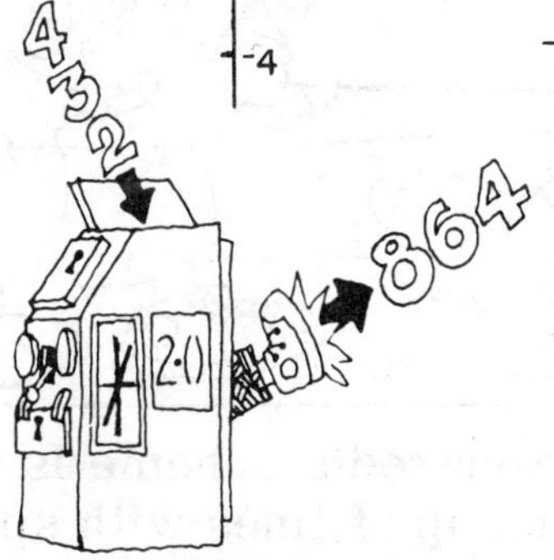

I.4.6

Assuming that g( ) has a straight line as its graph, here are some of the correspondence-arrows of g( ).

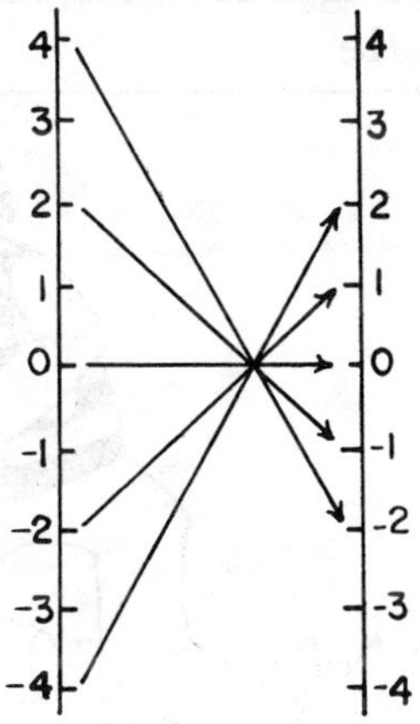

I.4.6 Continued:

Graph?

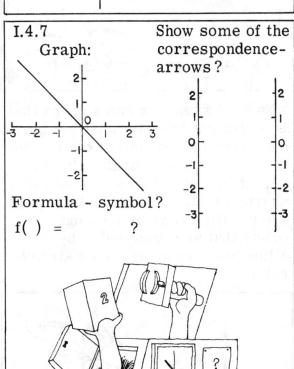

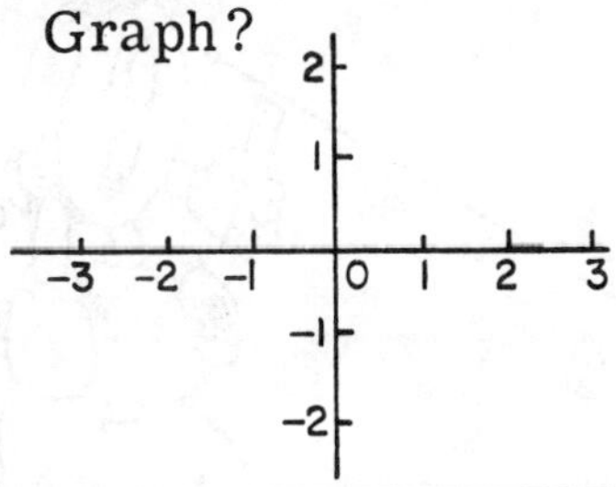

Formula - symbol?

g( ) = ?

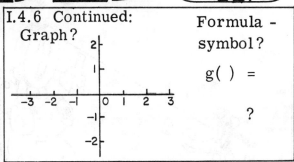

I.4.7

Graph:

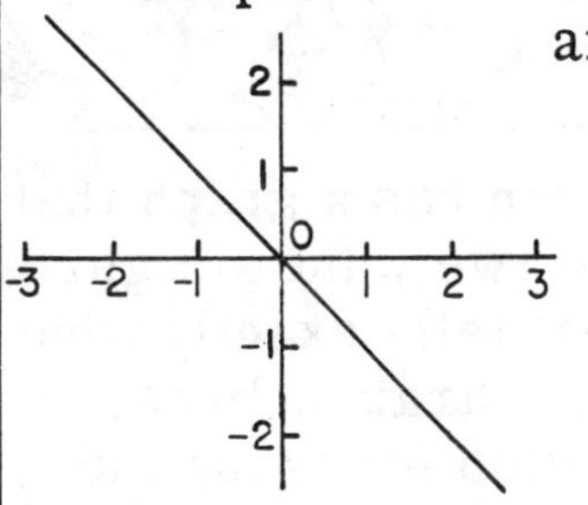

Formula - symbol?

f( ) = ?

Show some of the correspondence-arrows?

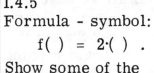

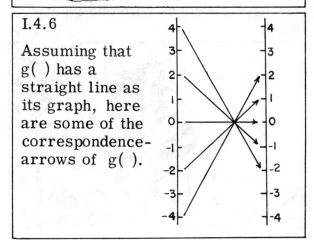

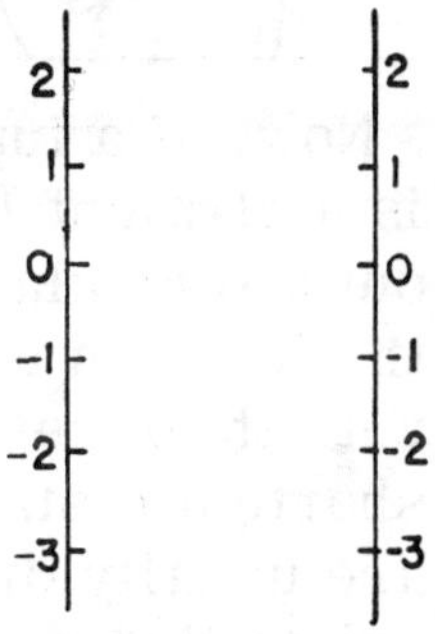

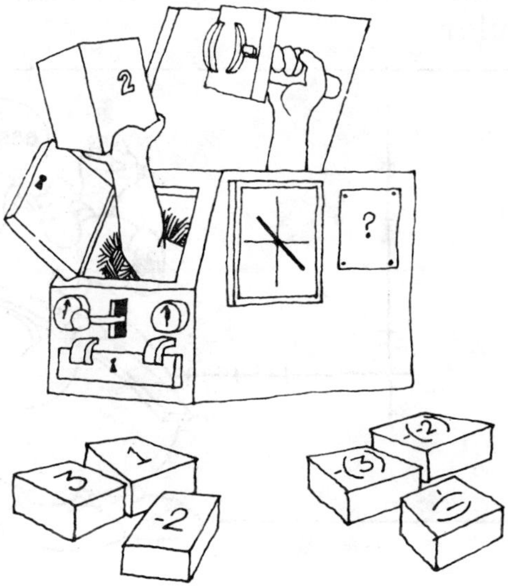

I·5 STRAIGHT LINES

Now, if a function has a graph that is a straight line, we want to figure out a formula that tells exactly what the function does. Straight lines, with their reputation for being the shortest distance between two points, are usually drawn by locating two points that are supposed to be on a line and then using a nice straight ruler.

But to find the formula for a function that has a particular line as its graph, it is handy to back up a bit and look at the job of constructing a line assuming that all we know at first is that one particular point is on the line - say (1, 3). If all we have is one point on the line, what is to be the direction of the line? Should it be steep or shallow? Should it point to the right or the left? Now there are lots of ways to prop up lines to give them definite directions.

Prof. McSquared's scheme is to do this propping up of lines with special triangular wedges with one right-angle.

In pursuit of a particular line, Mc-Squared picks the wedge

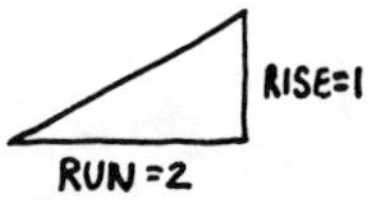

with a bottom edge that "runs" along length 2 , parallel to the x - axis , and a vertical side that "rises" length 1, parallel to the y - axis. He then pounds the wedge in from the right until the sharp corner (vertex) of the wedge sits on the point (1, 3).

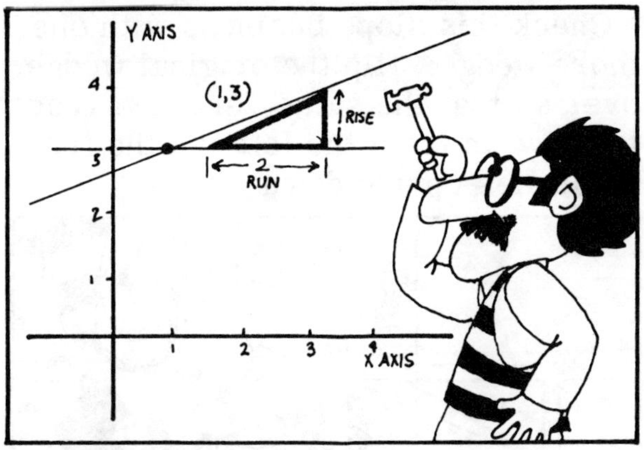

Now McSquared has got just the line he wants: the one lying on the slanted side of the triangular wedge.

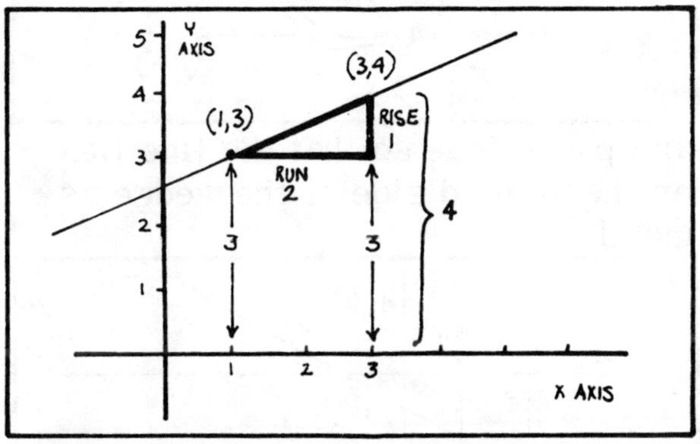

Once the wedge is in position and the line propped up in the desired direction, things are a bit more secure. For example, we can locate another point on the line by figuring out where the top vertex

of the triangle will sit. It is located directly above the x-axis at a point we can compute like this:

$$\left(\begin{array}{c}\text{X-COORDINATE OF}\\ \text{THE ORIGINAL POINT}\end{array}\right) + (\text{LENGTH OF RUN}) =$$
$$1 \quad + \quad 2 \quad = 3$$

and at height we can compute like this:

$$\left(\begin{array}{c}\text{y-COORDINATE OF}\\ \text{THE ORIGINAL POINT}\end{array}\right) + (\text{LENGTH OF RISE}) =$$
$$3 \quad + \quad 1 \quad = 4$$

So the point labeled (3, 4) is one more point on the line.

<u>The line will take the same direction</u> if Prof. McSquared takes a bigger triangular wedge in the form of a particular proportional(or similar) triangle with sides twice the length of the sides of the original triangle, i. e. ,

(new rise) = 2·(old rise) = 2·1 = 2

and (new run) = 2·(old run) = 2·2 = 4,

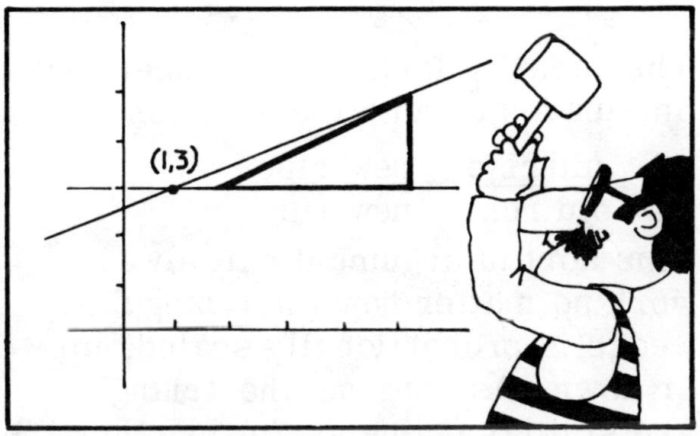

and pounds it into position.

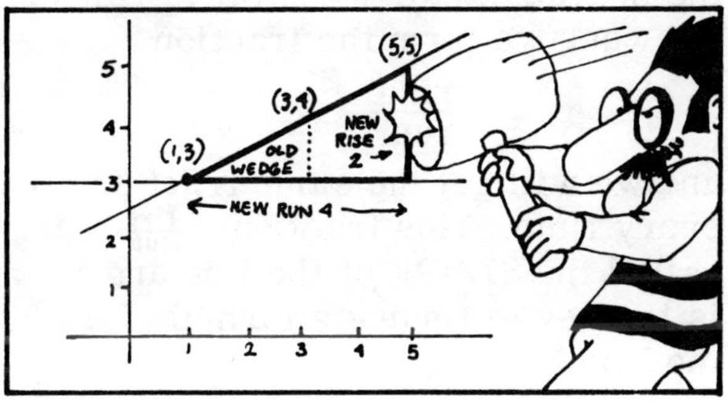

The reason that the line will still have the same direction is that the old wedge fits exactly in the sharp corner of the new bigger wedge - the slanted side of the old triangle will lie on the slanted side of the new triangle.

Mathematicians hunted around for some way to get hold of the notion of "direction" for a specific line that would be independent of which one of these similar wedges is used to prop up the line. Some way to do this had to be found that takes advantage of the fact that the sides of the new wedge are exactly twice the length of the sides of the old wedge and then gets rid of the factor 2 that will go along with "twice". This happens very neatly if we form the fraction $\frac{\text{rise}}{\text{run}}$ and compute like this:

$$\frac{2}{4} = \frac{(\text{NEW RISE})}{(\text{NEW RUN})} = \frac{\cancel{2}\cdot(\text{OLD RISE})}{\cancel{2}\,(\text{OLD RUN})} = \frac{(\text{OLD RISE})}{(\text{OLD RUN})} = \frac{1}{2} \; .$$

The "scaling factor" 2 cancels out and leaves us with the equation

$$\frac{1}{2} = \frac{\text{old rise}}{\text{old run}} = \frac{\text{new rise}}{\text{new run}} \; .$$

This sort of argument will always work no matter how the triangular wedge is proportionally scaled, up or down. As long as the triangles are similar (same as "proportional"), the scaling factor will always cancel out when we form the fraction

$$\frac{\text{rise}}{\text{run}}$$

and we will get the same fraction every time. This fraction $\frac{\text{rise}}{\text{run}}$ is called the SLOPE of the line and it is the key to finding a formula for the line.

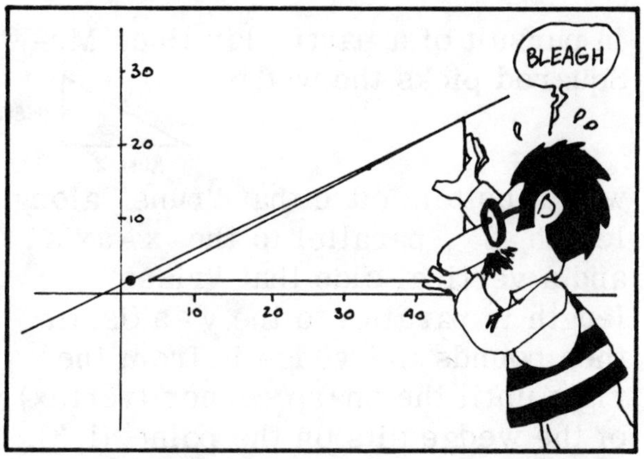

Check this slope business with one more wedge: flip the original wedge over so that the same vertex is ready to drive in from the left and the triangle lies upside down

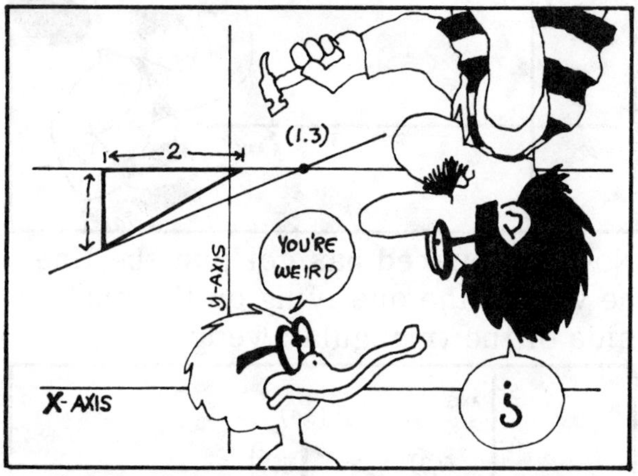

and pound it in so that the line lies on the slanted side of the wedge as usual.

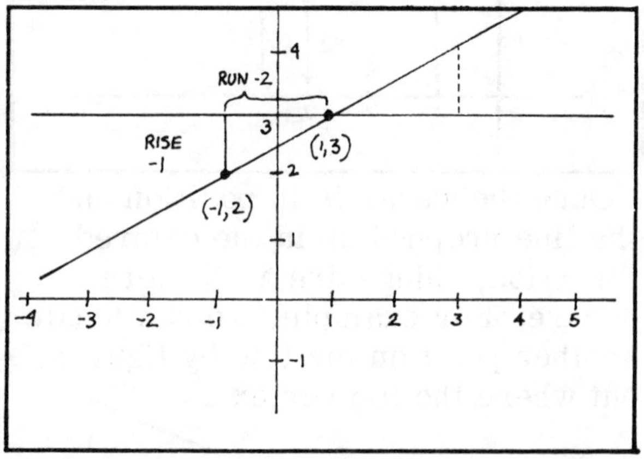

The "run" is considered negative in this new flipped triangle, since the triangle runs backward (to the left) from point (1, 3).

Furthermore, since the triangle doesn't "rise" either — it falls — the "rise" is considered negative as well.

So, in this case, the

"new run" $= (-1)\cdot(\text{ORIGINAL RUN}) = (-1)\cdot(2) = -2$

and the

"new rise" $= (-1)\cdot(\text{ORIGINAL RISE}) = (-1)\cdot(1) = -1$.

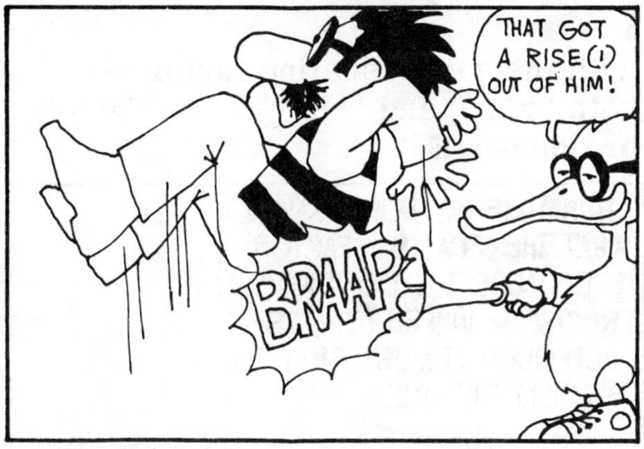

The line will again have the same direction, since the "flipped triangle" is proportional to the others. The slope concept still holds together since mathematicians have agreed to count the sides of the new flipped triangle as

(-1)•(sides of the original triangle)

and we can compute away to get

$$\frac{\text{new rise}}{\text{new run}} = \frac{-1}{-2} = \frac{(-1)\cdot(1)}{(-1)\cdot(2)} = \frac{(-1)\cdot(\text{original rise})}{(-1)\cdot(\text{original run})}$$

$$= \frac{\text{original rise}}{\text{original run}} = \frac{1}{2}.$$

The factor "-1" cancels out neatly as usual.

Finally, we can nail down one more point on the line — from the graph it is the point (-1, 2), figured out as follows: Its location above the x-axis is computed like this:

$$\left(\begin{matrix}\text{X-COORDINATE OF}\\ \text{THE ORIGINAL POINT}\end{matrix}\right) + \left(\begin{matrix}\text{LENGTH OF THE}\\ \text{NEW "RUN"}\end{matrix}\right) =$$

$$(1) \quad + \quad (-2) \quad = -1$$

and has height computed like this:

$$\left(\begin{matrix}\text{y-COORDINATE OF}\\ \text{THE ORIGINAL POINT}\end{matrix}\right) + \left(\begin{matrix}\text{length OF THE}\\ \text{NEW "RISE"}\end{matrix}\right) =$$

$$(3) \quad + \quad (-1) \quad = 2$$

These computations give the new point (-1, 2).

NOW, we can finesse a more general formula for finding any point on this line. (Recall - the line runs through the original point (1, 3) and is propped up from the right by the original wedge 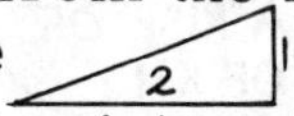).

Take any old point on the line and label it (x, y) for the moment. We want to figure out how x and y are related.

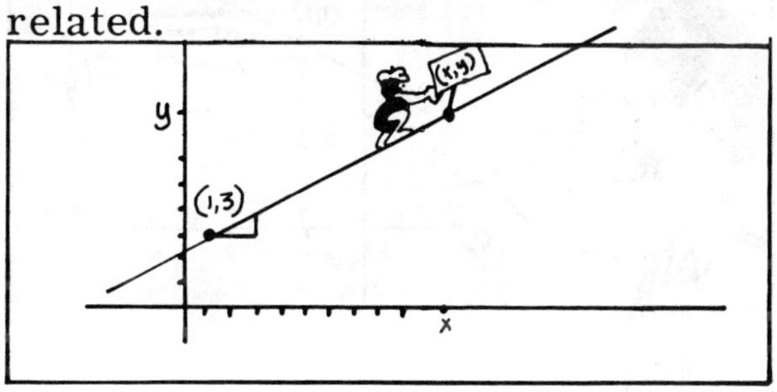

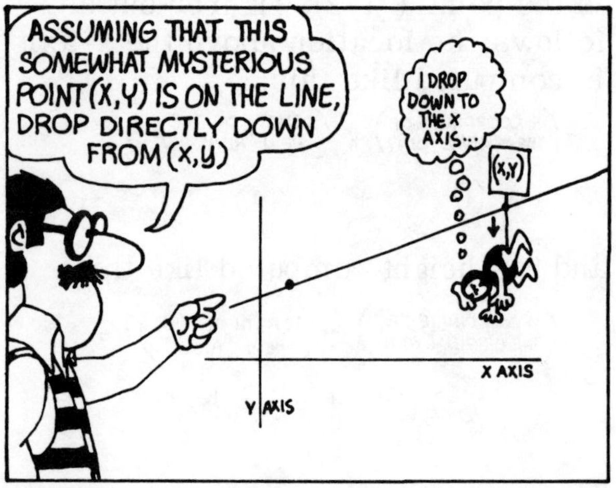

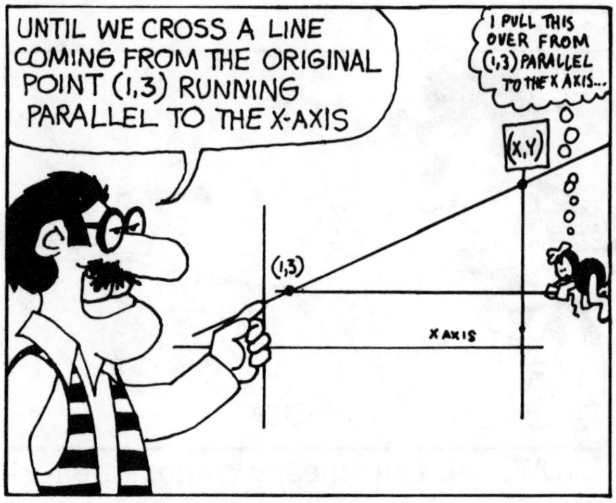

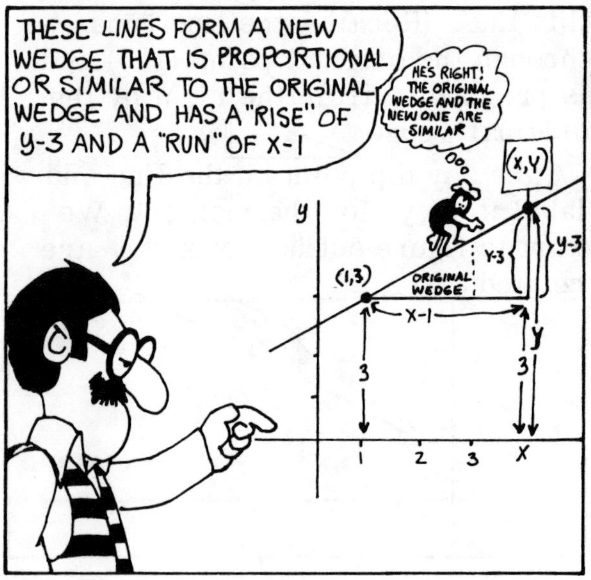

Now being "similar" or "proportional" again means that the lengths of the sides of the new triangle

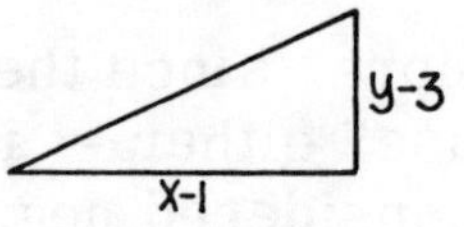

are just multiples of the lengths of the sides of the old triangle

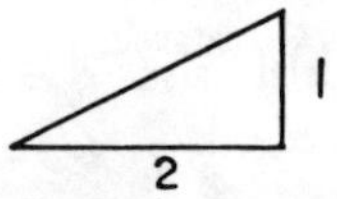

with the <u>same</u> scaling factor for both "rise" and "run", like "twice" or "one-half."

To tidy up a bit: multiply both sides of the equation by $(x - 1)$;

$$\frac{y-3}{\cancel{x-1}} \cdot \cancel{(x-1)} = \tfrac{1}{2} \cdot (x-1)$$

and cancel "$(x - 1)$" (which we can do even though we don't know "$(x - 1)$", as long as it isn't 0). Then we get

$$y - 3 = \tfrac{1}{2}(x - 1)$$

or (add 3) $\quad y = 3 + y - 3 = 3 + \tfrac{1}{2}(x - 1)$

or $\quad y = 3 + \tfrac{1}{2}(x - 1)$.

For any particular value of x , this formula tells what value y should take so that point (x, y) is on the line. Using wedges, we have seen that point $(3, 4)$ is on the line, so we should get 4 when we plug in 3 for x in the formula: $y = 3 + \tfrac{1}{2}(3 - 1) = 3 + \tfrac{1}{2}(2) = 4$; it checks!

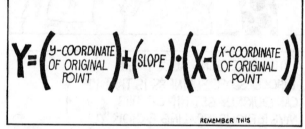

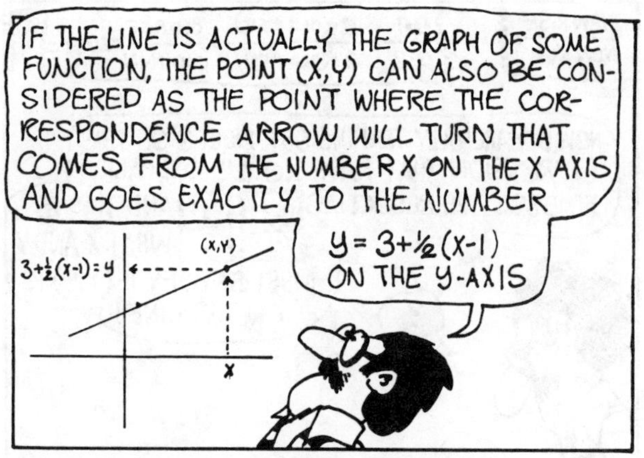

We haven't really checked this out completely - all the arguments worked only for points (x, y) on the line to the right of point (1, 3). If we were to work out a formula for points on the line to the left of (1, 3), we would get the same formula, but it is a hassle to do so and we skip the derivation and just declare it to be true "in a similar manner", as mathematicians say to get out of the hassle.

Since, in the formula for the line

$y=3+\frac{1}{2}(x-1)$: SLOPE $=\frac{1}{2}$,

(X COORDINATE OF THE ORIGINAL POINT) $=1$, AND (Y COORDINATE OF THE ORIGINAL POINT) $=3$,

WE CAN WRITE THE FORMULA AS THE SUPER-FORMULA

$$Y=\left(\text{y-COORDINATE OF ORIGINAL POINT}\right)+\left(\text{SLOPE}\right)\cdot\left(X-\left(\text{X-COORDINATE OF ORIGINAL POINT}\right)\right)$$

REMEMBER THIS

The reasoning we used will always end up with this SUPER-formula no matter what slope or original point we choose! So the super-formula gives us a way to find the formula for a line if we know just one point on the line and we know its slope: just "plug in" the information in the super-formula.

This super-formula is commonly known as

"THE POINT-SLOPE FORMULA for straight lines."

Here are all sorts of exercises about lines. Some of them are partly solved to help things along.

EXERCISES

I.5.1 Draw the line through $(-1, -2)$ with slope 3 and find the formula of the function that has the line as its graph.

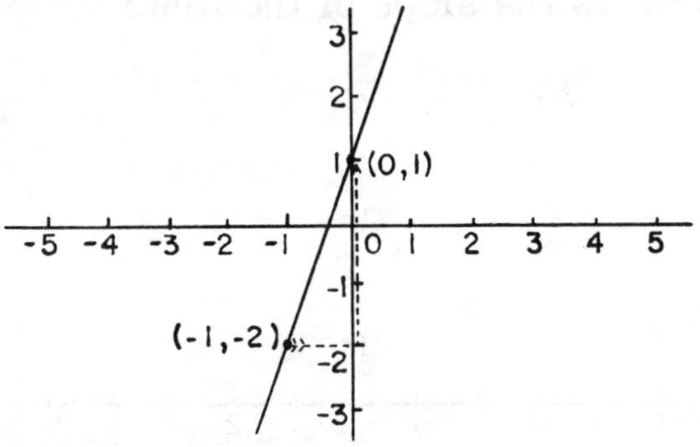

Answer: Finding the formula is easy - just plug in point $(-1, -2)$ and slope 3 in the super-formula:

$$y = (-2) + (3)\cdot(x - (-1))$$

or $y = -2 + 3\cdot(x+1)$.

then just convert this to a function-formula:

$$f(\ ) = -2 + 3\cdot((\ )+1).$$

The only tricky part is drawing the line. We need another point besides $(-1, -2)$. Since the slope is

$$3 = \frac{3}{1},$$

just use triangle: → which is one of the triangles that props up the line. Put the bottom vertex

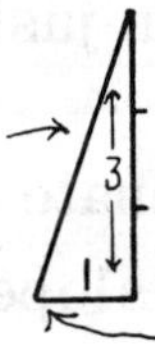

on $(-1, -2)$. Count over 1 from $(-1, -2)$ and then up 3 to find point $(0, 1)$, which will also be on the line. Then draw the line.

I.5.2 Draw the line through point $(2, 0)$ with slope $\frac{4}{5}$ and find the formula of the function that has the line as its graph.

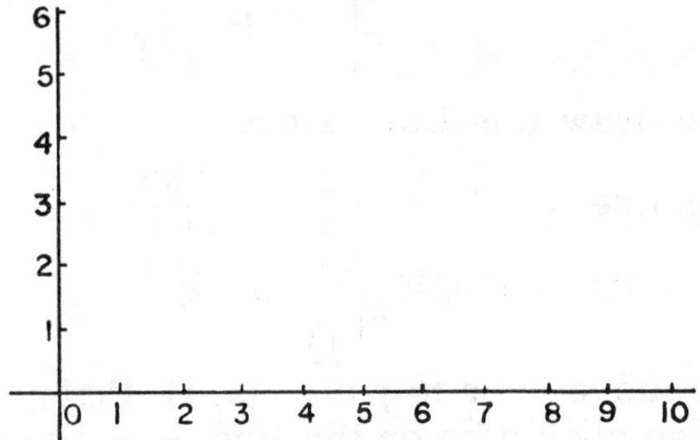

Formula?

I.5.3 Draw the line through point $(-2, -3)$ with slope $1\frac{1}{5}$ and find the formula of the function that has the line as its graph.

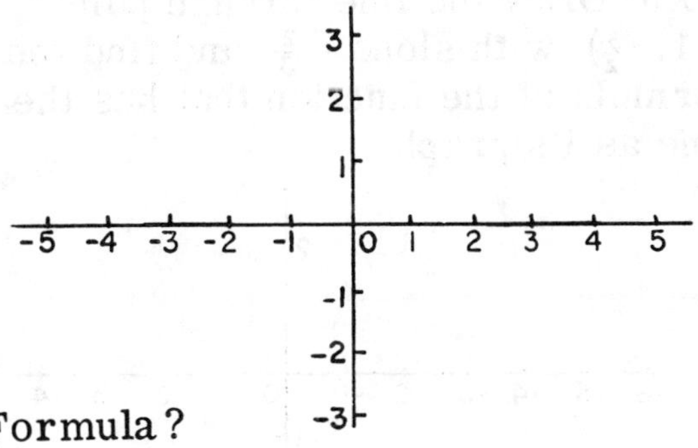

Formula?

I.5.4 Draw the line through point $(1, 3)$ with slope -2 and find the formula of the function that has the line as its graph.

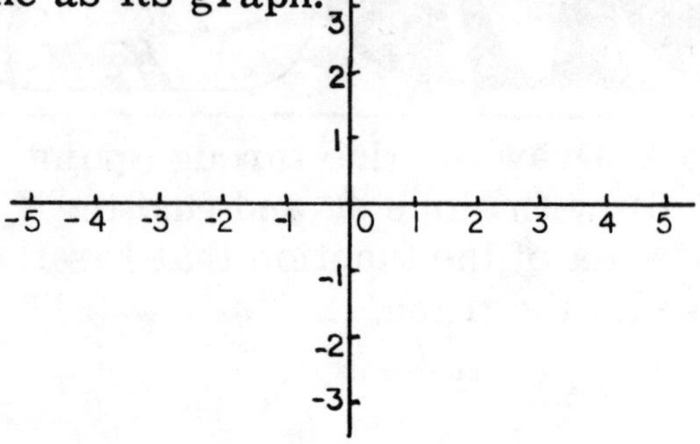

To draw this line, since

$$\text{slope} = -2 = \frac{-2}{1} = \frac{\text{rise}}{\text{run}},$$

just use triangle

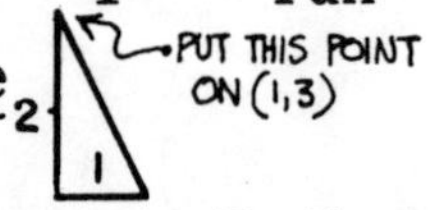
-2
1
PUT THIS POINT ON (1,3)

to move over to point $(2, 1)$ that also should be on the line.

NEGATIVE SLOPES ALWAYS mean that the line SLANTS UP TO THE LEFT!

Formula?

I.5.5 Draw the line through point $(-1, -\frac{1}{2})$ with slope $-\frac{2}{3}$ and find the formula of the function that has the line as its graph.

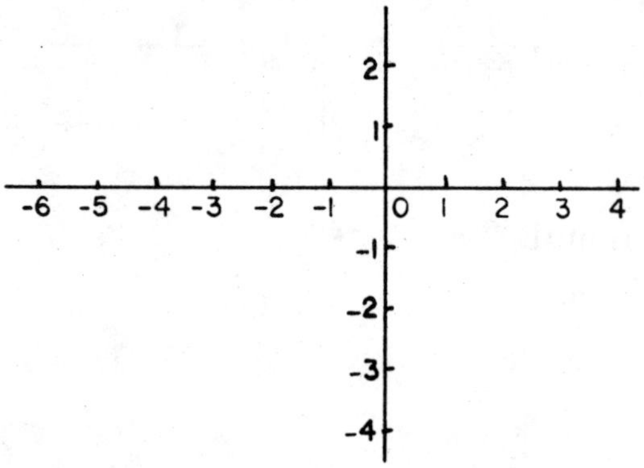

I.5.5 Cont'd. Formula?

I.5.6 Draw the line that has formula

$$y = -\tfrac{1}{3}x + 4 .$$

What is the slope of the line?

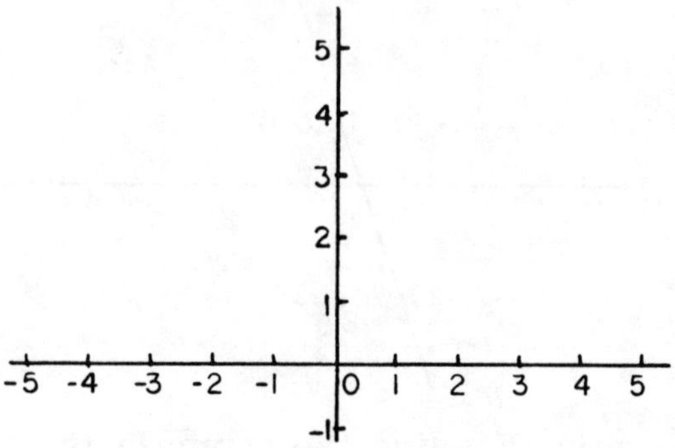

Answer: The slick trick here is to arrange the equation so that it fits the structure of the super-formula

$$y = \begin{pmatrix}\text{y-coord.}\\ \text{of the}\\ \text{point}\end{pmatrix} + (\text{slope})\cdot\left(x - \begin{pmatrix}\text{x-coord.}\\ \text{of the}\\ \text{point}\end{pmatrix}\right)$$

Now x appears only here, so try rearranging the original equation so that it looks like the super-formula:

$$y = 4 + (-\tfrac{1}{3})\cdot(x - 0)$$

and now we can just read off the information:

4 = (y-coordinate of the point) ,

$-\frac{1}{3} = (-\frac{1}{3})$ = slope , and

0 = (x-coordinate of the point).

From this we have both the point $(0, 4)$ and the slope $-\frac{1}{3}$ and this is enough information to draw the line.

Any function that looks like

$f(\ ) = m\cdot(\ ) + b \quad (y = mx + b\ )$

will have a straight line as its graph, where m and b are constant numbers. By pushing

$y = mx + b$

around to look like

$y = b + m\cdot(x - 0)$

and checking the structure of the super-formula we can figure out that

slope = m = the number that multiplies x

and the point (0, b) is on the line. Since this point (0, b) is also on the y-axis (where the x-coordinate of a point is always zero), (0, b) is usually given the official name "the y-intercept" of the line.

Note that we can always organize any formula for a straight line to look like

$y = mx + b$

by just adding together all constant numbers that don't multiply x to be "b."

I.5.7 Draw the line described by the formula

$y = -\frac{1}{2}x + 2.$

What is its slope?

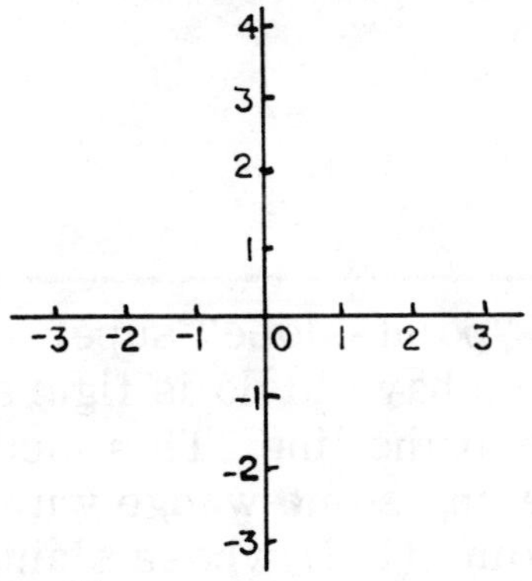

Slope?

I.5.7 Cont'd.

Once you figure out that a formula will have a straight line as a graph, it is a lot easier to draw the line by just finding a couple of points on the line instead of fooling around with point-slope formulas. The usual way to do this is to choose particular values for x to use in the formula so that the formula is easy to compute and the point is easy to put on the graph.

Here, with the formula

$y = -\frac{1}{2}x + 2$

(the same as $f(\ ) = -\frac{1}{2}\cdot(\ ) + 2$)

x = 0 is certainly easy to handle, since $f(0) = -\frac{1}{2}\cdot(0) + 2 = 2$, and we end up with one point on the line: $(0, f(0)) = (0, 2)$.

Another "easy" value for x is 2 (look ahead to some cancelling. . .)

$f(2) = -\frac{1}{2}\cdot(2) + 2 = -1 + 2 = 1$

and we get another point

$(2, f(2)) = (2, 1)$

on the line, and then drawing it is a snap.

I.5.8 Draw the line described by the formula

$y = \frac{3}{2} - 2x$.

What is its slope?

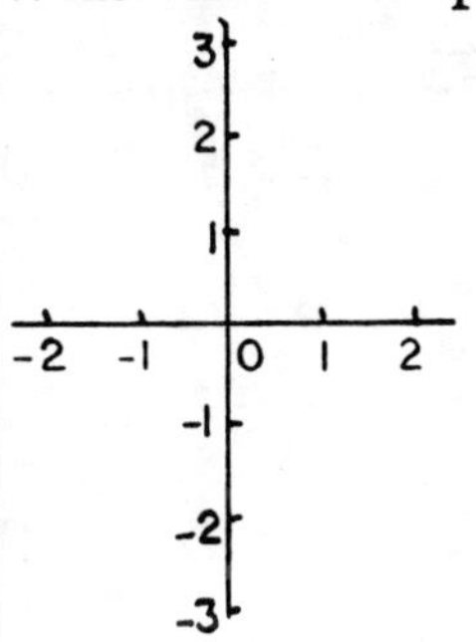

Slope?

I.5.9 (Hard!) Draw the line containing points (x, y) that satisfy the formula

$$2y = 3x - 2 \quad .$$

What is its slope?

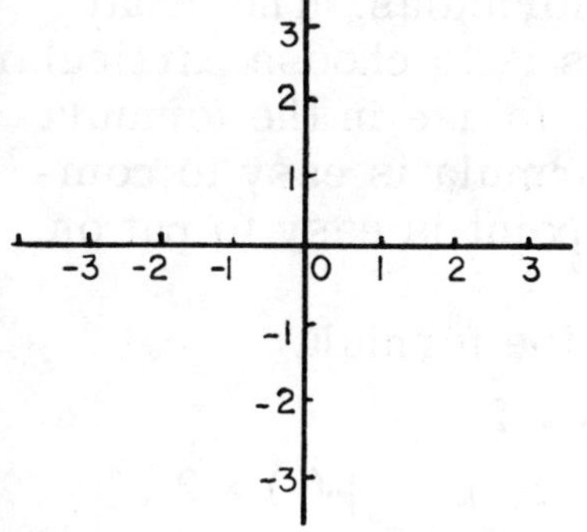

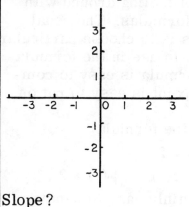

Slope?

I.5.10 (Harder!) Draw the line containing points (x, y) that satisfy the formula

$$2x = 3y - 2$$

What is its slope?

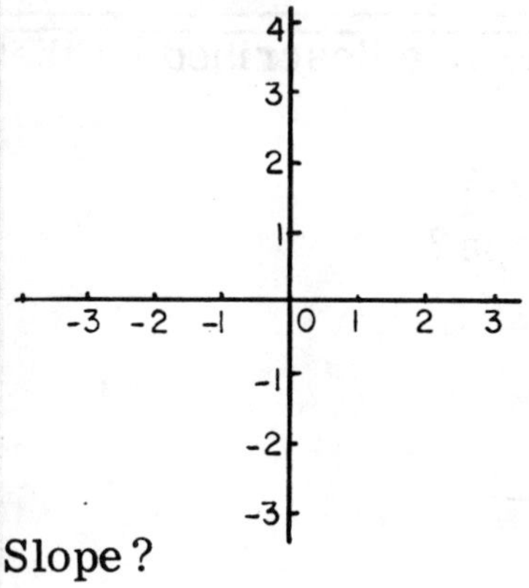

Slope?

NOW back to the problem of figuring out the symbol-formula for a function whose graph is the line through two prescribed points.

Suppose we want to find the formula for the function that gives the line through point (1, 3) and a second point (4, 5) .

To use the "point-slope" super-formula all we have to do is figure out the slope of the line. This means first discovering some wedge with its nose at point (1, 3) whose slanted side props up exactly this line. Then all we need is the rise and the run of this wedge to find the slope.

But this is easy! We just form a triangle by dropping down from (4,5), parallel to the y-axis, until we hit the line that comes from (1,3), parallel to the x-axis.

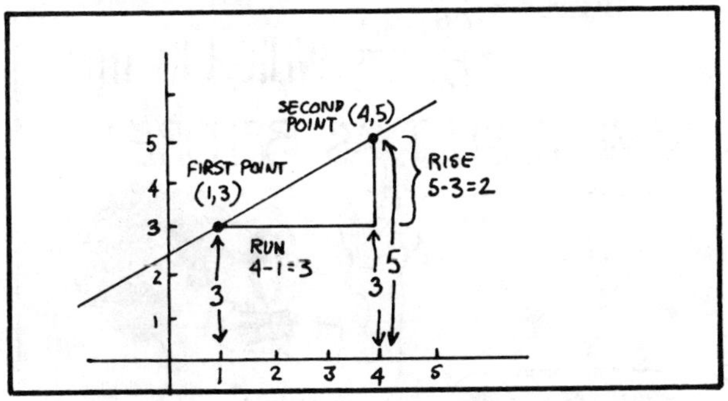

Then the "RISE" of this triangle is computed like this:

$$\left(\begin{array}{l}\text{y-COORDINATE}\\ \text{OF THE SECOND}\\ \text{POINT } (4,5)\end{array}\right) - \left(\begin{array}{l}\text{y-COORDINATE}\\ \text{OF THE FIRST}\\ \text{POINT } (1,3)\end{array}\right) =$$

$$5 \quad - \quad 3 \quad = 2$$

and the "RUN" of the triangle is computed like this:

$$\left(\begin{array}{l}\text{X-COORDINATE}\\ \text{OF THE SECOND}\\ \text{POINT } (4,5)\end{array}\right) - \left(\begin{array}{l}\text{X-COORDINATE}\\ \text{OF THE FIRST}\\ \text{POINT } (1,3)\end{array}\right) =$$

$$4 \quad - \quad 1 \quad = 3$$

So the slope is

$$\frac{\text{rise}}{\text{run}} = \frac{2}{3} .$$

If we put the whole thing together, the general formula for the slope of a line if we have two points on the line turns out to be

$$\text{SLOPE} = \frac{\text{RISE}}{\text{RUN}} = \frac{\left(\begin{array}{c}\text{y-COORDINATE}\\ \text{OF THE 2ND}\\ \text{POINT}\end{array}\right) - \left(\begin{array}{c}\text{y-COORDINATE}\\ \text{OF THE FIRST}\\ \text{POINT}\end{array}\right)}{\left(\begin{array}{c}\text{X-COORDINATE}\\ \text{OF THE 2ND}\\ \text{POINT}\end{array}\right) - \left(\begin{array}{c}\text{X-COORDINATE}\\ \text{OF THE FIRST}\\ \text{POINT}\end{array}\right)}$$

Now this is a mess to write, so mathematicians, in their constant search for conciseness, usually use some symbols to make this look neater (and, unfortunately, more mysterious).

To do this, they label the first point (1,3) with "(x_1, y_1)", using subscripts "$_1$."

So $1 = x_1$ (translated "x - one")
and $3 = y_1$ (translated "y- one")

The second point (4,5) is labeled with "(x_2, y_2)," using subscripts "$_2$."

So $4 = x_2$ (x - two)
and $5 = y_2$ (y - two) .

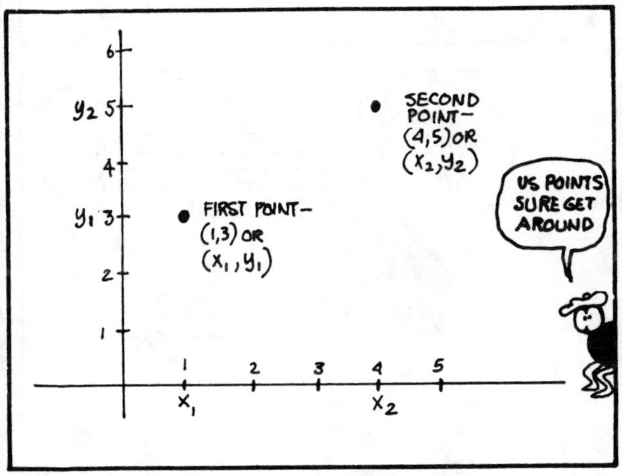

The little subscript numbers $_1$ and $_2$ are there to remind everyone which point the x's and y's refer to.

If we use this notation, the formula condenses down to

$$\text{SLOPE} = \frac{\text{RISE}}{\text{RUN}} = \frac{y_2 - y_1}{x_2 - x_1}$$

(y_2: 2ND POINT Y-COORD.; y_1: 1ST. POINT Y-COORD.; x_2: 2ND POINT X-COORD.; x_1: 1ST POINT X-COORD.)

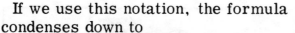

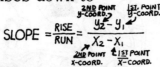

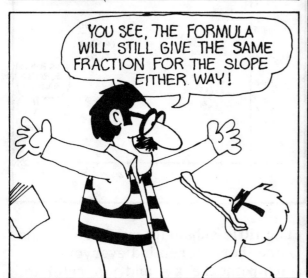

But the trouble again is which point shall we call the "original point" ? Once again, it turns out that it doesn't matter! If we use the "first point" (1, 3) in the role of the original point, we get, by collecting constant numbers

$$y = 3 + (\tfrac{2}{3})\cdot(x - 1) = 3 + \tfrac{2}{3}x - \tfrac{2}{3} = \tfrac{2}{3}x + \tfrac{7}{3}.$$

On the other hand, if we use the "second point" (4, 5) in the role of the "original point" we get

$$y = 5 + (\tfrac{2}{3})\cdot(x - 4) = 5 + \tfrac{2}{3}x - \tfrac{2}{3}\cdot 4 = \tfrac{2}{3}x + \tfrac{7}{3}.$$

We end up with the same formula

$$y = \tfrac{2}{3}x + \tfrac{7}{3}$$

in both cases! This is not exactly a general "proof" that this will happen no matter which two points we work with, but the same sort of computation always gives the same result for any two points.

Here are some more exercises involving lines and two points.

EXERCISES

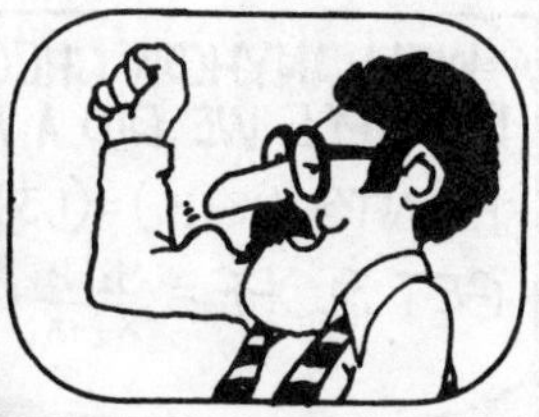

I.5.11 Find a formula for the line through points (-1, 4) and (1, 1).

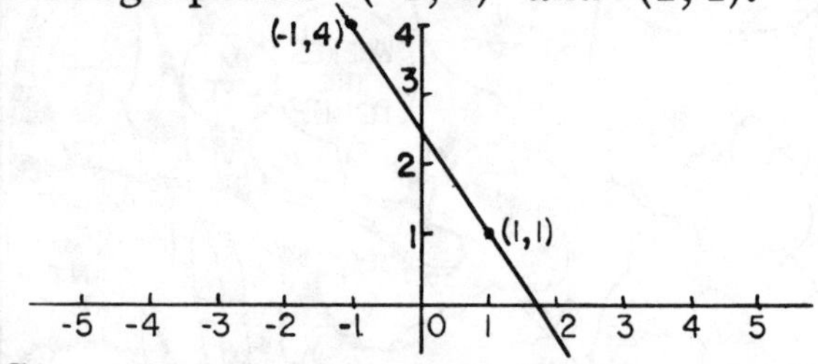

Since it turns out that the line slants up to the left the slope of the line should be a negative number. Check it out:

$$\text{slope} = \frac{(1, 1) - (-1, 4)}{(1, 1) - (-1, 4)} = \frac{1 - 4}{1-(-1)} = \frac{-3}{2} = -\frac{3}{2}$$

Formula?

I.5.12 Find a formula for the line through points (0, 1) and (2, 2) .

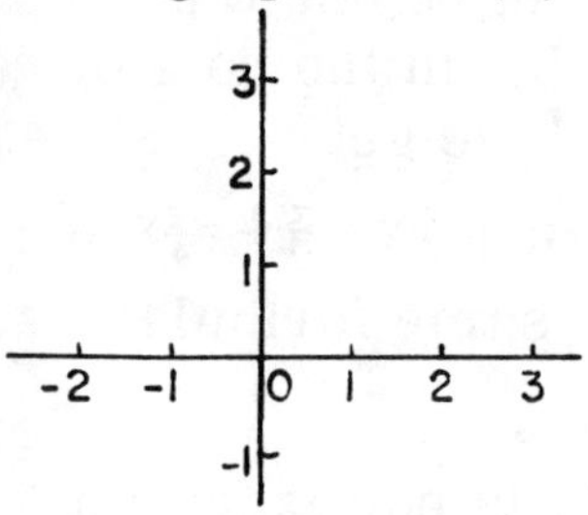

I.5.13 Find a formula for the line through points (-1, -2) and (1, 0).

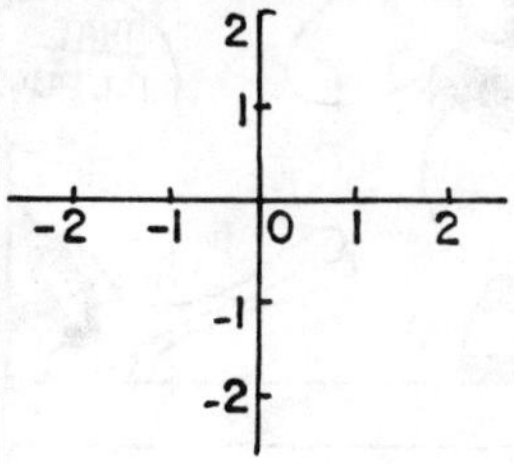

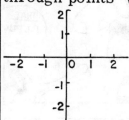

I.5.14 Find a formula for the line through points $(-3, \frac{2}{3})$ and $(-1, -\frac{1}{3})$.

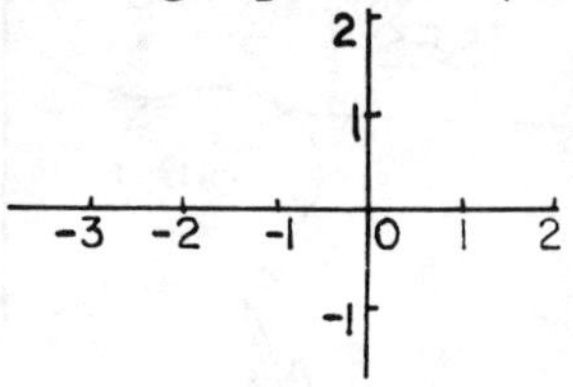

I.5.15 Does the point (4, -1) lie on the line that goes through points $(-1, \frac{3}{2})$ and $(1, \frac{1}{2})$?

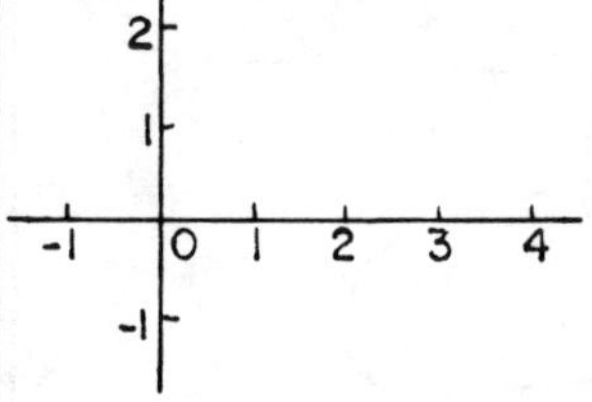

1.6 FURTHER FUNCTION FORMULAS FEATURING PARABOLAS & CUBICS

There are other graphs besides just straight lines which come from functions with handy formulas that tell exactly what the output should be for any particular input.

In the last section we showed how formulas that look like

$$f(\ ) = m\cdot(\ ) + b \quad (y = mx + b),$$

where m and b are constants, turn out to have graphs that are straight lines with slope m and go through point $(0, b)$.

In this section we mess around with formulas called POLYNOMIALS that look like

$$f(\ ) = (\ )^2 + 2\cdot(\ ) - 1$$ (which is the same as $y = f(x) = x^2 + 2x - 1$)

or $g(\ ) = (\ )^3 - 6\cdot(\ )^2 + 9\cdot(\ ) + 1$

(same as $y = g(x) = x^3 - 6x^2 + 9x + 1$)

or even

$$h(\ ) = \tfrac{1}{2}(\ )^5 - 3\cdot(\ )^4 + \tfrac{3}{5}(\ )^2 - 17\cdot(\ ) + 1$$

$(y = h(x) = \tfrac{1}{2}x^5 - 3x^4 + \tfrac{3}{5}x^2 - 17x + 1)$.

If Alfred is running a function with a formula like

$$f(\ ) = (\ )^2 + 2\cdot(\ ) - 1$$

we can tell exactly what he does with any given input x — for example, let's take the value 1 :

just put 1 in all the blanks;

$$f(1) = (1)^2 + 2\cdot(1) - 1$$

and then compute to get

$$f(1) = 1^2 + 2\cdot 1 - 1 = 2,$$

and so 2 is the output number. This tells us that a correspondence-arrow for f( ) goes from 1 to 2 and, more useful at the moment, the point $(1, 2) = (1, f(1))$ will be on the graph of f( ).

To get a further idea of what the graph of f( ) will look like, the usual thing to do is to choose other particular values of x more or less at random and use them to find a few more points $(x, f(x))$ on the graph. Then just connect them with a nice smooth curve and hope that we are about right.

To help get the right shape for the curve, here are a few pointers.

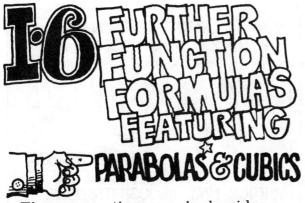

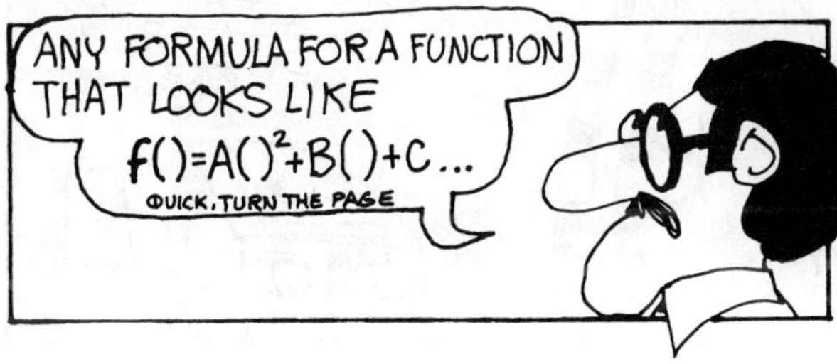

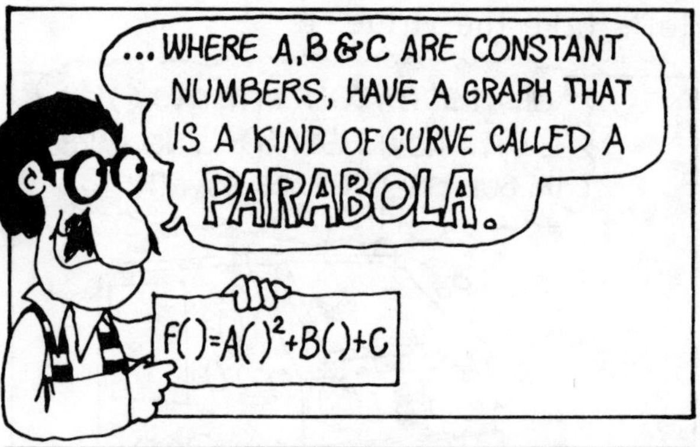
...WHERE A,B & C ARE CONSTANT NUMBERS, HAVE A GRAPH THAT IS A KIND OF CURVE CALLED A PARABOLA.
F( )=A( )²+B( )+C

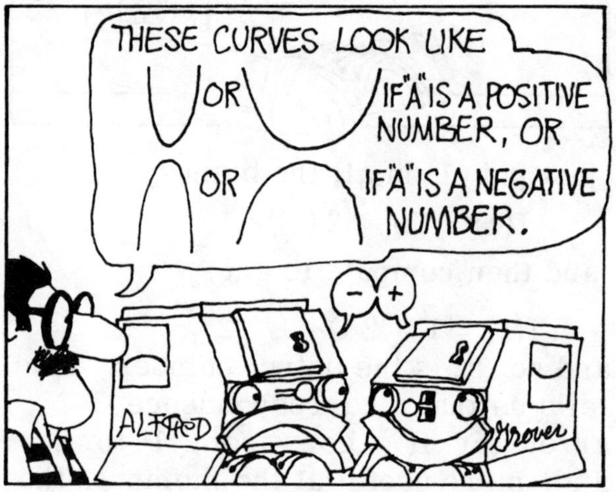
THESE CURVES LOOK LIKE
OR
IF "A" IS A POSITIVE NUMBER, OR
OR
IF "A" IS A NEGATIVE NUMBER.
ALFRED
Grover

IF |A| IS LARGE, LIKE A=6 OR A=-4, THE PARABOLA WILL BE THIN, LIKE (IF A>O) OR (IF A<O), AND IF |A| IS SMALL, LIKE A=¼ OR A=-⅓, THE PARABOLA WILL BE FAT, LIKE (IF A>O) OR (IF A<O).
ALFRED
Grover

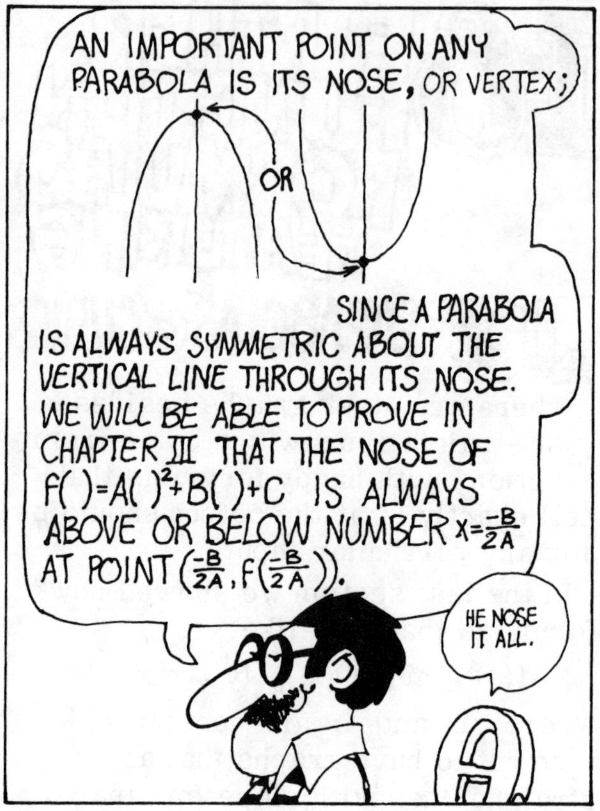
AN IMPORTANT POINT ON ANY PARABOLA IS ITS NOSE, OR VERTEX;
OR
SINCE A PARABOLA IS ALWAYS SYMMETRIC ABOUT THE VERTICAL LINE THROUGH ITS NOSE. WE WILL BE ABLE TO PROVE IN CHAPTER III THAT THE NOSE OF F( )=A( )²+B( )+C IS ALWAYS ABOVE OR BELOW NUMBER x=-B/2A AT POINT (-B/2A, F(-B/2A)).
HE NOSE IT ALL.

Now to draw the graph of a parabola like

$$f(\ ) = (\ )^2 + 2(\ ) - 1$$

it is handy to put down some general information about the shape of the parabola and the location of its nose first. Here is a check list:

$A = 1$ and $B = 2$ so

Up or down?	UP since $A = 1 > 0$
Fat or thin?	Medium. . . $A = 1$
Nose above?	$x = \frac{-B}{2 \cdot A} = \frac{-2}{2 \cdot 1} = -1.$

And so the nose is at point $(-1, f(-1))$ $= (-1, (-1)^2 + 2\cdot(-1) - 1) = (-1, -2).$

The next step is to compute a few more points on the graph. Start with choosing specific values of x near the "nose x-value", -1, like $x = 0$, $x = -2$, $x = 1$ and $x = -3$.

Next find "f(x)" in each case and make a list of the points (x, f(x)).

Points to go on the graph:

scratch work:

x - value	f(x)- value	scratch work
(0 ,	-1)	$f(0)=(0)^2+2\cdot(0)-1=-1$
(-2 ,	-1)	$f(-2)=(-2)^2+2\cdot(-2)-1=-1$
(1 ,	2)	$f(1)=(1)^2+2\cdot(1)-1=2$
(-3 ,	2)	$f(-3)=(-3)^2+2\cdot(-3)-1=2$

Now put in the points on the graph - don't forget the "nose point" (-1, -2)!

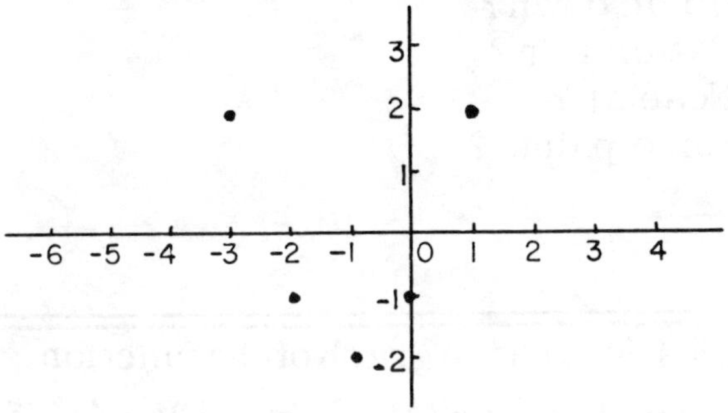

and draw a nice smooth curve through the points, and hopefully it is more or less the graph of the parabola.

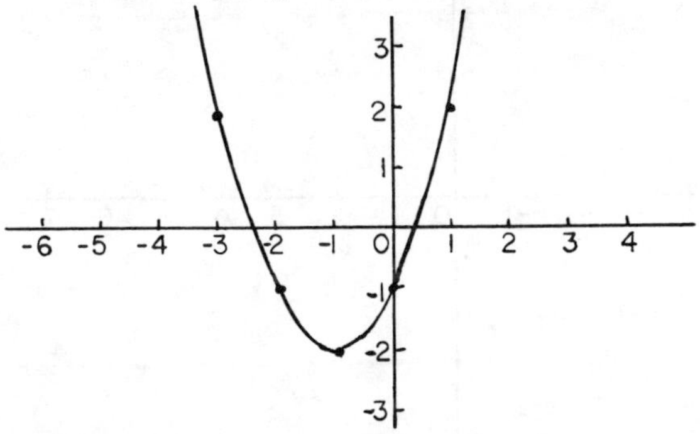

One important thing to remember in this follow-the-dots routine is that the order that one uses to move from dot to dot is always from left to right, with increasing x-coordinate values.

Here's an example that shows what happens if we don't hit the dots in the right order. Suppose we have a few points that we know are on the graph of some function, like

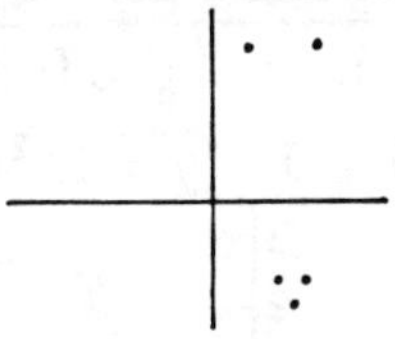

It is very tempting to draw the curve like this:

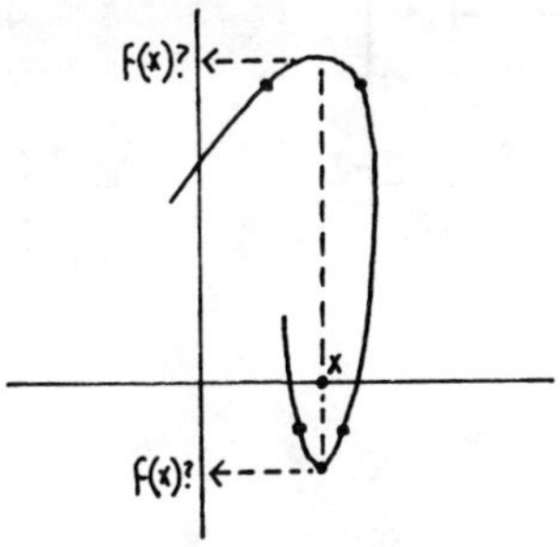

But this will never be right if the curve is the graph of a GENUINE function, since, from any particular value of x on the x-axis, there can be at most one correspondence-arrow from x by the RULES for functions. So there can be at most one dot on the graph above or below x. We rigged up an example with the graph above that shows that this rule is not followed for the curve that we just drew.

So the graph must look like

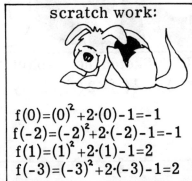

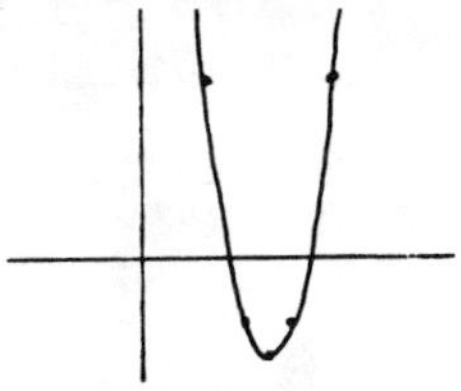

where there is no such problem.

Here are some exercises to help you practice on parabolas.

EXERCISES

I.6.1 Sketch the graph of the function

$f(\ ) = \frac{1}{2}(\ )^2 + 2(\ ) + 1 \quad (y = \frac{1}{2}x^2 + 2x + 1)$

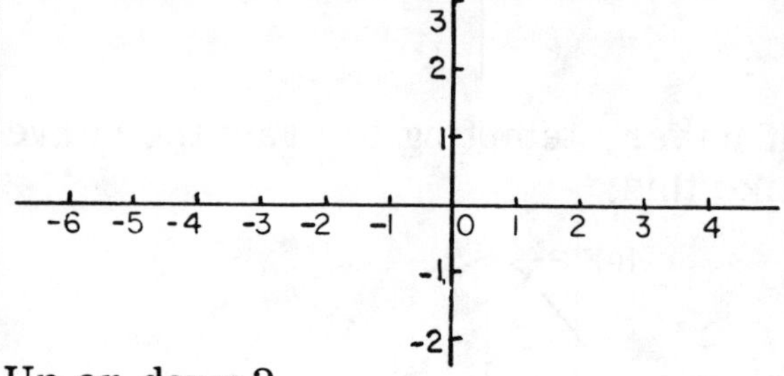

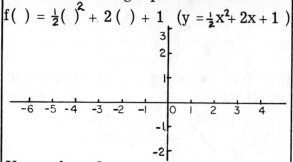

Up or down?
Fat or thin?
Nose at?
Some points?

I.6.3 Sketch the graph of the function

$f(\ ) = -\frac{1}{3}(\ )^2 - 2(\ ) - 1 \quad (y = -\frac{1}{3}x^2 - 2x - 1)$

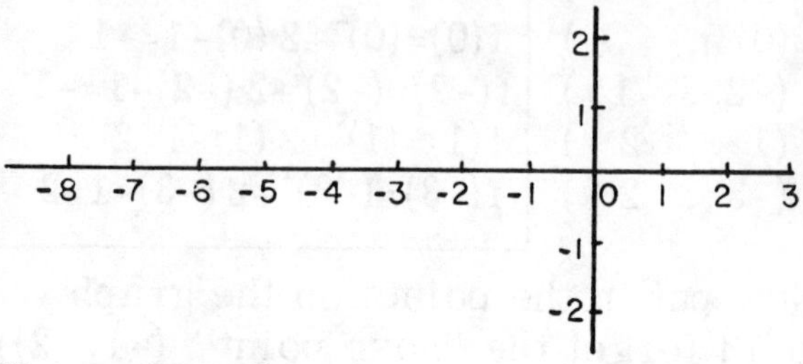

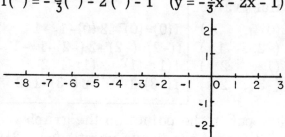

Up or down?
Fat or thin?
Nose at ?
Some points ?

I.6.2 Sketch the graph of the function

$f(\ ) = -2(\ )^2 + 4(\ ) + 1 \quad (y = -2x^2 + 4x + 1).$

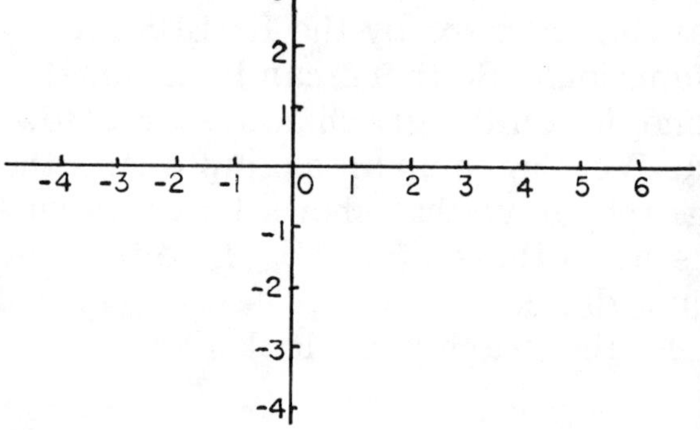

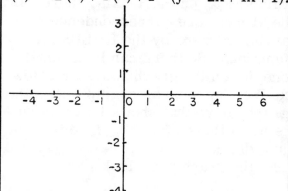

Up or down?
Fat or thin?
Nose at ?
Some points?

I.6.4 Sketch the graph of the function

$f(\ ) = -12(\ ) + 4(\ )^2 + 5 \quad (y = -12x + 4x^2 + 5)$

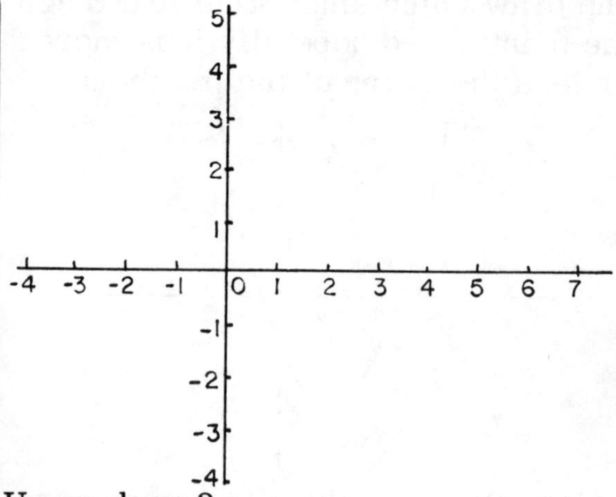

Up or down?
Fat or thin?
Nose at ?
Some points?

If g( ) has a formula-symbol that looks like

$$g(\ ) = A\cdot(\ )^3 + B\cdot(\ )^2 + C\cdot(\ ) + D$$

(which stands for the same function as

$$y = g(x) = Ax^3 + Bx^2 + Cx + D\ )$$

where A , B , C and D are constant numbers, then g( ) has a graph that is a kind of curve that is called a CUBIC . Cubics look like

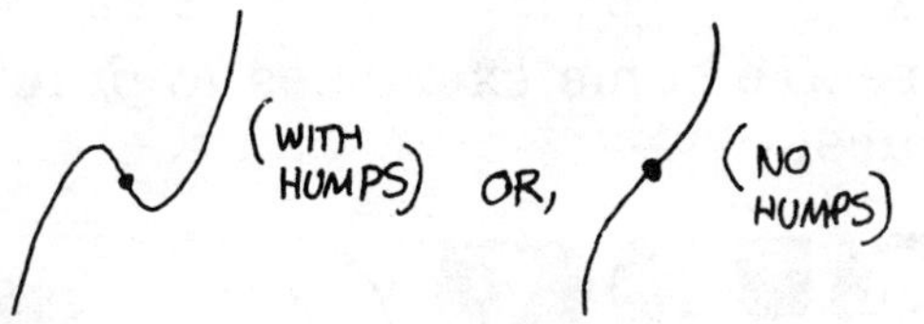

if A is a positive number or

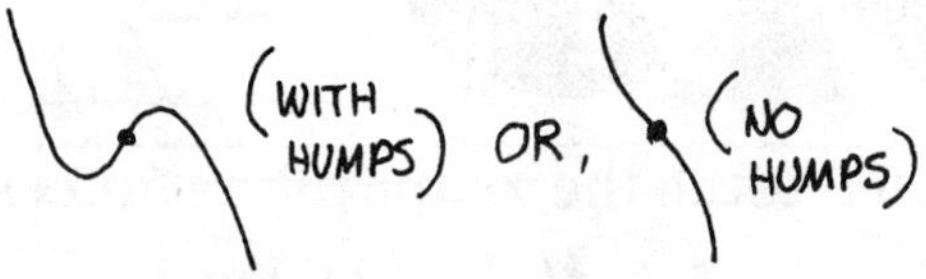

if A is a negative number.

A HANDY POINT TO KNOW ON THE CURVE IS THE POINT WHERE (IF A>0) THE CURVE STOPS DOING
AND STARTS DOING
IT IS CALLED THE "POINT OF INFLECTION".

This point is always above (or below)

$$x = -\frac{B}{3\cdot A}$$

at point $(-\frac{B}{3\cdot A}, g(-\frac{B}{3\cdot A}))$. On all the different kinds of cubics the "point of inflection" is circled:

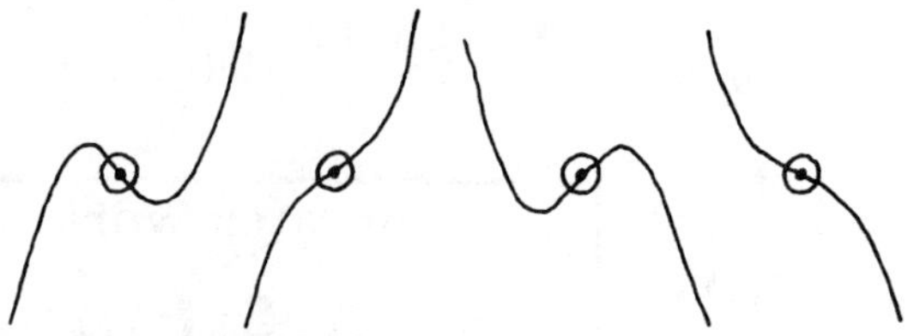

These "points of inflection" are kind of the center of action of any cubic since whatever the cubic does on one side of the point it repeats, upside down and backwards, on the other side of the point.

A graph will have humps if

$B^2 - 3AC > 0$.

To draw the graph of a cubic like

$$g(\ ) = (\ )^3 - 6\cdot(\ )^2 + 9\cdot(\ ) + 1\ ,$$

first get an idea of its shape: A = 1 in this case, which is positive, so it will look like

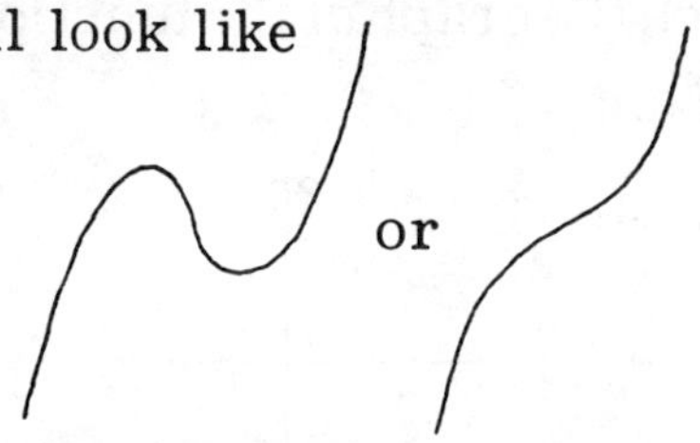

Next, since A = 1 and B = -6 , the "point of inflection" is above

$$x = \frac{-B}{3A} = \frac{-(-6)}{3\cdot 1} = 2\ .$$ Also, since

$$g(2) = (2)^3 - 6\cdot(2)^2 + 9\cdot(2) + 1$$
$$= 8 - 24 + 18 + 1 = 3\ ,$$

the point of inflection is actually

$$(2, g(2)) = (2, 3)\ .$$

$B^2 - 3AC = (-6)^2 - 3\cdot1\cdot9 = 9 > 0$, so there will be humps.

Now just find a few more sample points on the graph - choose sample values of x near the "point-of-inflection-x" = 2.

With

$$g(\ ) = (\ )^3 - 6\cdot(\ )^2 + 9\cdot(\ ) + 1$$

If $x =$	Then $f(x) =$	Scratch Work
(3	, 1)	$g(3)=3^3-6\cdot3^2+9\cdot3+1= 1$
(1	, 5)	$g(1)=1^3-6\cdot1^2+9\cdot1+1 = 5$
(4	, 5)	$g(4)=4^3-6\cdot4^2+9\cdot4+1 = 5$
(0	, 1)	$g(0)=0^3-6\cdot0^2+9\cdot0+1 = 1$

Finally, put in the points and draw the curve:

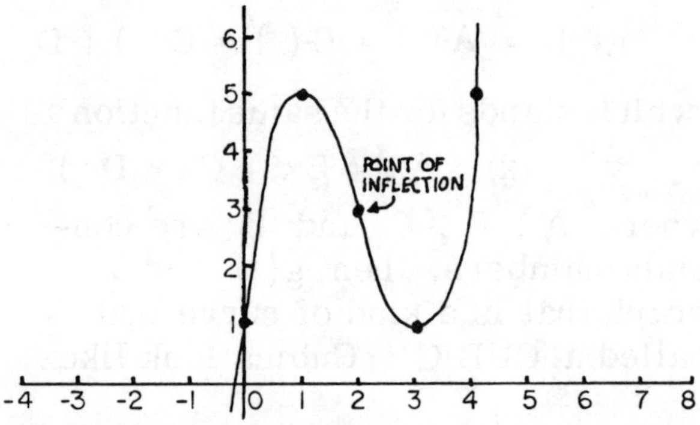

Here are some exercises to practice cubics.

EXERCISES

I.6.5 Sketch the graph of the function $f(\ ) = -(\ )^3 - 6\cdot(\ )^2 - 9\cdot(\ ) - 3$.

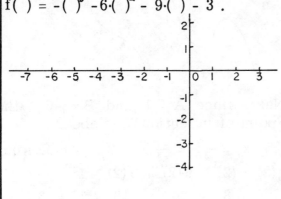

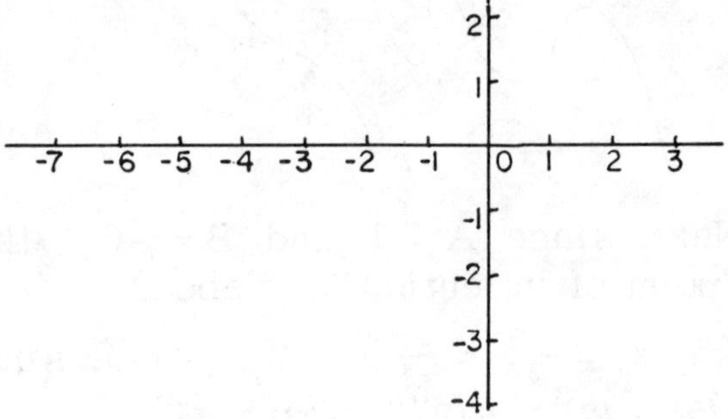

I.6.6 Sketch the graph of the function $f(\ ) = \frac{1}{2}(\ )^3 + \frac{3}{2}(\ )^2 + 2\cdot(\ ) + 2$.

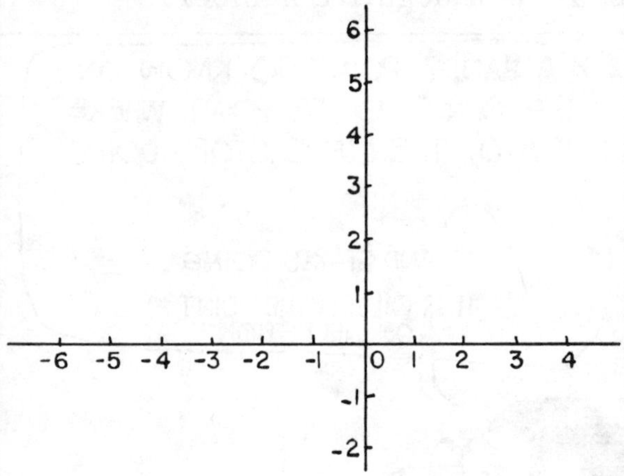

I.6.7 Sketch the graph of the function
$f(\ \ ) = \frac{1}{2}(\ \ )^3 - 3\cdot(\ \ )^2 + \frac{9}{2}(\ \ ) + 1$.

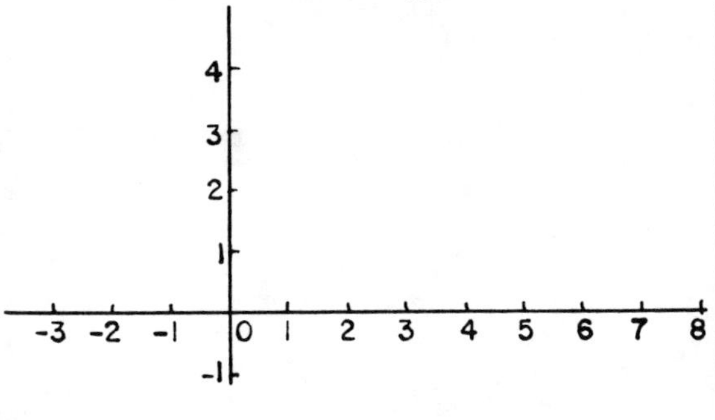

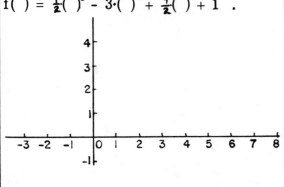

I.6.8 Sketch the graph of the function
$f(\ \ ) = -(\ \ )^3 + 3\cdot(\ \ ) + 2$.

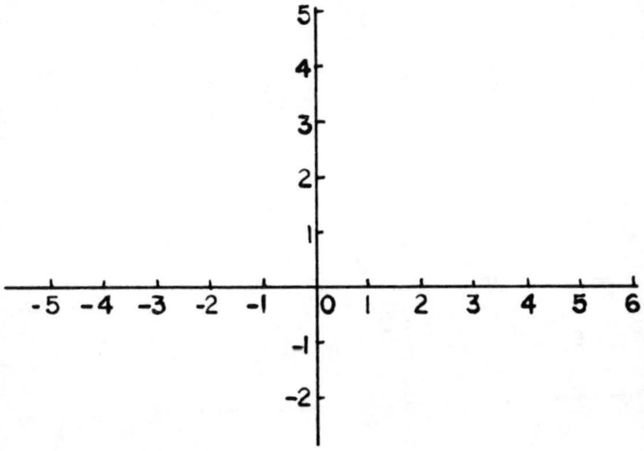

Good Vibes
f()
ALFRED

EXTRA EXERCISES & EXCURSIONS

Extra exercises are given for many sections in chapter 1, and they are numbered, for example, like EI.3.1 to indicate that they belong to section I.3. Selected answers and solutions can be found at the end of the book.

For Section I.3.

Exercise EI.3.1. Sample.
(a) If x satisfies $2<x\leq 3$, what inequality will x^2 satisfy?
(b) If x satisfies $-3<x<2$, what inequality will x^2 satisfy?

Solution: (a) Any x's satisfying $2<x\leq 3$ will be positive, so we can multiply $2<x\leq 3$ by positive x to get $2x < x^2 \leq 3x$. To find an interval for the values of x^2, we can multiply inequality $2<x$ by positive 2 to get $4 = 2\cdot 2 < 2x$ and multiply inequality $x\leq 3$ by positive 3 to get $3x \leq 3\cdot 3 = 9$.

So, for x's satisfying $2<x\leq 3$, we have
$4 < 2x$,
$2x < x^2 \leq 3x$ and
$3x \leq 9$, or, putting these together, we get

$$4 < 2x < x^2 \leq 3x \leq 9,$$

so x^2 will satisfy $4 < x^2 \leq 9$.

We have shown that if we start with an inequality like $2<x\leq 3$ where none of the numbers are negative, then we can square each number to obtain a new inequality that has the same form: $4 < x^2 \leq 9$. This only works when none of the numbers are negative, as we will see in (b).

(b) Suppose x is in the interval described by $-3<x<2$. If we just square each term as we did above and try to construct an inequality of the same form, we would write

$$9 = (-3)^2 < x^2 < 2^2 = 4,$$

which is certainly NOT true; 9 is not less than 4.

Instead, we can reason as follows: If $-3 < x < 2$, then $-3 < x < 2 < 3$, so $-3 < x < 3$ too. The results from section I.3 can be used to conclude that $|x| < 3$. Now $|x|$ is positive or zero, $(0 \leq |x|)$ so we can use the reasoning in part (a) to conclude that

$$0 \leq x^2 = |x|\cdot|x| < 3^2 = 9,$$

so $0 \leq x^2 < 9$.

Exercise EI.3.2.
(a) If x satisfies inequality $-2 < x < 3$, what inequality does x^2 satisfy?
(b) If x satisfies inequality $-2 \leq x < 3$, what inequality does x^2 satisfy?
(c) If x satisfies inequality $-2 < x \leq 3$, what inequality does x^2 satisfy?

Exercise EI.3.3. Hard!
If x satisfies inequality $-3 < x < -2$, what inequality does x^2 satisfy?

Exercise EI.3.4.
If x satisfies inequality $2 \leq x < 3$, what inequality does x^3 satisfy?

Exercise EI.3.5. $(\text{Hard})^2$.
If x satisfies inequality $-3 < x < -2$, what inequality does x^3 satisfy?

Exercise EI.3.6. Sample.
If x satisfies $1 < x < 4$,
(a) what inequality will $3x+x^2$ satisfy?
(b) What inequality will x^2-3x satisfy?
(c) What inequality will $|3x+x^2|$ satisfy?

Solution: From section I.3 of the book, we know that if $1<x<4$, then $3<3x<3 \cdot 4=12$. From exercise EI.3.1, we know that $1 < x^2 < 4^2=16$.

It turns out that we can 'add' inequalities to get new inequalities in the following way:

We start with $3<3x$ and $1<x^2$. Using what we know about adding numbers to inequalities, we can add 1 to $3<3x$ to get $4=3+1<3x+1$. We can add $3x$ to $1<x^2$ to get $3x+1<3x+x^2$. These two results give $4<3x+1<3x+x^2$, so $4<3x+x^2$. We have 'added' the inequalities

$$\begin{array}{l} 3 < 3x \quad \text{and} \\ \underline{1 < x^2} \quad \text{to get} \\ 4 < 3x+x^2. \end{array}$$

Similarly, we can 'add' inequalities $3x<12$ and $x^2<16$ to get $3x+x^2<12+16=28$. This reasoning allows us to obtain the final result without the intermediate steps. We know that if $1<x<4$, then

$$\begin{array}{rcccl} & 3 < & 3x & < 12 & \text{and} \\ & \underline{1 <} & \underline{x^2} & \underline{< 16}, & \\ \text{so (add)} & 4 < & 3x+x^2 & < 28. & \end{array}$$

When we 'add' inequalities like this, we have to be sure that all the inequalities that are added go 'in the same direction.' For example, suppose we try to figure out what inequality x^2-3x will satisfy if $1<x<4$. We already know that $1<x^2<16$ and $3<3x<12$. So, from the results in section I.3 of the book, $-3 > -3x > -12$.

We have to reverse the direction of this inequality to use the method of adding inequalities described above. We can write it as $-12 < -3x < -3$.

Now we can add the inequalities for x^2 and $-3x$: We know that if $1<x<4$, then

$$-12 < -3x < -3 \quad \text{and}$$
$$1 < x^2 < 16\ ,$$

so (add) $-11 < x^2-3x < 13$.

Finally, to find out what inequality $|3x+x^2|$ will satisfy if $1<x<4$, we have already shown that $4<3x+x^2<28$. So

$$-28 < 4 < 3x+x^2 < 28,$$

and the results of section I.3 allow us to conclude that

$$|3x+x^2|<28.$$

Exercise EI.3.7.
If x satisfies $2 < x < 3$,
(a) what inequality will $7x+2x^2$ satisfy?
(b) What inequality will $7x-2x^2$ satisfy?
(c) What inequality will $|7x-2x^2|$ satisfy?

Exercise EI.3.8.
If $|2x-3|<1$,
(a) what inequality will x satisfy?
(b) What inequality will x^2 satisfy?
(c) What inequality will $4x+x^2$ satisfy?
(d) What inequality will $|4x+x^2|$ satisfy?

EXCURSION.

Although $0<3<5$, when we compare $\frac{1}{3}$ and $\frac{1}{5}$ on the number line, we note that $\frac{1}{3}$ is <u>greater</u> than $\frac{1}{5}$

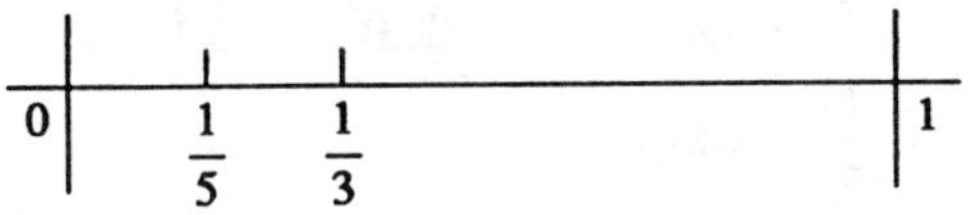

because $\frac{1}{3}$ is on the right of $\frac{1}{5}$ on the usual number line.

This sort of result is true in general: If $0 < a < b$, then $\frac{1}{b} < \frac{1}{a}$ (NOT $\frac{1}{a} < \frac{1}{b}$). Let's see why this will be true.

With $0 < a < b$, both a and b will be positive, so $a \cdot b$ will be positive and $\frac{1}{ab}$ will be positive too. From section I.3 of the book, We can multiply inequality $a<b$ by positive $\frac{1}{ab}$ and the '<' sign will still point in the same direction: The following inequality will be true:

$$(\tfrac{1}{ab})\cdot a < (\tfrac{1}{ab})\cdot b.$$

We can cancel here: $\frac{1}{ab}a = \frac{a}{ab} = \frac{1}{b}$ and $\frac{1}{ab}b = \frac{b}{ab} = \frac{1}{a}$. So

$$\frac{1}{b} = (\tfrac{1}{ab})\cdot a < (\tfrac{1}{ab})\cdot b = \frac{1}{a}\ ,$$

which is the inequality that we were trying to establish:

$$\text{If } 0 < a < b, \text{ then } \frac{1}{b} < \frac{1}{a}\ .$$

Exercise EI.3.9.

If $3 < a$, what inequality will $\frac{1}{a}$ satisfy?

Exercise EI.3.10.

If $0 < x < 5$, what inequality will $\frac{1}{x}$ satisfy?

Exercise EI.3.11.

Show that $|a-b| = |b-a|$.

For Section I.4.

Exercise EI.4.1.

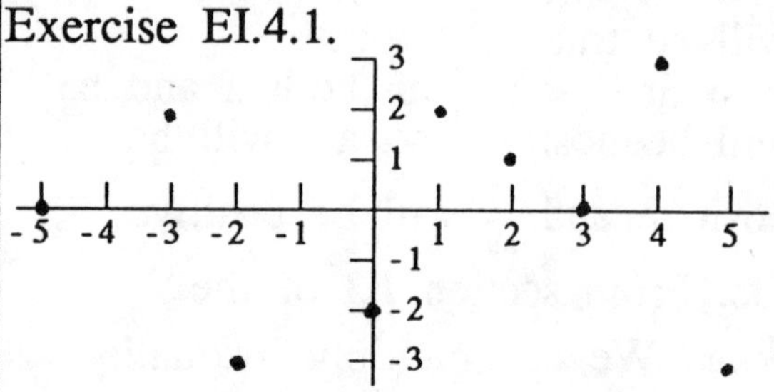

(a) Label the dots.
(b) If the dots represent a function, what is its domain? What is the function's range?

$f(-5) =$? $\quad f(5) =$?
$f(0) =$? $\quad f(3) =$?
$f(2) =$?
$f(-3) =$?

(c) Put in the appropriate arrows to show what corresponds to what in this function.

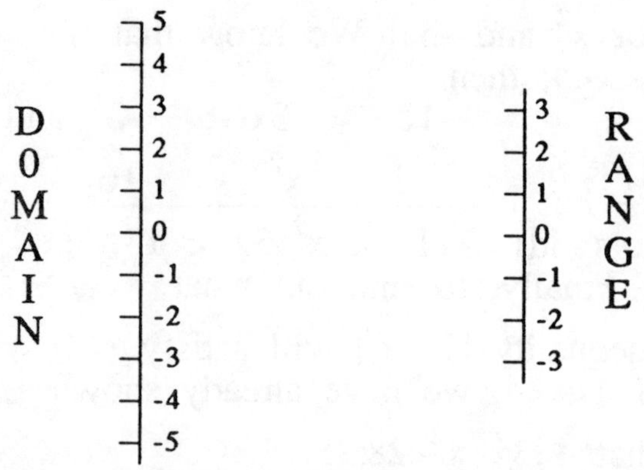

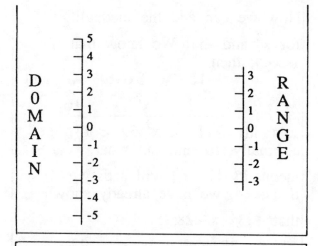

Exercise EI.4.2.

If f( ) is the correspondence shown by the following arrows,

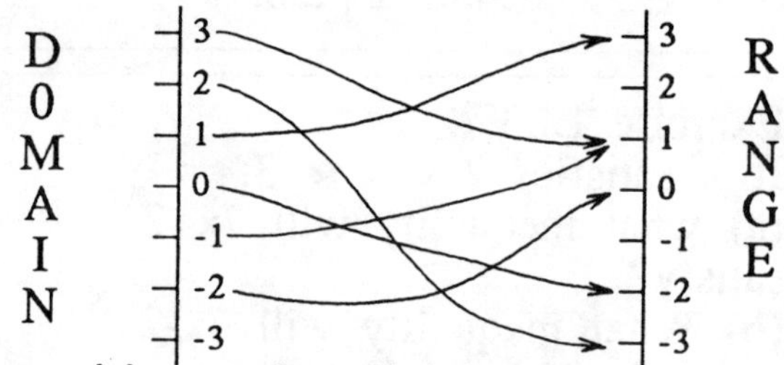

draw it's graph and label the dots on the graph.

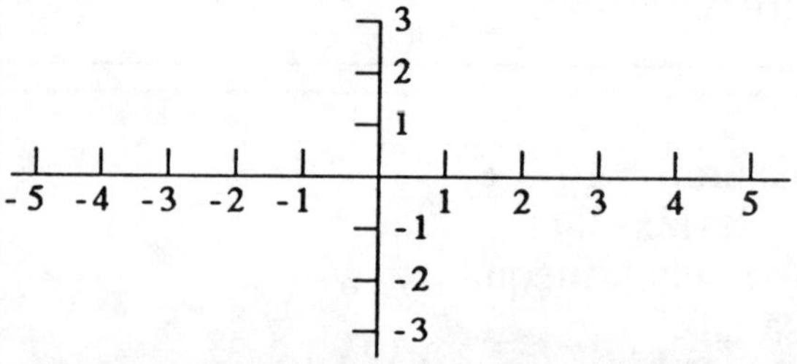

Exercise EI.4.3.

If the domain for f( ) is

{-2,-1,-1/2,0,1,2}

and f(-2)=0, f(-1)=-2, f(-1/2)=2, f(0)=3,f(1)=-1 and f(2),

(a) Put in the appropriate arrows to show what corresponds to what in this function.

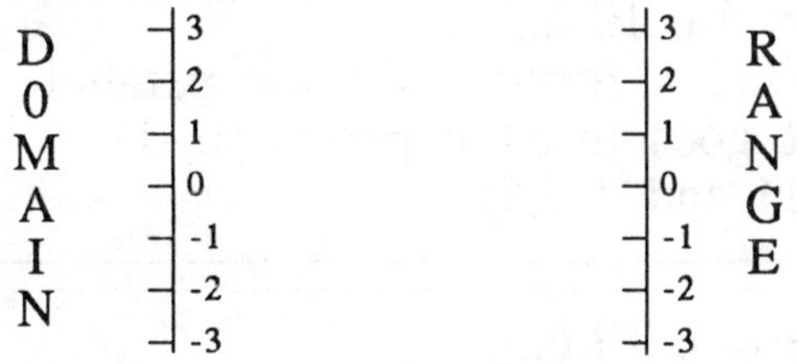

(b) Draw the graph of the function.

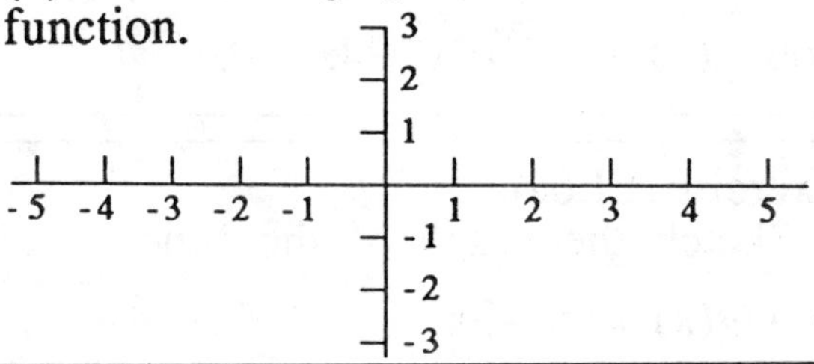

Exercise EI.4.4.

If the graph of f( ) is shown in the diagram below,

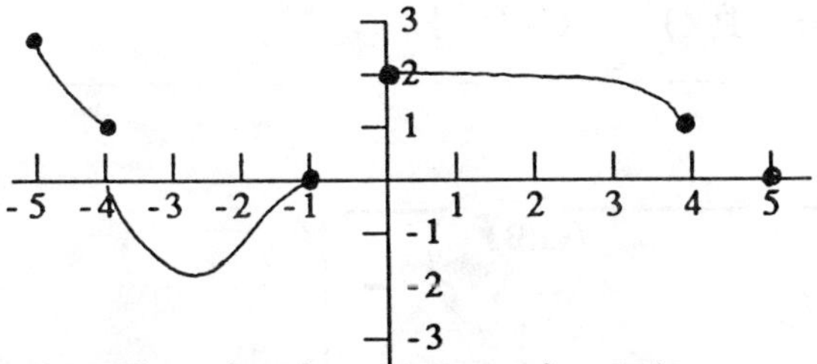

(a) What is the (approximate) domain of f( )? What is the (approximate) range for f( )?

(b) What is

f(-5) = ? f(-3) = ?

f(-1) = ? f(0) = ?

f(1) = ? f(4) = ?

(c) f(?) = 5/2. f(?) = 1.

f(?) = -1. f(?) = 0.

f(??) = 0.

Exercise EI.4.5.

Here are some of the correspondence-arrows for a function with a simple formula.

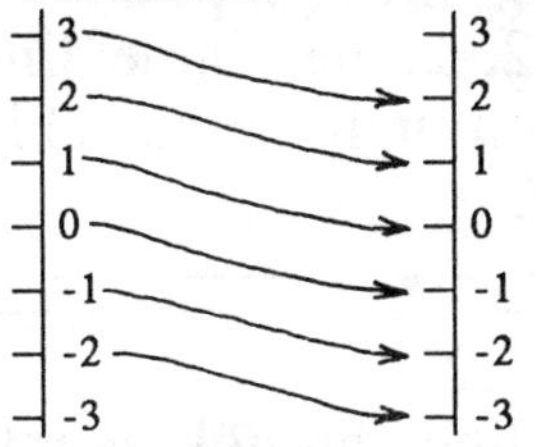

Sketch its graph.

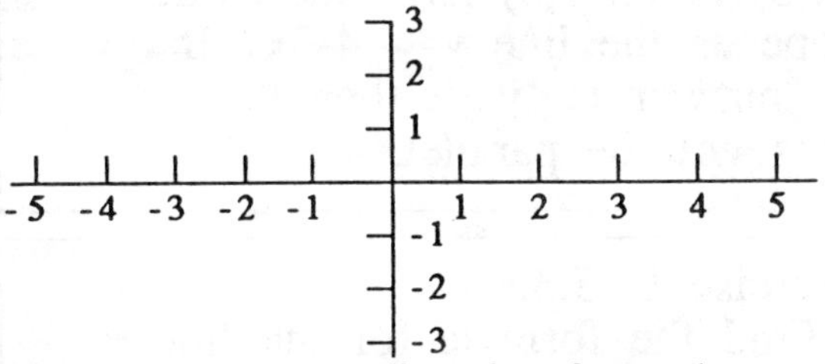

Can you guess what its formula might be?

Exercise EI.4.6.

Here are some of the correspondence-arrows for a function with a simple formula.

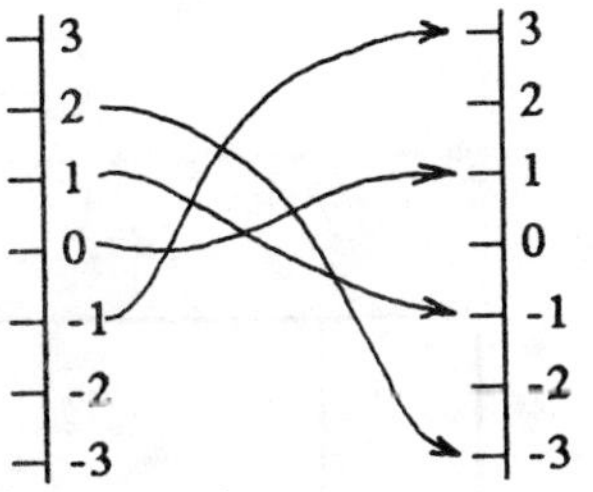

Sketch its graph.

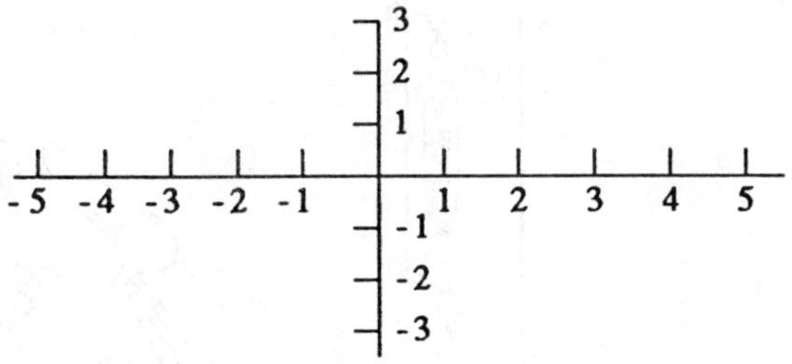

Can you guess what its formula might be?

For Section I.5.

Exercise EI.5.1.
Is the line through points (0,3) and (-2,-1) the same line as the one through points (11,25) and (-7,-11)?

Exercise EI.5.2.
Does the line through points (3,-3) and (-1,3) have the same slope as the line $y = 4-3x$? If the answer is 'yes', then the lines will be parallel.

Exercise EI.5.3.
Find the formula for the line through point (-2,-1) that is parallel to the line through points (-1,8) and (2,2).

For Section I.6.

Exercise EI.6.1
Sketch the graph of the function $y = f(x) = 2+2x- \frac{1}{2} x^2$.

Exercise EI.6.2.
Find a formula for the parabola that goes through points (0,-1), (1,1) and (-3,5)

Exercise EI.6.3.
Sketch the graph of the function $g(x) = -x^3+9x^2-24x+20$.

Exercise EI.6.4.
Sketch the graph of the function $g(x) = x^3+3x^2-1$.

Exercise EI.6.5.
Sketch the graph of the function $f(x) = x^3-3x-1$.

EMC2

CHAPTER 2

LIMITS

Chapter II is all about how mathematicians came up with a somewhat slippery answer to the question. . .

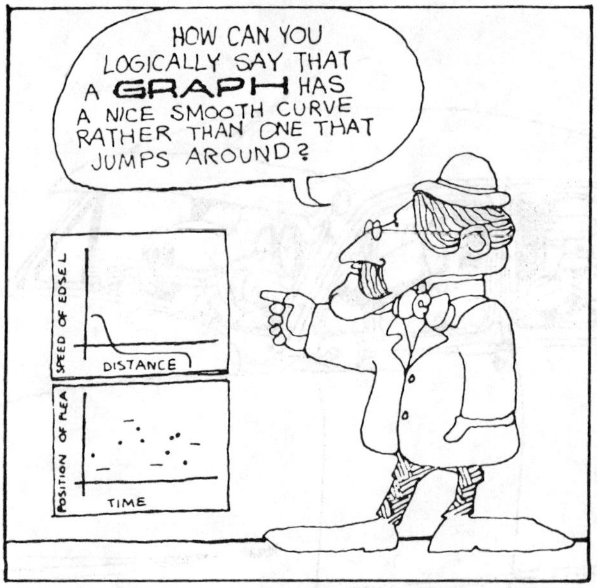

II·1 THE SMOOTHNESS PROBLEM

If Alfred is running a function whose domain-set and range-set are sets of numbers, we can draw the function's graph easily enough by just keeping the dots where the correspondence-arrows turn.

Now Alfred's graph is a mess—nothing but a whole bunch of disorganized dots! Not at all like the graphs that show things that happen in physics, chemistry, economics, etc. These sorts of graphs almost always turn out to be nice smooth curves.

But the rules for functions say nothing about whether or not the dots on a graph make a nice smooth curve. So mathematicians set out to find some way to describe this "smoothness" logically in such a way that they could actually PROVE that a function has a graph that is a smooth curve.

At first glance, it doesn't look like much of a problem. Suppose Grover is running a function whose graph looks nice and smooth.

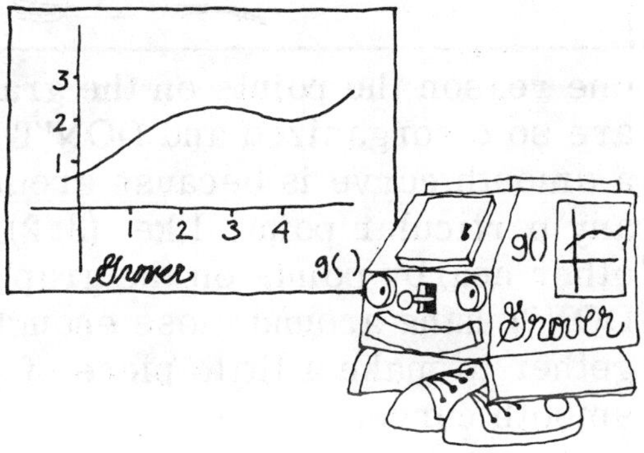

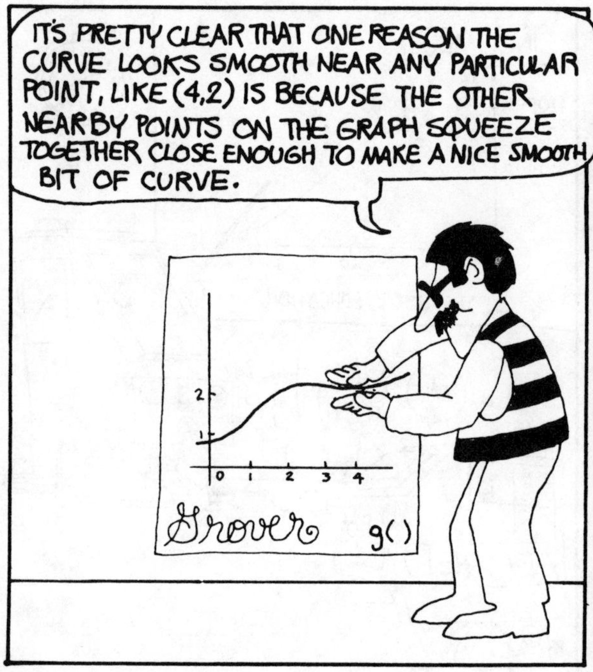

If we look back at Alfred's function from the same point of view,

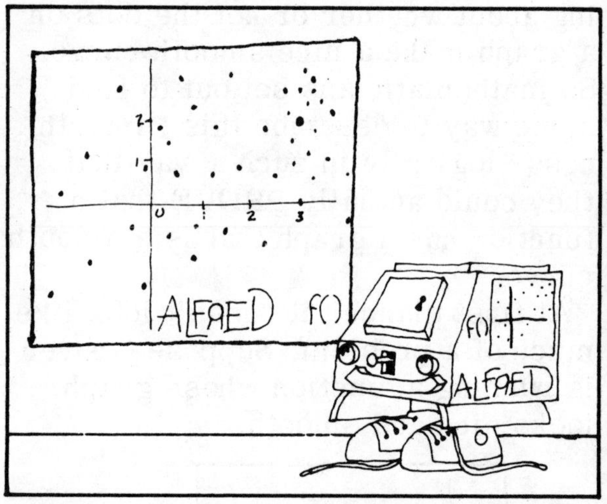

one reason the points on the graph are so disorganized and DON'T make a smooth curve is because around any particular point, like (3, 2), the other nearby points on the graph DON'T hang around close enough together to make a little piece of smooth curve.

But...

before we let McSquared sum things up like this, we ought to look at Grover's graph again to be really sure that it is smooth and squeezed together no matter how close anybody looks.

Things look good for a while (except to Good Vibes) . . .

but it turns out that there is a slight problem.

II·2 LIMIT MACHINES

Check all the fine print in this one!

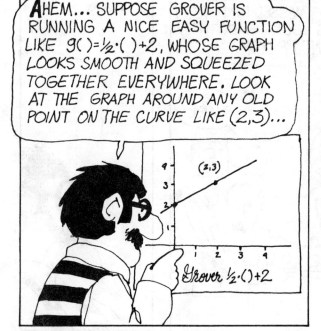

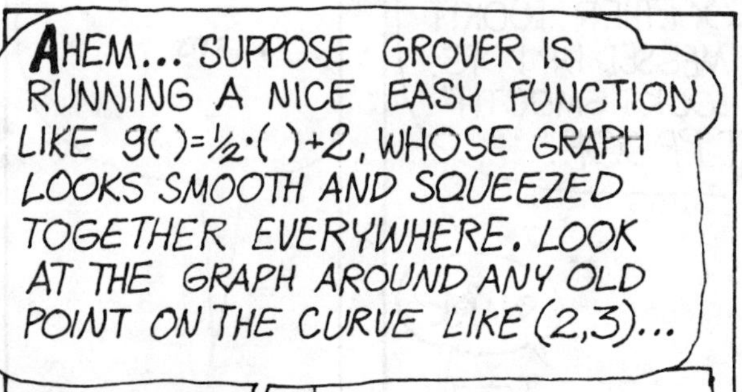

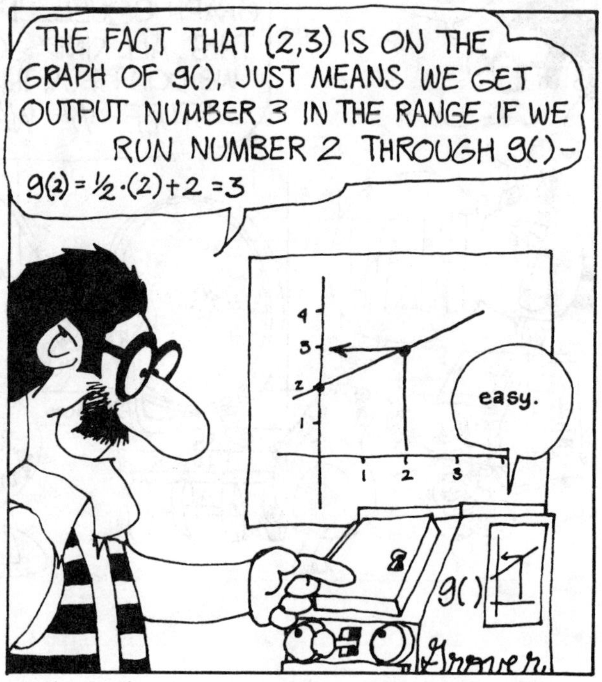

Now one special thing that happens only when a curve is smooth like Grover's graph is that if we run a number through g( ) that is close to 2 , like 2.2 , it looks like the output of g( ) should be close to

$$g(2) = \tfrac{1}{2}(2) + 2 = 3.$$

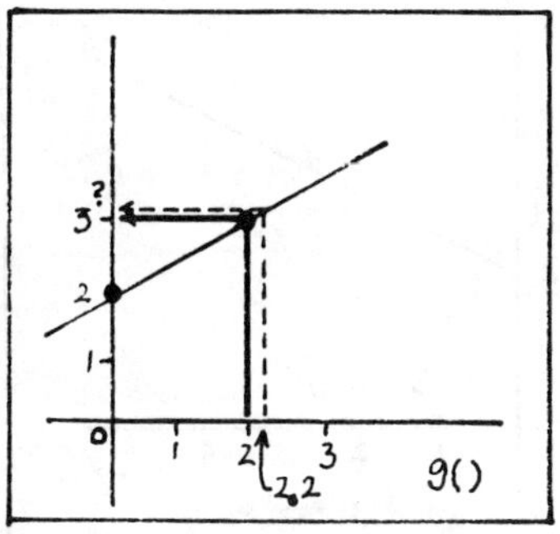

Check it -

$$g(2.2) = \tfrac{1}{2}\cdot(2.2) + 2 = 1.1 + 2 = 3.1 ,$$

which IS pretty close to 3 !

In fact, for the graph of the function

$$g(\ ) = \tfrac{1}{2}\cdot(\ ) + 2$$ to be "smooth,"

the closer x gets to 2 , the closer g(x) should be to 3 . So g(x) should approach 3 as its LIMIT NUMBER as x goes to 2 since <u>the points where the correspondence-arrows turn will just slide down the smooth graph</u> until they get to (2, 3).

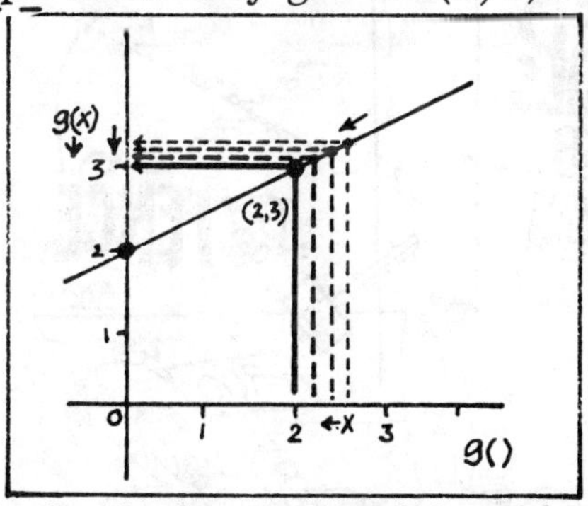

This is the <u>basic</u> <u>intuitive</u> <u>idea</u> of limit, but nobody can prove anything from an intuitive idea.

So mathematicians decided to get into this limit business by the back door. They first figured out a nice way to talk about "closeness." Here's how it goes:

Suppose we want the output g(x) of g( ) to be so close to g(2) = 3 that the output is within some particular ERROR-TOLERANCE of 3 .

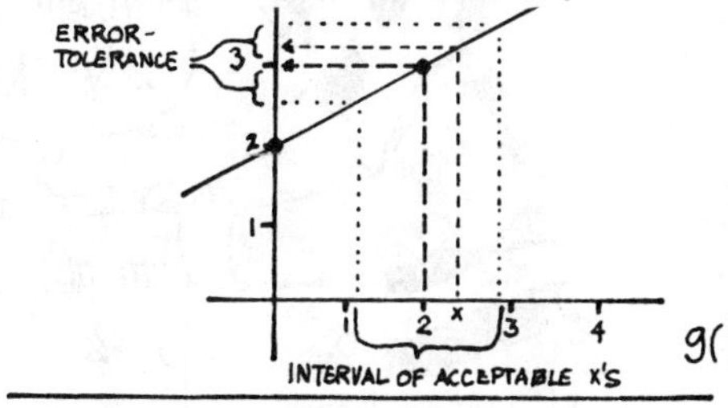

It looks like there should be a whole <u>interval</u> on the x-axis around 2 where, if we run any input number "x" <u>from</u> <u>this</u> <u>interval</u> through g( ), the output g(x) should be close enough to 3 to be within the error-tolerance!

This sort of thing won't necessarily happen if a function's graph isn't a smooth curve.

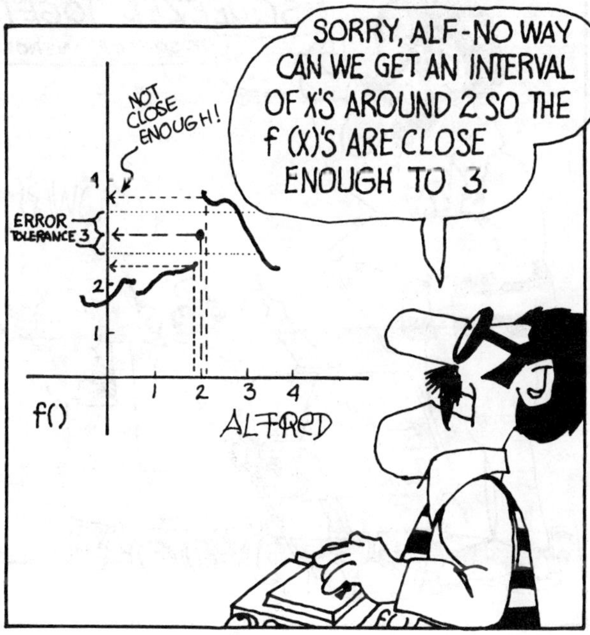

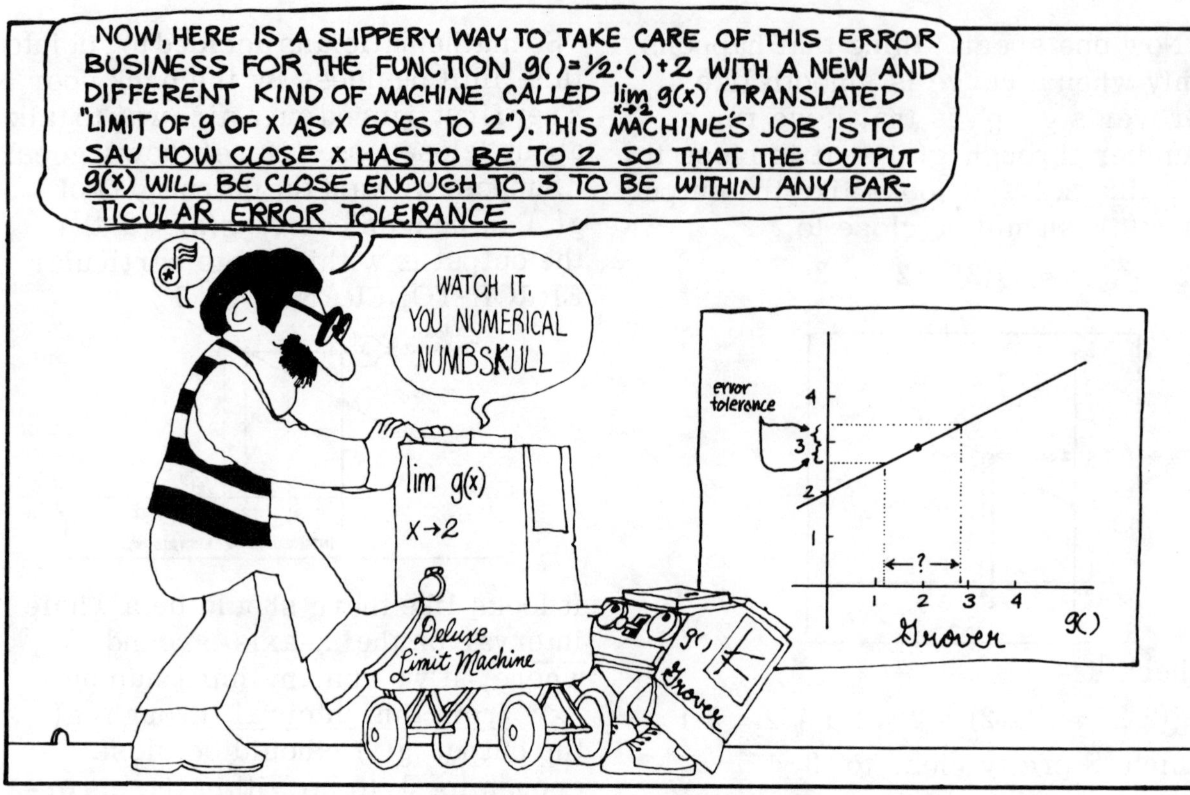

NOW, HERE IS A SLIPPERY WAY TO TAKE CARE OF THIS ERROR BUSINESS FOR THE FUNCTION g( ) = 1/2·( ) + 2 WITH A NEW AND DIFFERENT KIND OF MACHINE CALLED lim g(x) x→2 (TRANSLATED "LIMIT OF g OF X AS X GOES TO 2"). THIS MACHINE'S JOB IS TO SAY HOW CLOSE X HAS TO BE TO 2 SO THAT THE OUTPUT g(x) WILL BE CLOSE ENOUGH TO 3 TO BE WITHIN ANY PARTICULAR ERROR TOLERANCE
WATCH IT, YOU NUMERICAL NUMBSKULL
lim g(x) x→2
Deluxe Limit Machine
g( ) Grover
error tolerance
4
3
2
1
?
1 2 3 4
Grover g( )

USING THIS FINE NEW LIMIT-MACHINE, I WILL FENCE IN THE GRAPH OF g( ) AROUND (2,3) SO YOU WILL SEE THAT IT IS REALLY SMOOTH AND SQUEEZED TOGETHER THERE!
(er, SORT OF, ANYHOW)
GROWR
g( ) Grover
lim g(x) x→2
FULL X AND EVEN Y RATED DETAILS LATER
POST NO BILLS
SNEAK PREVIEW
FENCE
WHAZZAT!
PROO

But first we have to see what this new machine does. To begin with, McSquared, in his usual arbitrary way, tosses the number $\frac{1}{2}$ into the limit-machine.

This means that he is CHALLENGING this limit-machine to say HOW CLOSE x has to be to 2 so that GROVER'S output g(x) will be no more than distance $\frac{1}{2}$ away from 3.

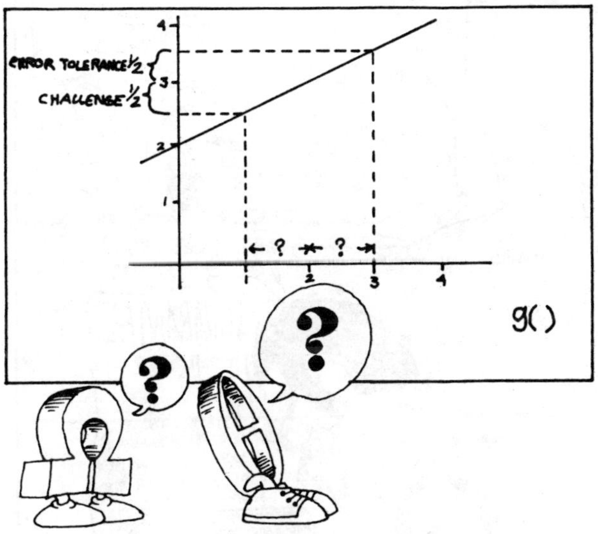

The limit-machine then mysteriously answers the challenge "how close" by producing the number 1 (with some sort of guarantee attached) to indicate that x must be within distance 1 of 2 for output g(x) to be no more than the requested error-tolerance $\frac{1}{2}$ away from 3.

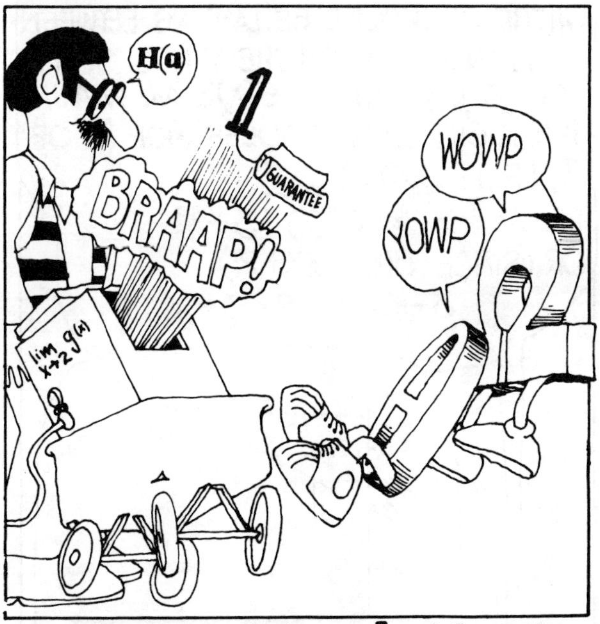

To be within distance 1 of 2 on the x-axis means that a number "x" has to be in the interval (1, 3) running from $2 - 1 = 1$ to $2 + 1 = 3$. So such x's that are within distance 1 of 2 must satisfy the inequality

$$2 - 1 < x < 2 + 1 .$$

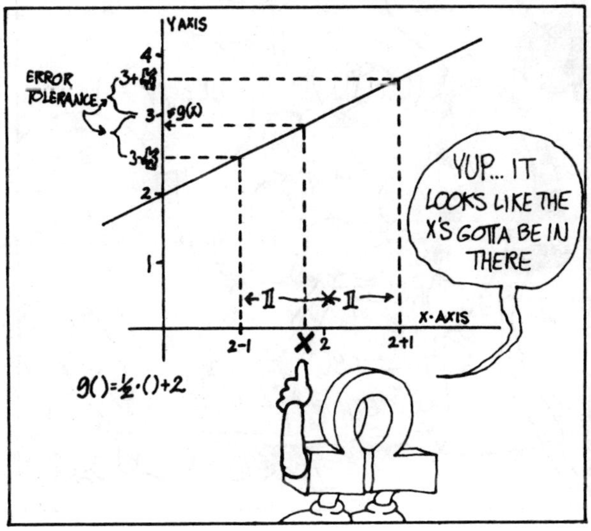

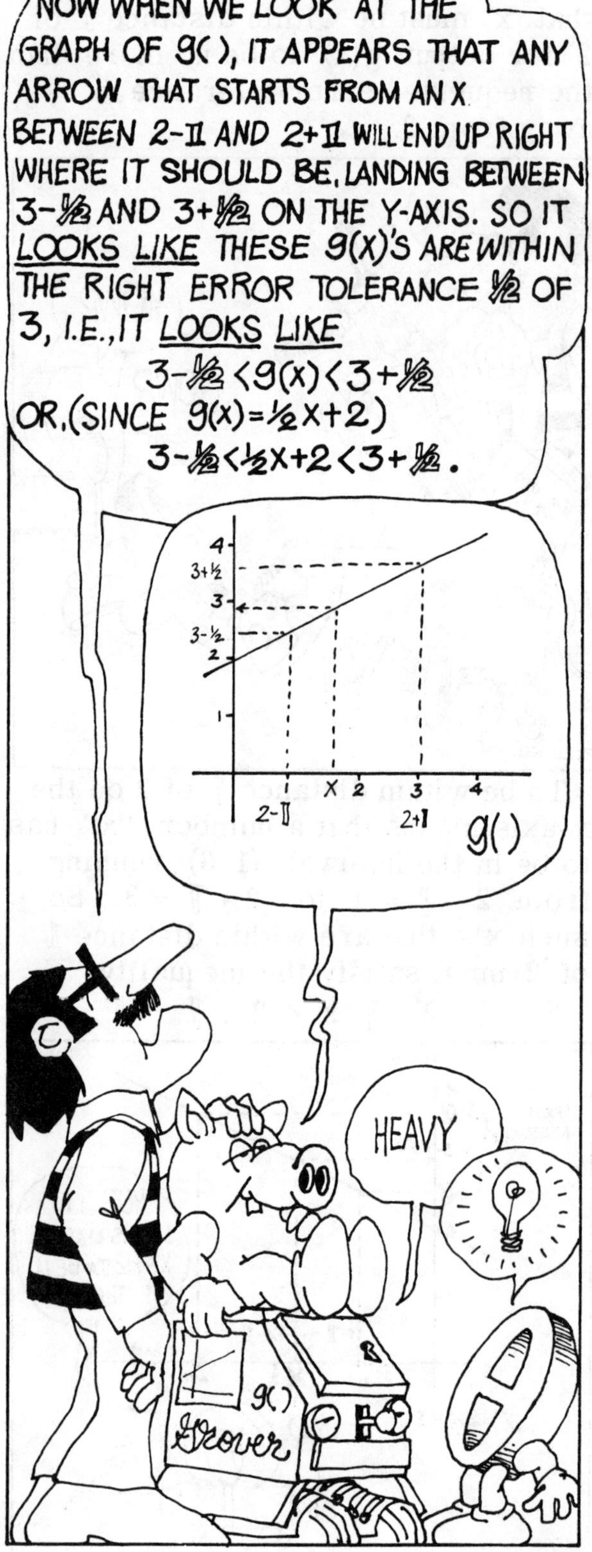

But it isn't enough that things LOOK GOOD any more. Somebody has got to PROVE that if x satisfies

$$2 - \mathbb{I} < x < 2 + \mathbb{I}$$

then, for those x's ,

$$3 - \tfrac{1}{2} < \tfrac{1}{2} \cdot x + 2 < 3 + \tfrac{1}{2} \quad .$$

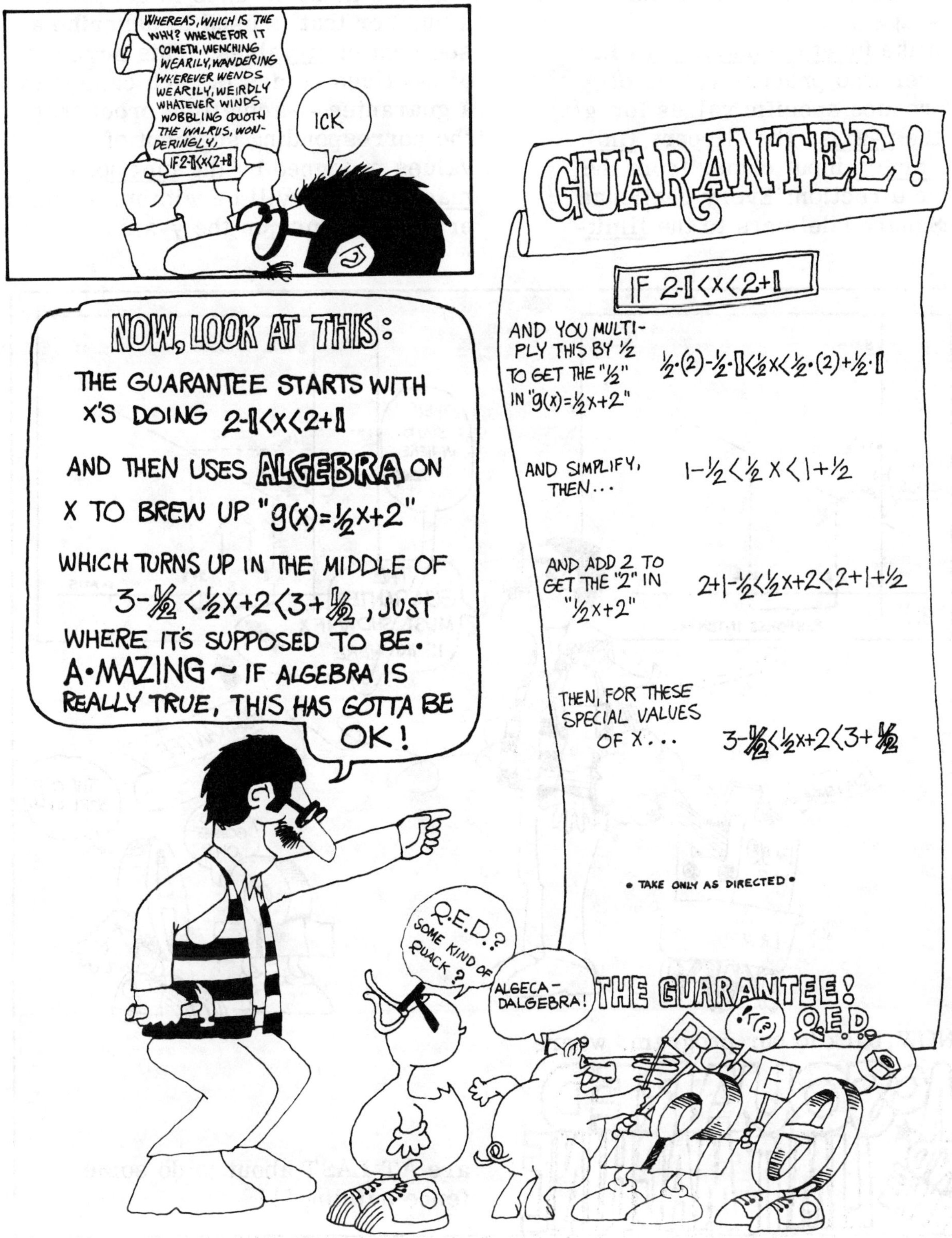

WHEREAS, WHICH IS THE WHY? WHENCE FOR IT COMETH, WENCHING WEARILY, WANDERING WHEREVER WENDS WEARILY, WEIRDLY WHATEVER WINDS WOBBLING QUOTH THE WALRUS, WONDERINGLY,
IF 2-1<x<2+1
ICK
NOW, LOOK AT THIS:
THE GUARANTEE STARTS WITH X'S DOING 2-1<x<2+1
AND THEN USES ALGEBRA ON X TO BREW UP "g(x)=½x+2"
WHICH TURNS UP IN THE MIDDLE OF 3-½<½x+2<3+½, JUST WHERE IT'S SUPPOSED TO BE.
A•MAZING ~ IF ALGEBRA IS REALLY TRUE, THIS HAS GOTTA BE OK!
GUARANTEE!
IF 2-1<x<2+1
AND YOU MULTIPLY THIS BY ½ TO GET THE "½" IN "g(x)=½x+2"
½·(2)-½·1<½x<½·(2)+½·1
AND SIMPLIFY, THEN...
1-½<½x<1+½
AND ADD 2 TO GET THE "2" IN "½x+2"
2+1-½<½x+2<2+1+½
THEN, FOR THESE SPECIAL VALUES OF X...
3-½<½x+2<3+½
• TAKE ONLY AS DIRECTED •
Q.E.D.? SOME KIND OF QUACK?
ALGECA-DALGEBRA!
THE GUARANTEE!
PROOF
Q.E.D.

We pause briefly for a mathematical message . . .

Unlike function-machines like Grover who process values of x to produce specific values for g(x) on the y-axis, the slippery limit-machine's input comes from the other direction: Every challenge McSquared delivers to the limit-machine must produce in response a number that is used to describe a segment of x-values. This segment of x-values comes in each case with a guarantee — actually a proof that the corresponding segment of y-values produced by the function-machine g( ) will be within a certain error tolerance on the y-axis.

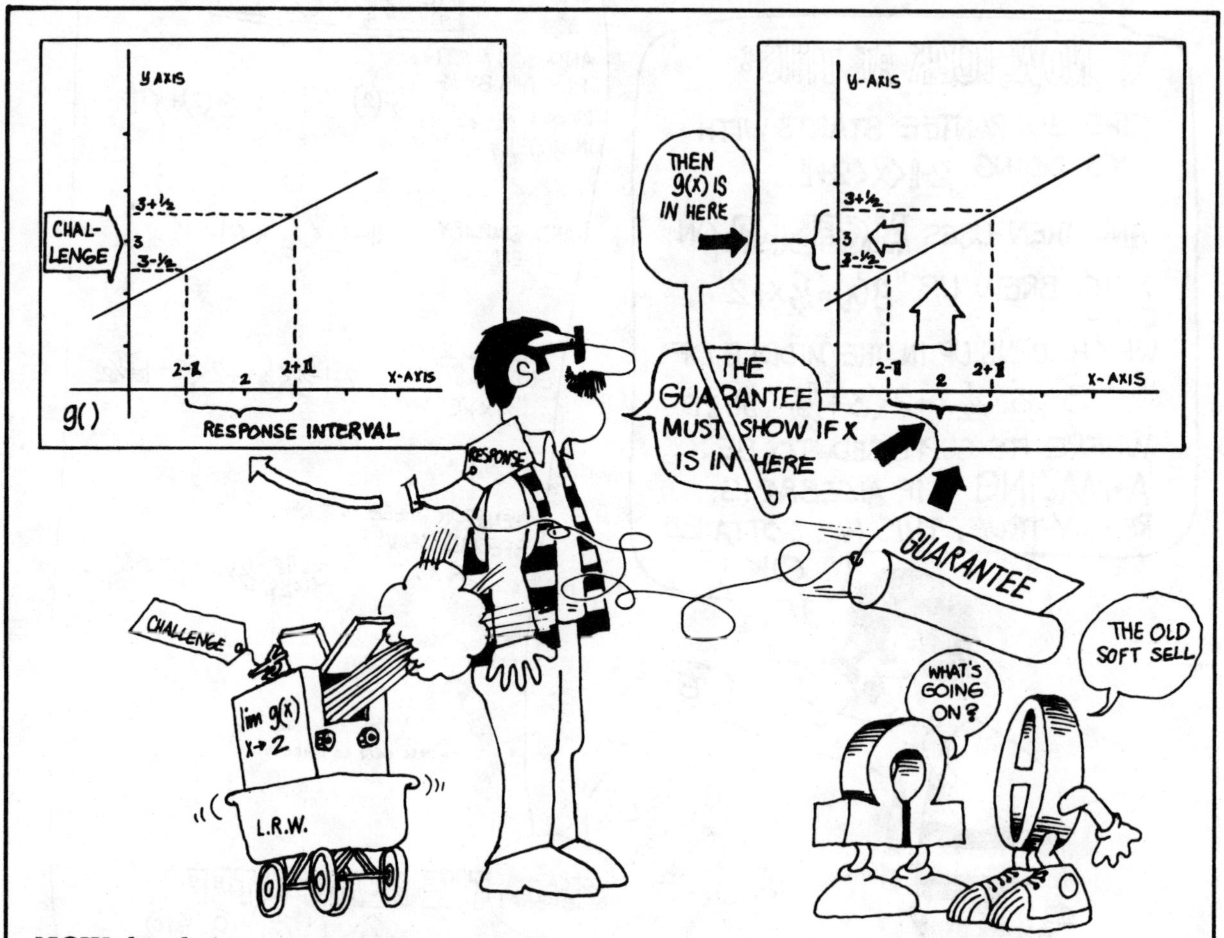

NOW, back to our program, where

McSQUARED AND THE LIMIT MACHINE

are AT LAST about to do some fence building!!!

NOW THAT IT IS REALLY TRUE THAT IF X IS BETWEEN 1 AND 3, THEN g(X) GOTTA BE BETWEEN 3-½ AND 3+½. I CAN SHOW THAT THE GRAPH OF g() IS TRAPPED BETWEEN THESE 2 FENCES NEAR (2,3)!
SHORE AIN'T MY KIND OF FENCE– WANNA BUY A HOT WATCH?
THAT'S A FENCE?
3+½
3
3-½
2
1
2
3
?
lim g()
x→2
g()
Deluxe Limit Machine
FENCE

POW!
ZAP
3+½
3
3-½
2
2-1
2+1
FIRST, DO A LITTLE DELICATE TRIMMING ON THESE FENCES, SO THEIR END POINTS ARE RIGHT ABOVE THE POINTS 2-1 AND 2+1
CAREFUL! VIOLENCE IS A NO NO IN SCHOLARLY COMIX
lim g(x)
x→2
Deluxe Limit Machine

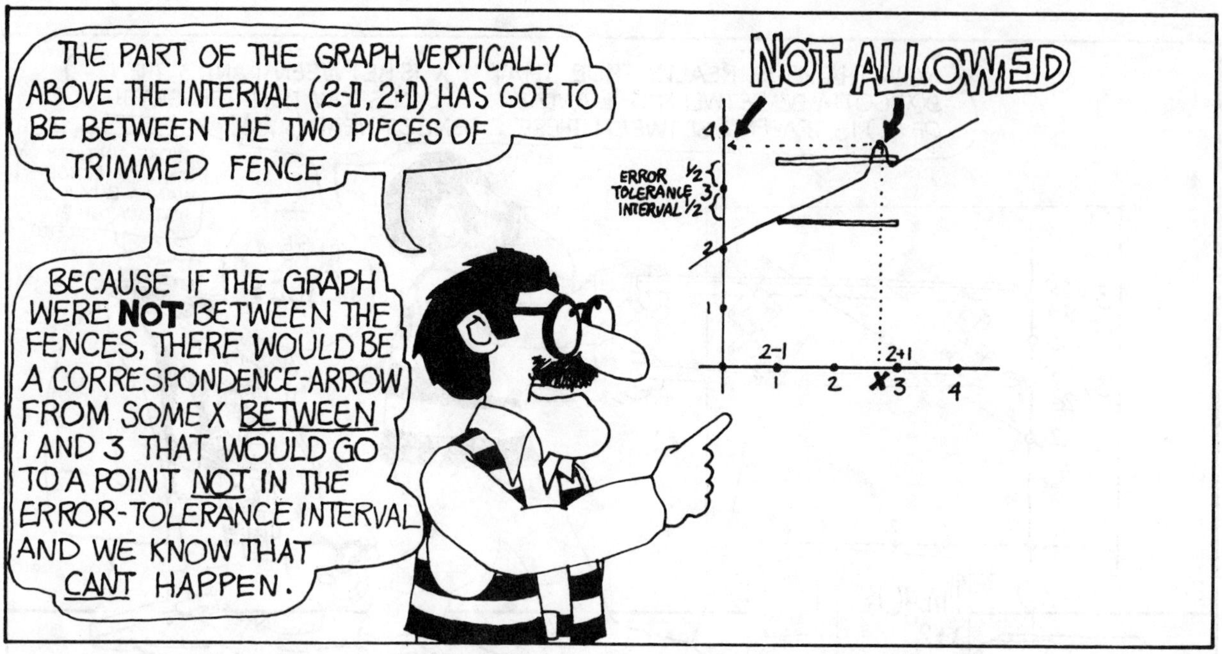
THE PART OF THE GRAPH VERTICALLY ABOVE THE INTERVAL (2-1, 2+1) HAS GOT TO BE BETWEEN THE TWO PIECES OF TRIMMED FENCE
BECAUSE IF THE GRAPH WERE **NOT** BETWEEN THE FENCES, THERE WOULD BE A CORRESPONDENCE-ARROW FROM SOME X BETWEEN 1 AND 3 THAT WOULD GO TO A POINT NOT IN THE ERROR-TOLERANCE INTERVAL AND WE KNOW THAT CAN'T HAPPEN.
NOT ALLOWED
ERROR TOLERANCE INTERVAL
½ 3 ½
4
2
1
2-1
2+1
1 2 X 3 4

WITH THOSE TWO PIECES OF FENCE FIRMLY IN PLACE, WE NOW FENCE IN THE GRAPH A BIT MORE CLOSELY BY DUMPING ERROR-TOLERANCE CHALLENGE ¼ INTO THE OLD LIMIT-MACHINE HERE.
HA! OUT COMES RESPONSE ½, WITH A GENUINE AND CORRECT GUARANTEE -- SHOWING THAT ANY X'S WITH DISTANCE ½ OF 2 (SITTING IN THE "RESPONSE INTERVAL" 2-½ < x < 2+½) GIVES g(x)'S IN 3-¼ < g(x) < 3+¼
½
GUARANTEE! IF 2-½<x<2+½ THEN (TIMES ½) 1-¼<½x<1+¼ AND... ADD 2 3-¼<½x+2<3+¼ GUARANTEED
DON'T FENCE ME IN...
BLAP.
¼
lim g(x) x→2
Deluxe Limit Machine
PROOF
PROOF
PROOF
PROOF!

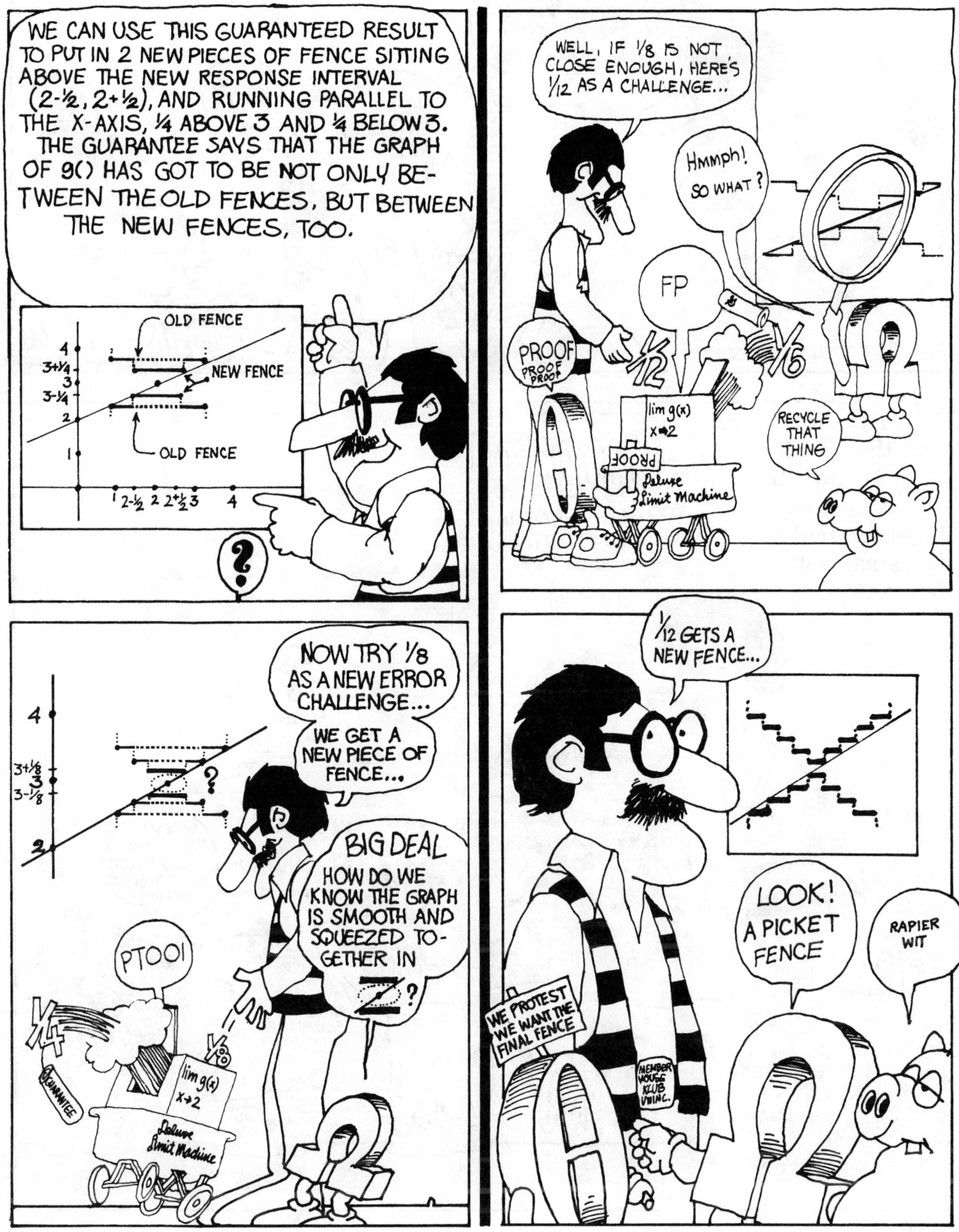

WE CAN USE THIS GUARANTEED RESULT TO PUT IN 2 NEW PIECES OF FENCE SITTING ABOVE THE NEW RESPONSE INTERVAL (2-½, 2+½), AND RUNNING PARALLEL TO THE X-AXIS, ¼ ABOVE 3 AND ¼ BELOW 3. THE GUARANTEE SAYS THAT THE GRAPH OF g() HAS GOT TO BE NOT ONLY BETWEEN THE OLD FENCES, BUT BETWEEN THE NEW FENCES, TOO.
OLD FENCE
NEW FENCE
OLD FENCE
4
3+¼
3
3-¼
2
1
1
2-½
2
2+½
3
4
?
WELL, IF 1/8 IS NOT CLOSE ENOUGH, HERE'S 1/12 AS A CHALLENGE...
Hmmph! SO WHAT ?
FP
PROOF PROOF PROOF
1/12
1/6
lim g(x) x→2
PROOF
Deluxe Limit Machine
RECYCLE THAT THING
NOW TRY 1/8 AS A NEW ERROR CHALLENGE...
WE GET A NEW PIECE OF FENCE...
4
3+1/8
3
3-1/8
2
?
BIG DEAL
HOW DO WE KNOW THE GRAPH IS SMOOTH AND SQUEEZED TOGETHER IN
?
PTOOI
1/4
GUARANTEE
1/8
lim g(x) x→2
Deluxe Limit Machine
1/12 GETS A NEW FENCE...
LOOK! A PICKET FENCE
RAPIER WIT
WE PROTEST WE WANT THE FINAL FENCE
MEMBER VOUGE KLUB VBINC.

LOW DISCOUNT 1/2643
RANTEE
YOUIRREVERENTYRANNICAL WIERDOVERLYPUNKIDS MONSTEROWDYCLOWNS
HE'S MAD
HE'S FRUSTRATED... MOST ILLOGICAL
WOꟻ
ALL FIDEL SAID WUZ "WOꟻ"
VOCAL INFIDEL
Good Vibes

Still no one believes that the graph of g( ) is smooth and squeezed together inside the smallest fence . . .
NOW LOOKIT, YOU GUYS, YA GOTTA BELIEVE IT WITH THESE
BROWLR RARF SNARK
GLUMF...
1/20 1/50
lim g(x) x→2
YOU KICKED THETA BARA MY FWIEND
Deluxe Limit Machine
?
(2,3)
CLOSE-UP BY Grover
WATCH IT WITH THE ROUGH STUFF- THIS IS ME BUDDY, MALICIOUS MELVIN
Grr.
STILL NOT ENUFF
GIT THE MICRO-SCOPE
GOOD VIBES

GLICH
FOOM
FRANG
ROWF
lim g(x)
X→2
Deluxe Limit Machine
MECHANICAL COCKROACH
GOOD VIBES
HE'LL NEVER GET THERE
THAT'S A PROOF?
NO LOGIC THERE
HE KICKED ME
PROOF
GET TO THE POINT!
YOU KICKED HIM
SIGH
YAH
OFF THE DEEP END, MATE

I COULD CONTINUE TO RUN MORE ERROR CHALLENGE NUMBERS THROUGH $\lim_{x \to 2} g(x)$ AND CUSTOM-FIT MORE AND MORE FENCE, BUT THERE IS AN INFINITE NUMBER OF POSSIBLE CHALLENGE NUMBERS AND... AND...
...#!★ I CAN'T PROVE g() IS SQUEEZED TOGETHER!
HMM
PUT HIM OUT TO PASTURE
HE FINALLY FLIPPED
HE'S ABOUT TO DO SOMETHING
GOOD VIBES
HEH, SNARKLE, GUFFAW, I CLAIM THE GRAPH IS SQUEEZED TOGETHER AROUND (2,3) ANYHOW (EVEN THOUGH I CAN'T PROVE IT). LET'S SEE YOU GUYS TRY TO PROVE THAT IT'S **NOT** SMOOTH AND SQUEEZED TOGETHER.
ANYBODY CARE TO CHALLENGE ME WITH AN ERROR-TOLERANCE YOU THINK I CAN'T HANDLE?
∞
$\lim_{x \to 2} g(x)$
GOOD VIBES
D.L.M.
COP OUT!
BUCK PASSER
NOPE, THAT # $\lim_{x \to 2} g(x)=2$ WILL JUST PUT OUT A GUARANTEED 1/500
( )
TRY 0?
1/1000 MAYBE?
GOOD VIBES
D.L.M.
HAW! AS LONG AS I HAVE OL' $\lim_{x \to 2} g(x)$ AROUND WITH HIS GUARANTEED RESPONSES, I CAN CUSTOM-FIT MORE FENCE NO MATTER HOW CLOSELY YOU LOOK-SO I SAY THE GRAPH OF g(x) IS SMOOTH AND SQUEEZED TOGETHER NEAR (2,3) SINCE YOU CAN'T SHOW IT ISN'T. HAH!
SNEAK
HE THINKS HE'S RIGHT JUST CAUSE WE CAN'T PROVE HIM WRONG

NOBODY KNOWS whether or not this legislation really means that the graph of g( ) is "smooth and squeezed together" near (2, 3) ; no one has ever gotten close enough to check the "final infinitesimal" (there isn't any, alas...) to see whether or not this does the trick. We will inquire further into the effectiveness of this legislation later in section II-6 and in a startling INTERGALACTIC encounter in the epilogue.

EXERCISES

II.2.1 When McSquared put $1/8$ into  , out came $1/4$. Check the guarantee, i.e. argue from IF to THEN using algebra.

IF $2 - 1/4 < x < 2 + 1/4$,

THEN $3 - 1/8 < \frac{1}{2}x + 2 < 3 + 1/8$.

II.2.2 When  suggested $1/1000$ as an error-tolerance challenge number, R.D. said that $\lim_{x \to 2} g(x)$ would just put out a guaranteed $1/500$. Check to see that this is really guaranteed.

II·3 ε's AND δ's

Mathematicians have used this tricky argument by legislation to set up a procedure for dealing with any function f( ) whose graph is supposed to be smooth and squeezed together near some point (a, f(a)).

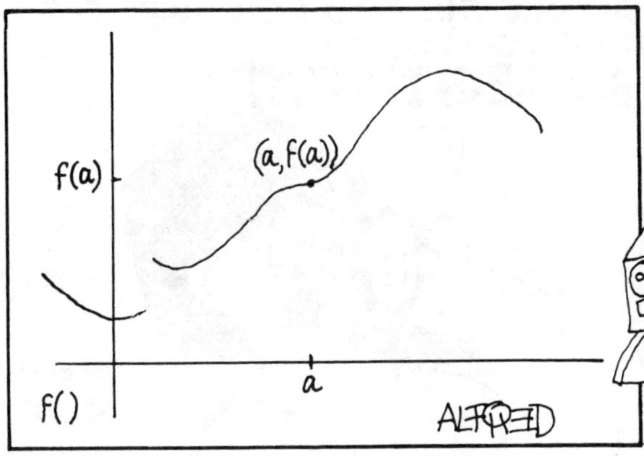

We don't worry about whether or not the graph is REALLY smooth and squeezed together near (a, f(a)) (nobody will ever know.) Instead, WE take over and set the rules: If we can construct a limit-machine that gives a guaranteed response interval for every error-tolerance challenge number, we LEGISLATE that our ability to construct such a limit-machine for f( ) is sufficient justification for us to say that f( )'s graph is smooth and squeezed together near (a, f(a)).

Assume that a limit-machine is constructable for f( ) near a point (a, f(a)). (By legislation, this means that f( )'s graph is officially smooth and squeezed together near (a, f(a)).)

First the limit-machine (called $\lim_{x \to a} f(x)$) gets challenged with a specific positive error-tolerance challenge number. The official mathematical symbol representing any such specific error-tolerance challenge number is "ϵ" (called "epsilon").

Now, this time, when McSquared dumps ϵ into the limit-machine (instead of an actual number)

it stands for a typical challenge to the limit-machine asking it to say HOW CLOSE x has to be to a on the x-axis so that ALFRED'S output f(x) will be close enough to f(a) on the y-axis to be within the particular representative error-tolerance distance ϵ.

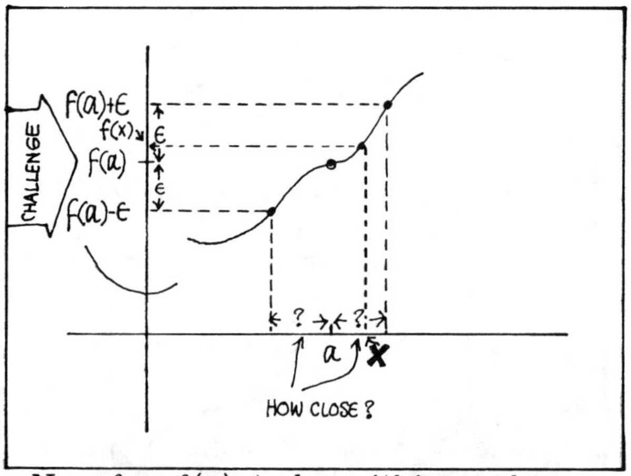

Now for f(x) to be within such an error-tolerance ϵ of f(a) simply means that f(x) must lie within a segment of the y-axis defined by the inequality $f(a) - \epsilon < f(x) < f(a) + \epsilon$.

In response to tossing in this symbol ϵ , the limit-machine puts out another symbol " $\delta_{(\epsilon)}$ " (called "delta-sub-epsilon").

This symbol $\delta_{(\epsilon)}$ stands for a distance: it is the answer to the challenge "HOW CLOSE . . . ?"

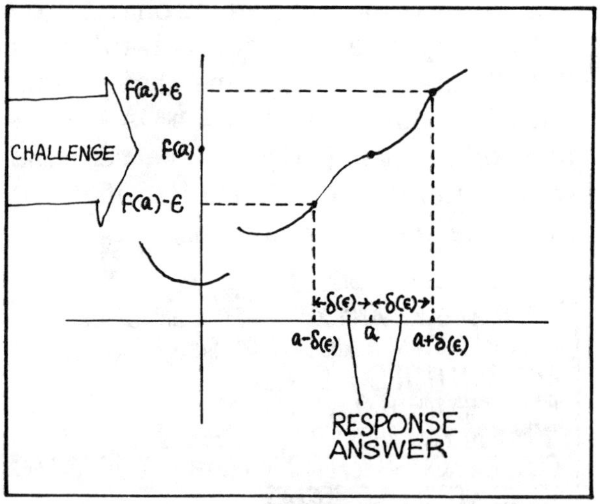

When the limit-machine produces $\delta_{(\epsilon)}$, it CLAIMS that IF x is within distance $\delta_{(\epsilon)}$ of a on the x-axis, THEN ALFRED'S output f(x) will be close enough to f(a) on the y-axis to be within the requested representative error-tolerance ϵ .

If x is to be within distance $\delta_{(\epsilon)}$ of a on the x-axis, then x has to be in the interval

$$(a - \delta_{(\epsilon)} , a + \delta_{(\epsilon)})$$

running from a- $\delta_{(\epsilon)}$ to a+ $\delta_{(\epsilon)}$. So such x's will have to satisfy the inequality

$$a - \delta_{(\epsilon)} < x < a + \delta_{(\epsilon)} .$$

This symbol " $\delta_{(\epsilon)}$ " does NOT stand for " δ" times "(ϵ) "; the blank ( ) in $\delta_{(\ )}$ is like the blank in "f( )". Function f( ) waits for a number , like 13 , and produces another number called "f(13)". $\delta_{(\ )}$ is the symbol for the function of the LIMIT-MACHINE, which waits for an actual error-tolerance challenge, like $1/2$, and is then supposed to produce another number to be called " $\delta_{(1/2)}$." As yet we don't know what this number should be, since we have no explicit FORMULA for $\delta_{(\ )}$.

To sum things up: when challenged with a representative error-tolerance " ϵ ," the limit-machine that belongs with function f( ) at x = a is supposed to produce a representative response with symbol " $\delta(\epsilon)$." The limit-machine then CLAIMS that

This is what the machine CLAIMS is true and eventually it is supposed to produce a guarantee to actually PROVE IT. But of course a limit-machine can't even start to prove anything without knowing exactly

(1) the formula for the specific function in question ,

(2) the number "a" on the x-axis,

and

(3) a FORMULA for $\delta(\epsilon)$,

since an actual PROOF can only be constructed if we know this information.

Let's see if we can figure out how to construct a limit-machine to go with the specific function

$$g(\ ) = \tfrac{1}{2}(\ ) + 2$$

that Grover was running in the last section when the point "a" is 2 .

Now, in this case,

$g(a) = g(2) = \tfrac{1}{2}\cdot(2) + 2 = 3,$

and $g(x) = \tfrac{1}{2}(x) + 2,$.

So the limit-machine's problem is to figure out precisely what the distance $\delta(\epsilon)$ has to be (in response to any specific error-tolerance challenge number ϵ) . . .

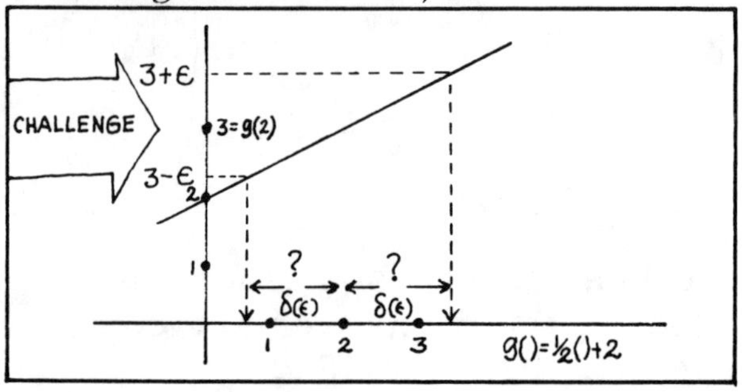

so that it can guarantee that IF we take any x such that

$$\overset{a}{\overset{\downarrow}{2}} - \delta_{(\epsilon)} < x < \overset{a}{\overset{\downarrow}{2}} + \delta_{(\epsilon)}$$

THEN, for those x's ,

$$\underset{g(2)}{\underset{\uparrow}{3}} - \epsilon < \underbrace{\tfrac{1}{2}(x) + 2}_{g(x)} < \underset{g(2)}{\underset{\uparrow}{3}} + \epsilon .$$

But, because ϵ stands for any one of an infinite number of possible error-tolerance challenge numbers, the guarantee can't be shown for the individual response to ALL actual challenge numbers — as McSquared unfortunately found out in the last section. The best that a "proof" of the guarantee can do is to find a formula for $\delta_{(\ )}$ that tells exactly what $\delta_{(\epsilon)}$ should be for ANY challenge ϵ and then use ALGEBRA to get from

"IF $2 - \delta_{(\epsilon)} < x < 2 + \delta_{(\epsilon)}$"

to " THEN $3 - \epsilon < \frac{1}{2}(x) + 2 < 3 + \epsilon$."

If the right formula for $\delta_{(\ )}$ is found, then WHEN AN ACTUAL ERROR CHALLENGE NUMBER IS SUBSTITUTED FOR ϵ , ALGEBRA SAYS THAT THE REASONING WILL ALWAYS BE VALID ! !

In the last section, each time McSquared put an error-tolerance challenge number into  ,

the number that came out was twice the number that was dumped in. Check back - when McSquared challenged the limit machine with $\frac{1}{2}$, out came  , and the guarantee used algebra to show that 1 actually worked! So, in this case, with the error tolerance challenge of $\frac{1}{2}$, the response 1 is twice $\frac{1}{2}$, and proved to be guaranteeable. Later McSquared tried challenging with $\frac{1}{4}$ and got response $\frac{1}{2}$ as the limit-machine's $\delta_{(1/4)}$, again twice the challenge $\frac{1}{4}$. (Notice how the blank in $\delta_{(\ )}$ gets filled in with the specific value of ϵ !) So maybe a good GUESS for a FORMULA for $\delta_{(\ )}$ is to make the response $\delta_{(\epsilon)}$ always TWICE the value of the error challenge ϵ, i.e.,

let $\delta_{(\ )} = 2(\ )$

or, more conventionally,

let $\delta_{(\epsilon)} = 2(\epsilon)$.

NOW THAT WE HAVE A TENTATIVE FORMULA FOR δ(ε), LET'S SEE IF WE CAN USE ALGEBRA TO COMPLETE THE ★ GUARANTEE ! ★
★ IF WE TAKE ANY X SUCH THAT.. 2−δ(ε)<X<2+δ(ε)
★ NOW, SUBSTITUTE 2ε FOR δ(ε) (SINCE δ(ε)=2ε)....... 2−2ε<X<2+2ε
(NOW USE ALGEBRA ON THIS X TO BREW UP g(x)=½x+2)
★ MULTIPLY BY ½ TO GET THE "½" IN "g(x)=½x+2"......... ½·2−½·2ε<½X<½·2+½·2ε
★ ... AND SIMPLIFY, SO........ 1−ε<½X<1+ε
★ AND ADD 2 TO GET THE "2" IN g(x)=½(x)+2......... 2+1−ε<½x+2<2+1+ε
★★★ THEN............. 3−ε<½X+2<3+ε
OR
3−ε<g(x)<3+ε !

THE GUARANTEE WORKS! SINCE THIS BIT OF ALGEBRA IS GENUINELY OK FOR ANY ε, ANY TIME WE CHALLENGE WITH ANY ACTUAL ERROR CHALLENGE-NUMBER (NOT JUST THE SYMBOL ε), ALGEBRA LETS US "PLUG IN" TWICE THAT NUMBER IN PLACE OF δ(ε) (SINCE NOW δ()=2·()) AND THE GUARANTEE HAS TO HOLD! ALGEBRA SAYS IT IS SO!
IT WORKS
THE MANAGEMENT

So mathematicians have declared that the following RITUAL is an officially acceptable "proof."

OFFICIAL MATHEMATICAL JARGON

PROOF that the graph of

$$g(\ ) = \tfrac{1}{2}(\ ) + 2$$

is smooth and squeezed together near $(2, g(2))$.

1. For any $\epsilon > 0$, let $\delta(\epsilon) = 2\epsilon$.

   (This is the same as constructing the limit machine "$\lim_{x \to 2} g(x)$" that puts out two times the number "ϵ" it is challenged with.)

2. Guarantee: IF $2 - \delta(\epsilon) < x < 2 + \delta(\epsilon)$

   (Substitute 2ϵ for $\delta(\epsilon)$) $2 - 2\epsilon < x < 2 + 2\epsilon$

   (Multiply by $\tfrac{1}{2}$) $1 - \epsilon < \tfrac{1}{2}x < 1 + \epsilon$

   (Add $+2$) $3 - \epsilon < \tfrac{1}{2}(x) + 2 < 3 + \epsilon$

   or, substituting back for $g(\ )$: THEN $g(2) - \epsilon < g(x) < g(2) + \epsilon$.

   (And this is all that is needed in the guarantee!)

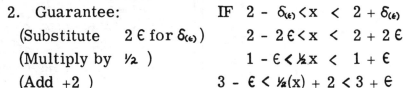

Now Professor McSquared will put all this theory about general functions f( ) and points "a" into an Official Mathematical Definition! Watch out for the new word "continuous," which is used instead of "smooth and squeezed together." ("Squeezed together" is definitely NOT official.)

OFFICIAL MATHEMATICAL DEFINITION

IF f() IS A FUNCTION,

1. $\lim_{x \to a} f(x)$ EXISTS AND EQUALS $f(a)$ **IF**, FOR EVERY CHALLENGE NUMBER $\epsilon > 0$, THERE IS A RESPONSE NUMBER $\delta(\epsilon) > 0$ SUCH THAT, **IF** X SATISFIES THE CONDITION $a - \delta(\epsilon) < x < a + \delta(\epsilon)$ **THEN**, FOR THESE VALUES OF X, $f(a) - \epsilon < f(x) < f(a) + \epsilon$.

2. f() IS SAID TO BE <u>CONTINUOUS</u> AT $x = a$ IF $\lim_{x \to a} f(x)$ EXISTS AND EQUALS $f(a)$.

(CONFIDENTIAL AND UNOFFICIAL TRANSLATION)

IF 

IS A FUNCTION MACHINE,

1. A GUARANTEED LIMIT MACHINE EXISTS FOR FUNCTION f() NEAR $(a, f(a))$ IF THERE IS A MACHINE 

THAT HAS THE PROPERTY THAT WHENEVER YOU DUMP IN AN ERROR-TOLERANCE CHALLENGE NUMBER WITH THE SYMBOL $\epsilon > 0$, 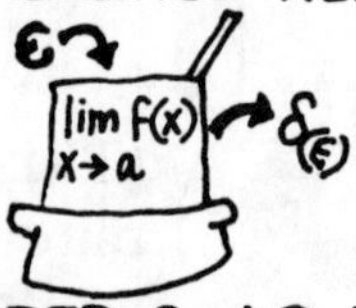

OUT COMES ANOTHER NUMBER $\delta(\epsilon) > 0$ THAT IS <u>GUARANTEED</u> AS FOLLOWS: IF YOU TAKE ANY VALUE OF X SATISFYING $a - \delta(\epsilon) < x < a + \delta(\epsilon)$ AND RUN IT THROUGH 

, THEN THE RESULT $f(x)$ SATISFIES $f(a) - \epsilon < f(x) < f(a) + \epsilon$.

2. THE GRAPH OF f() IS SMOOTH AND SQUEEZED TOGETHER NEAR $(a, f(a))$ IF 

EXISTS, READY TO DO ITS ϵ, $\delta(\epsilon)$ THING.

Mathematicians often sum this up by saying just "the limit of f(x) as x goes to a IS f(a)" or, in jargon

$$\lim_{x \to a} f(x) = f(a) \quad ,$$

which gets us back to the intuitive idea that (for smooth f( )) as x gets closer to a , f(x) approaches f(a) as its LIMIT number.

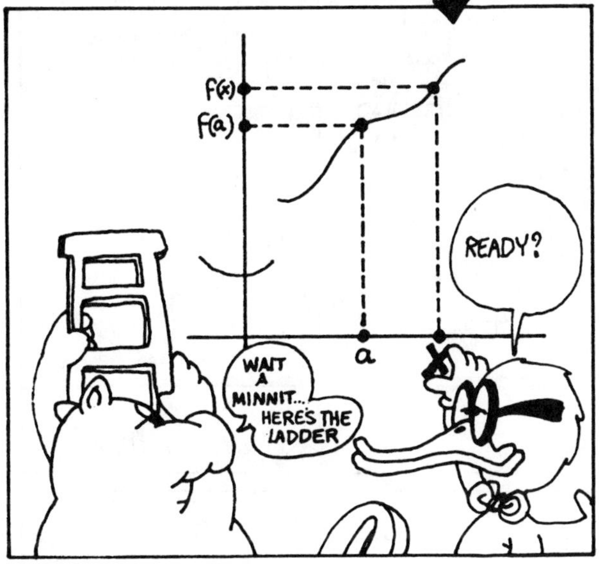

But we can never get in close enough to actually watch the final arrival of x at a and, at the same time, see the arrival of f(x) at f(a) ! ! ! Instead, we have to be content with the official LOGICAL meaning of

$$\lim_{x \to a} f(x) = f(a) \quad ,$$

which is always ONLY that an appropriate limit machine exists. Then, with this guaranteed limit-machine, we CAN show that we can get f(x) "as close to f(a) as is ever necessary" (the same as "within any ϵ error tolerance") by getting x "close enough" to a (just make the distance from x to a less than $\delta(\epsilon)$). Somehow this isn't quite the same as the intuitive idea, but it is the best that we can do.

After all that heavy stuff, let's go back and have a look at a couple of properties of the official definition of limits and continuity.

Challenge intervals have to be genuine intervals of positive length, so ϵ should be greater than zero. $\delta_{(\epsilon)}$ must be greater than zero too! This is because any function h( ), whether or not its graph is squeezed together near (a, h(a)), has only ONE point on its graph above (or below) x = a, namely (a, h(a)). (Check the rules for functions: There is only one correspondence arrow from "a" in the domain.) For any challenge ϵ, we could always "fence in" the POINT (a, h(a)) with "one-point" fences at the coordinates

$$(a,\ h(a)+\epsilon) \text{ and } (a,\ h(a)-\epsilon).$$

But this wouldn't prove anything about whether or not the graph of h( ) is actually "squeezed together" NEAR (a, h(a)). So, since $\lim_{x\to a} h(x)$ is trying to show that the graph of h( ) is squeezed together NEAR (a, h(a)), we LEGISLATE that $\lim_{x\to a} h(x)$ must always produce a response-interval

$$(a-\delta_{(\epsilon)},\ a+\delta_{(\epsilon)})$$

with positive length, i.e., in our official mathematical definition we must have $\delta_{(\epsilon)} > 0$ for ANY challenge number $\epsilon > 0$.

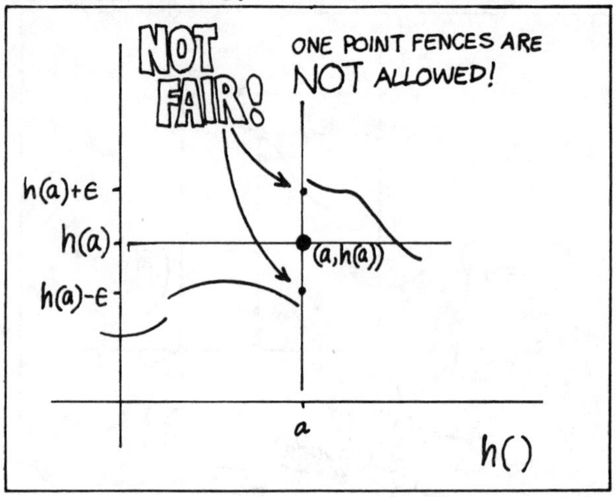

Let's have a look at what would happen if a challenge interval of zero length were allowed. If some $\delta_{(0)}$ were produced of positive size, the guarantee on such a $\delta_{(0)}$ would force the graph of the function above the interval $(a-\delta_{(0)}, a+\delta_{(0)})$ to be a flat line. This would make the argument applicable for flat-line (constant) functions only, and that would restrict the definition of continuity so badly that it wouldn't be of any use. (As before, if we let $\delta_{(0)} = 0$, we would end up saying nothing at all about the function being continuous.) So we have to have $\epsilon > 0$ to get anywhere.

EXERCISES

II.3.1 If $f(\ ) = 3(\ )-1$, $\lim_{x\to 2} f(x)$ puts out $\delta_{(\epsilon)}$ with formula $\delta_{(\epsilon)} = \frac{1}{3}\epsilon$. Check the guarantee; i.e., letting "a"=2, reason from

IF $2-\delta_{(\epsilon)} < x < 2+\delta_{(\epsilon)}$

to THEN $f(2) - \epsilon < f(x) < f(2) + \epsilon$.

II.3.2 If $f(\ ) = -3(\ )+5$, $\lim_{x\to 1} f(x)$ puts out $\delta_{(\epsilon)} = \frac{1}{3}\epsilon$. Check the GUARANTEE. (Don't forget that multiplying inequalities by a negative number flips the inequality signs!)

II.3.3 Using the same function as II.3.2 , suppose that $\lim_{x\to 1} f(x)$ puts out $\delta_{(\epsilon)} = \frac{1}{6}\epsilon$ instead. Show that this can still be guaranteed. (Hint: use the fact that $\frac{1}{2}\epsilon < \epsilon$ since $\epsilon > 0$.)

IF $1 - \frac{1}{6}\epsilon < x < 1 + \frac{1}{6}\epsilon$ (NOW USE ALGEBRA TO WORK DOWN FROM HERE)

THEN $2 - \epsilon < -3(x)+5 < 2 + \epsilon$.

Hoping that you have worked out the guarantee all right, this shows that there need NOT be a unique value for $\delta_{(\epsilon)}$ — there can be more than one way to program a limit-machine to show that the same function is continuous at the same point!

I·4 (SECRET) COMPUTATIONS

McSquared will try to solve the following problem: He will try to PROVE that the function

$$f(\ ) = 4 - 2(\ )$$

is continuous at $x = 1$, which is the same as trying to show that

$$\lim_{x \to 1} (4 - 2(x))$$

exists and equals $f(1) = 4 - 2(1) = 2$.

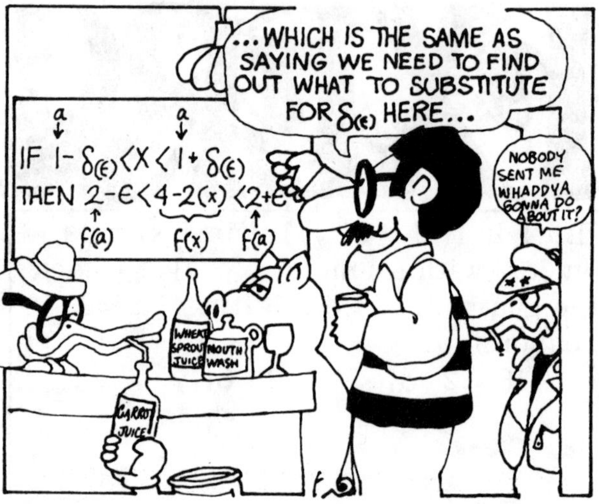

The secret computation starts with the challenge interval $(2-\epsilon, 2+\epsilon)$ on the range of f( ) and figures out where it "bounces" back down to a patch on the domain of f( ).

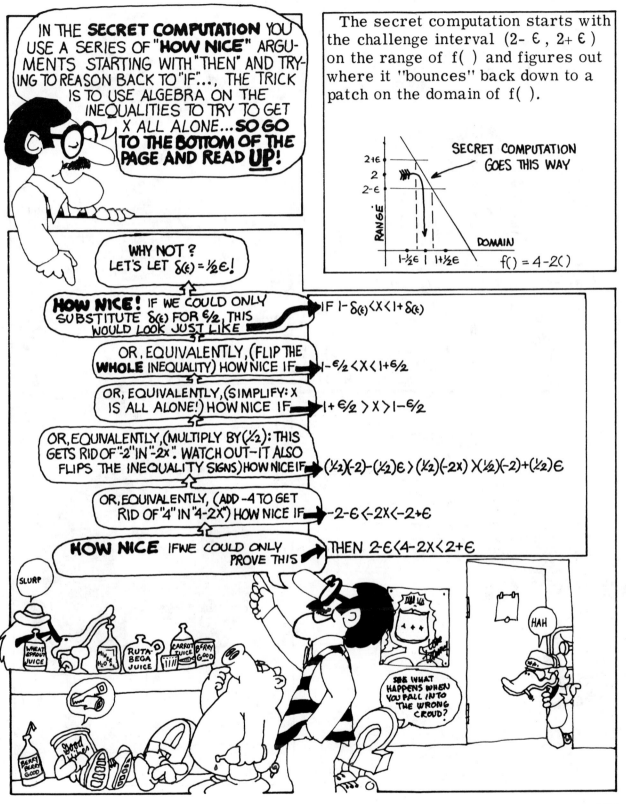

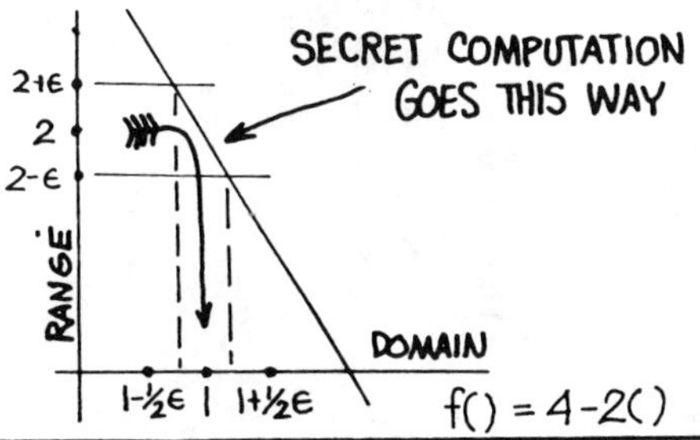

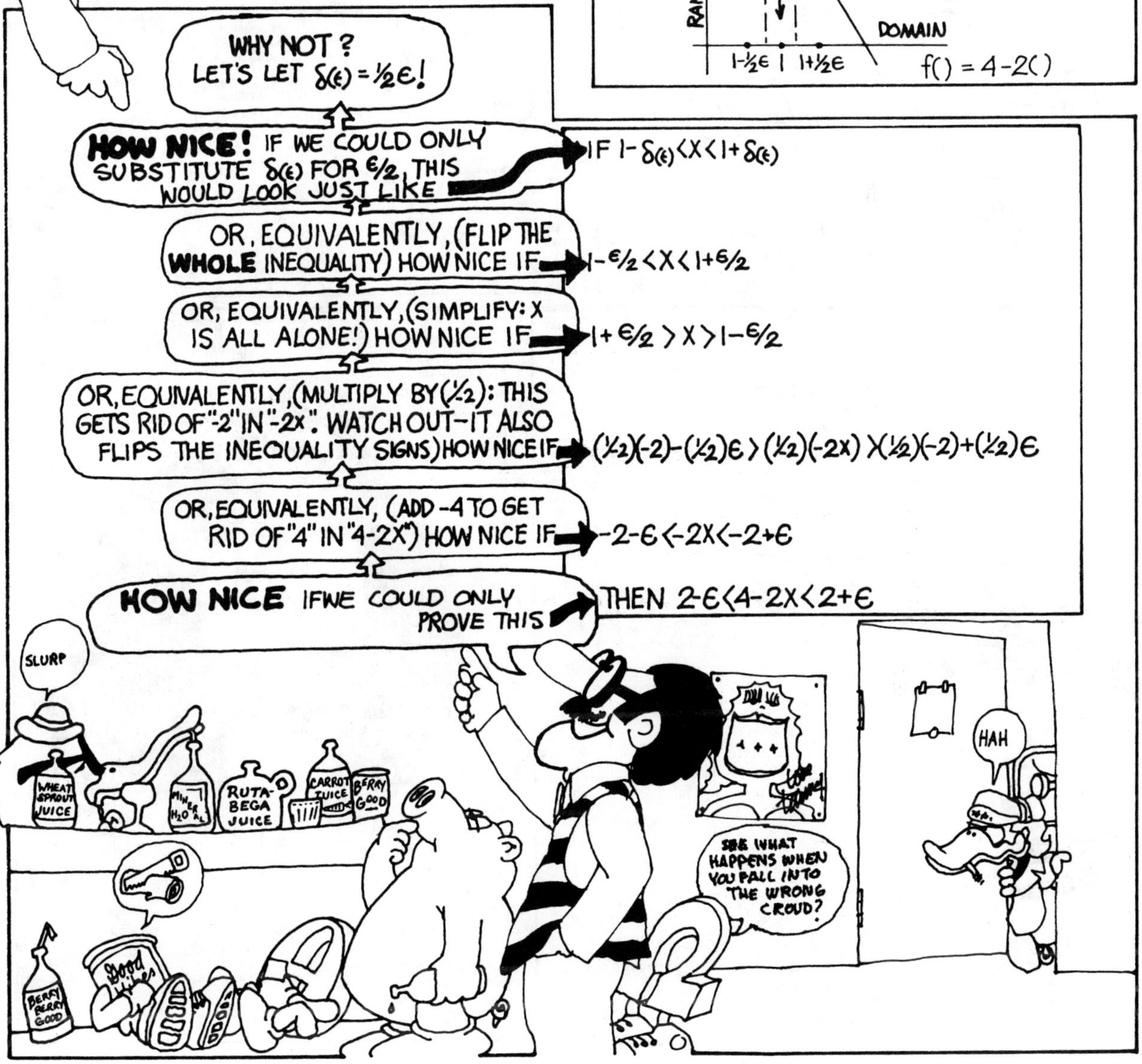

SO, YOU SEE, THIS SECRET COMPUTATION IS REALLY A HANDY METHOD TO...
SNARK
ANTI-DIET SQUAD
RAID
ANTI-DIET SQUAD
GROWF YOU #@ RUBBERDUCK MOTORCYCLEHIPPIECREEP-FREAK PSYCHOCOMMIE WISEGUY-WIERDO RATFINK!
QUICK-HIDE THE TOFU..
WHAZZAMATTER CHUCKLES, YA CAN'T TAKE A YUK?
AHEM... BACK TO LIMIT MACHINES...
TRASH

BUT WE CAN ARRANGE THAT $\delta(\epsilon)=\tfrac{1}{2}(\epsilon)$ BY JUST PROGRAMMING $\lim_{x\to 1}(4-2x)$ WITH THE FORMULA $\delta()=\tfrac{1}{2}()$! THIS CONSTRUCTS THE LIMIT-MACHINE — JUST WHAT WE SET OUT TO DO! CHECK THE GUARANTEE — IT OUGHT TO GO THROUGH IF THE SECRET COMPUTATION IS REVERSIBLE

OFFICIAL PROOF THAT $\lim_{x\to 1}(4-2(x))$ EXISTS AND EQUALS 2.

1. FOR ANY $\epsilon>0$, LET $\delta(\epsilon)=\tfrac{1}{2}\epsilon$
2. IF $1-\tfrac{1}{2}\epsilon < x < 1+\tfrac{1}{2}\epsilon$ (NOW TRY TO "BREW UP" $f(x)=4-2(x)$ BY USING ALGEBRA ON THIS INEQUALITY)

MULTIPLY BY (-2) TO GET "$-2x$" IN $f(x)=4-2x$. $-2+\epsilon > -2(x) > -2-\epsilon$

ADD 4 TO GET THE "4" IN $f(x)=4-2x$ THEN $2+\epsilon > 4-2(x) > 2-\epsilon$

Guaranteed!

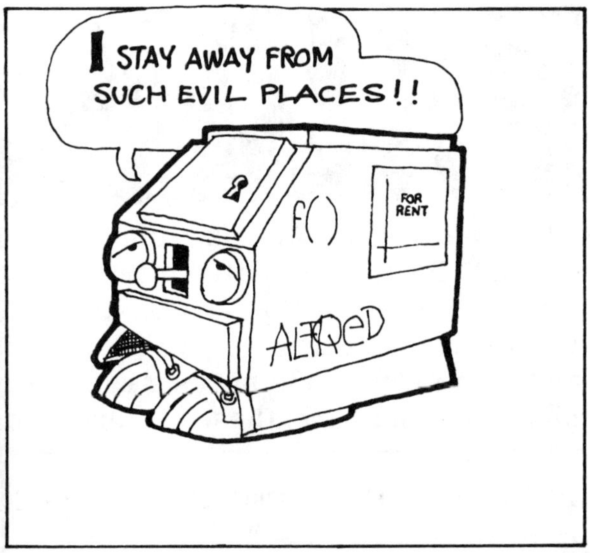

Although the Secret Computation reasons from the challenge in the RANGE of f( ) BACK to the response interval in the DOMAIN, the Official Proof of the guarantee must always proceed the other way: from x in the response interval (in the domain) to certain f(x)'s (in the range), hopefully landing within the challenge interval. This is the same direction as the arrows in the original definition of functions.

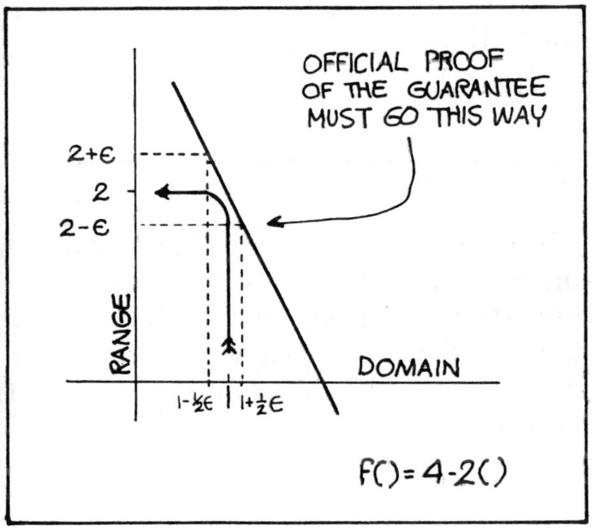

EXERCISES

Prove that the following functions are continuous at the appropriate value of "a." By the legislated official mathematical definition, this means that you have to show $\lim_{x \to a} f(x)$ exists and equals $f(a)$ by constructing a suitable limit machine that gives a formula for what $\delta_{(\epsilon)}$ should be in terms of ϵ, and then proving the necessary guarantee. Secret computations should be hidden behind parentheses. Here is one worked out:

PROBLEM: Prove that $f(\ ) = 3(\ ) - 2$ is continuous at $x = 2$.
Translation: Prove that $\lim_{x \to 2}(3(x) - 2)$ exists and equals $f(2) = 3(2) - 2 = 4$.

1. For any $\epsilon > 0$, let $\delta_{(\epsilon)} = \frac{1}{3}\epsilon$.

2. If $\qquad 2 - \frac{1}{3}\epsilon < x < 2 + \frac{1}{3}\epsilon$
(Now brew up $f(x) = 3(x) - 2$)
(times 3) $\qquad 6 - \epsilon < 3(x) < 6 + \epsilon$
(add -2) $\qquad 4 - \epsilon < 3(x) - 2 < 4 + \epsilon$.
and this is just what had to be guaranteed!

Secret Computation: Start with THEN and work back towards IF. Try to get x all alone between $2 - (?)$ and $2 + (?)$. Why 2 ? It's the number here: $\lim_{x \to 2}(3(x) - 2)$

IF $\quad 2 - \delta_{(\epsilon)} < x < 2 + \delta_{(\epsilon)}$

HOW NICE if $\quad \delta_{(\epsilon)} = \frac{1}{3}\epsilon$!!!

(times $\frac{1}{3}$) $\quad 2 - \frac{1}{3}\epsilon < x < 2 + \frac{1}{3}\epsilon$

(add +2) $\quad 6 - \epsilon < 3(x) < 6 + \epsilon$

THEN $\quad 4 - \epsilon < 3(x) - 2 < 4 + \epsilon$

Notice that the actual inequalities in both the secret computation and the official proof are exactly the same! Only the explanations of how to go from one step to the next are different. After you practice a few proofs, you usually don't even have to bother writing down the secret computation at the side. You can just write the inequalities in the official guarantee from the bottom up!

II.4.1 Prove that $f(\ ) = \frac{1}{2}(\ ) + 3$ is continuous at $x = 4$.

II.4.2 Prove that $f(\ ) = -\frac{1}{9}(\ ) + 5$ is continuous at $x = -3$.

II.4.3 Prove that $f(\ ) = 27 - \frac{1}{10}(\ )$ is continuous at $x = 5$.

You may have noticed that one formula that seems to always work is to let $\delta_{(\epsilon)} = \frac{1}{|\text{slope of the line}|} \cdot (\epsilon)$.

This actually WILL be O.K., but for straight lines only! Check it: From the picture,

$$(\text{slope of the line}) = \frac{\text{rise}}{\text{run}} = \frac{\epsilon}{\delta_{(\epsilon)}},$$

and you can solve for $\delta_{(\epsilon)}$ to get

$$\delta_{(\epsilon)} = \frac{1}{\text{slope}} \cdot (\epsilon) \quad !$$

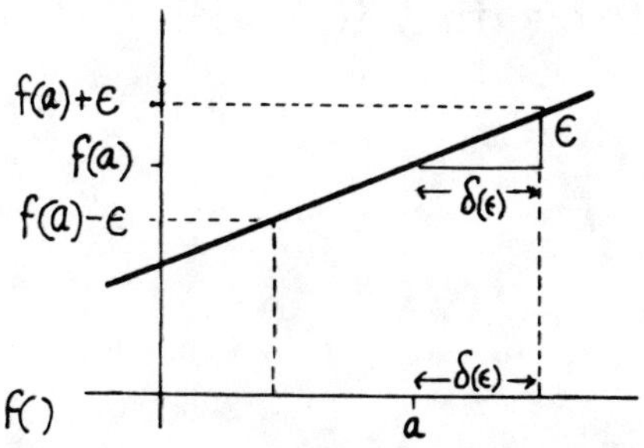

(But remember that other different $\delta_{()}$'s may also work as well - just as long as you can come up with a guarantee that goes through.)

II.4.4 Prove that f( ) = 3 - 18( ) is continuous at x = -6.

II.4.5 Prove that f( ) = $-\frac{2}{5}$( ) + $\frac{7}{5}$ is continuous at x = $-\frac{13}{2}$.

II.4.6 Prove that f( ) = 55( ) + 44 is continuous at x = 11.

II.4.7 Prove that f( ) = 2( ) + 7 is continuous at x = a for ANY real number "a".

II.4.8 Prove that f( ) = ( ) (f(x) = x) is continuous at x = a for any real number "a".

II·5 MORE SECRET COMPUTATIONS ON PARABOLAS!

FEATURING EVEN MORE ELABORATE LIMIT MACHINES!

Once we get away from simple functions with graphs that are straight lines, it gets considerably more difficult to figure out a formula for $\delta_{(\epsilon)}$ in terms of ϵ in order to be able to properly program a limit machine.

For example, suppose we try to show that

$$f(\ ) = (2\cdot(\ )^2 - 3\cdot(\ ) + 1)$$

is continuous at $x = 2$, which means somehow coming up with a proof that

$$\lim_{x \to 2} (2x^2 - 3x + 1)$$

exists and equals $f(2)$.

The function

$$f(x) = 2x^2 - 3x + 1$$

is a parabola, looking like $\cup$ since the coefficient of "x^2" is greater than zero (in this case "2"). The "nose" x-value is

$$x = \frac{-(-3)}{2\cdot 2} = \frac{3}{4}$$

(from page 62).

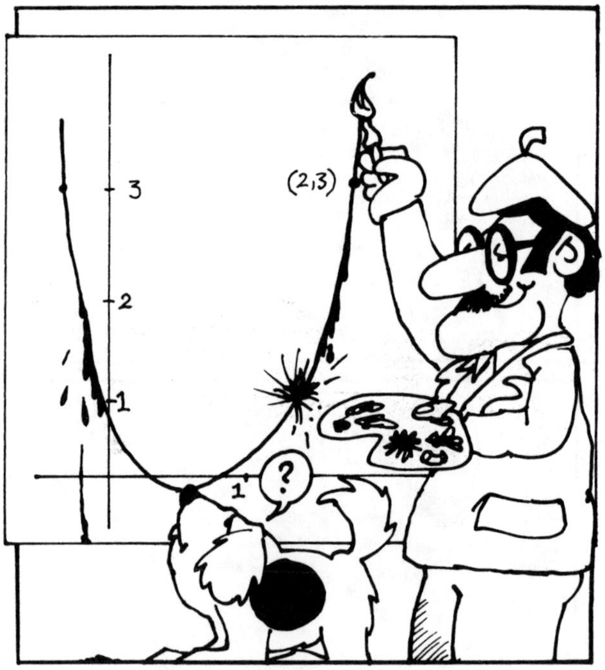

We want to show that the graph of $f(\ )$ is smooth and squeezed together near $(2, f(2))$— since

$$f(2) = 2\cdot(2)^2 - 3\cdot(2) + 1 = 3\ ,$$

this is the point $(2, 3)$.

This time, if we challenge with the error-tolerance interval

$$(\underset{f(2)}{3}-\epsilon, \underset{f(2)}{3}+\epsilon)\ ,$$

we find that this interval bounces back down to the x-axis in two patches* instead of one. However, the patch on the left doesn't count— trying to prove that $\lim_{x\to 2} f(x)$ exists means we are only interested in patches near 2 on the x-axis.

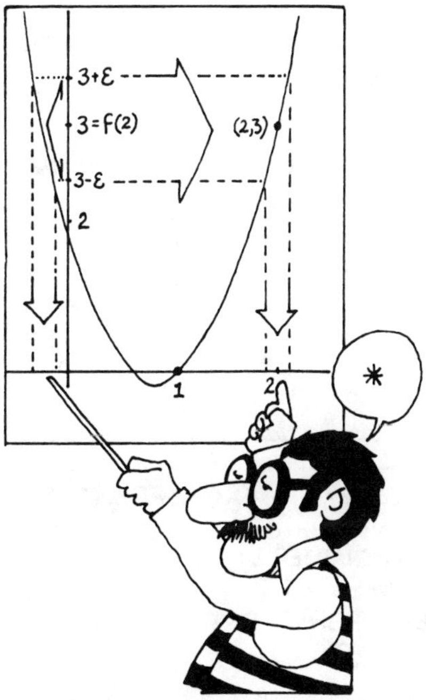

3+ε
3=F(2)
3-ε
2
(2,3)
1
2
*

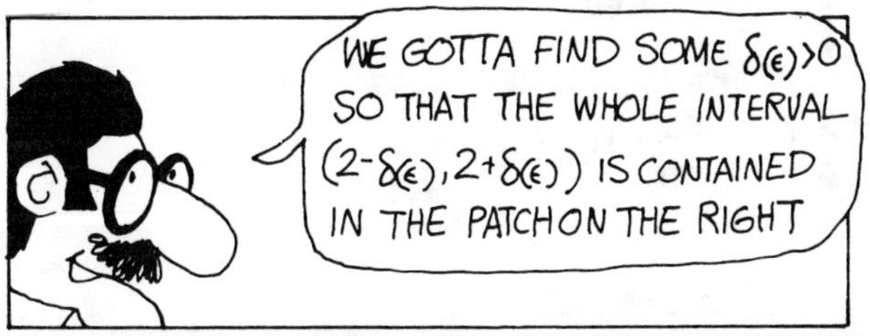

WE GOTTA FIND SOME δ(ε)>0 SO THAT THE WHOLE INTERVAL (2-δ(ε), 2+δ(ε)) IS CONTAINED IN THE PATCH ON THE RIGHT

NOW, TRY A LITTLE CALCULATING USING SOME MORE SECRET COMPUTATIONS!
ARGGH! NOT MORE SECRET COMPUTATIONS!
TEST PATTERN
PUT COIN IN SLOT
BRONK
ZAP
FLEEP

SOMEHOW WE HAVE TO FIND A FORMULA FOR δ(ϵ) SO WE CAN SUBSTITUTE IT IN "IF 2-δ(ϵ) < X < 2+δ(ϵ)" AND USE ALGEBRA ON THIS INEQUALITY TO EVENTUALLY PROVE THAT "THEN f(2)-ϵ < f(x) < f(2)+ϵ".
FIRST CHANGE "2-δ(ϵ) < X < 2+δ(ϵ)" TO AN EXPRESSION INVOLVING ABSOLUTE VALUES. USE THE STUFF FROM CHAPTER I (1.3), PAGE 25 ABOUT ABSOLUTE VALUES THAT SAYS YOU CAN GO BACK AND FORTH BETWEEN 2-δ(ϵ) < X < 2+δ(ϵ) AND (SUBTRACT 2) -δ(ϵ) < X-2 < δ(ϵ) TO GET EQUIVALENT |X-2| < δ(ϵ).
SINCE f(2)=2(2)²-3(2)+1 = 8-6+1 = 3, AND f(x)=2x²-3x+1, THE EXPRESSION "f(2)-ϵ < f(x) < f(2)+ϵ" FIRST TRANSLATES TO 3-ϵ < 2x²-3x+1 < 3+ϵ AND THEN, TO USE ABSOLUTE VALUES, SUBTRACT 3: -ϵ < 2x²-3x+1-3 < ϵ AND SIMPLIFY TO GET -ϵ < 2x²-3x-2 < ϵ AND FINALLY, WITH ABSOLUTE VALUES, |2x²-3x-2| < ϵ.
SO, NOW THE SECRET COMPUTATION HAS TO TELL US HOW TO GET SOME FORMULA FOR δ(ϵ) SO THAT WE CAN SUBSTITUTE THE FORMULA IN "IF |X-2| < δ(ϵ)" AND USE ALGEBRA TO SHOW THAT, FOR VALUES OF X SATISFYING THIS, "THEN |2x²-3x-2| < ϵ" MUST HOLD.
IN THE SECRET COMPUTATION WE REASON BACKWARD AS USUAL FROM "HOW NICE IF |2x²-3x-2| < ϵ," HOPEFULLY ENDING UP WITH "HOW NICE IF |X-2| < (SOMETHING INVOLVING ϵ)." WE CAN THEN DISCOVER WHAT (SOMETHING INVOLVING ϵ) SHOULD BE, NEXT WE CAN LEGISLATE IT TO BE δ(ϵ), AND FINALLY WE CAN RUN THE "HOW NICE" ARGUMENT BACKWARD TO GET AN OFFICIAL PROOF (WHICH MUST GO FROM "IF" TO "THEN").
BUT THERE'S SOME TROUBLE ALONG THE WAY. LET'S SEE WHAT HAPPENS.
NOTES

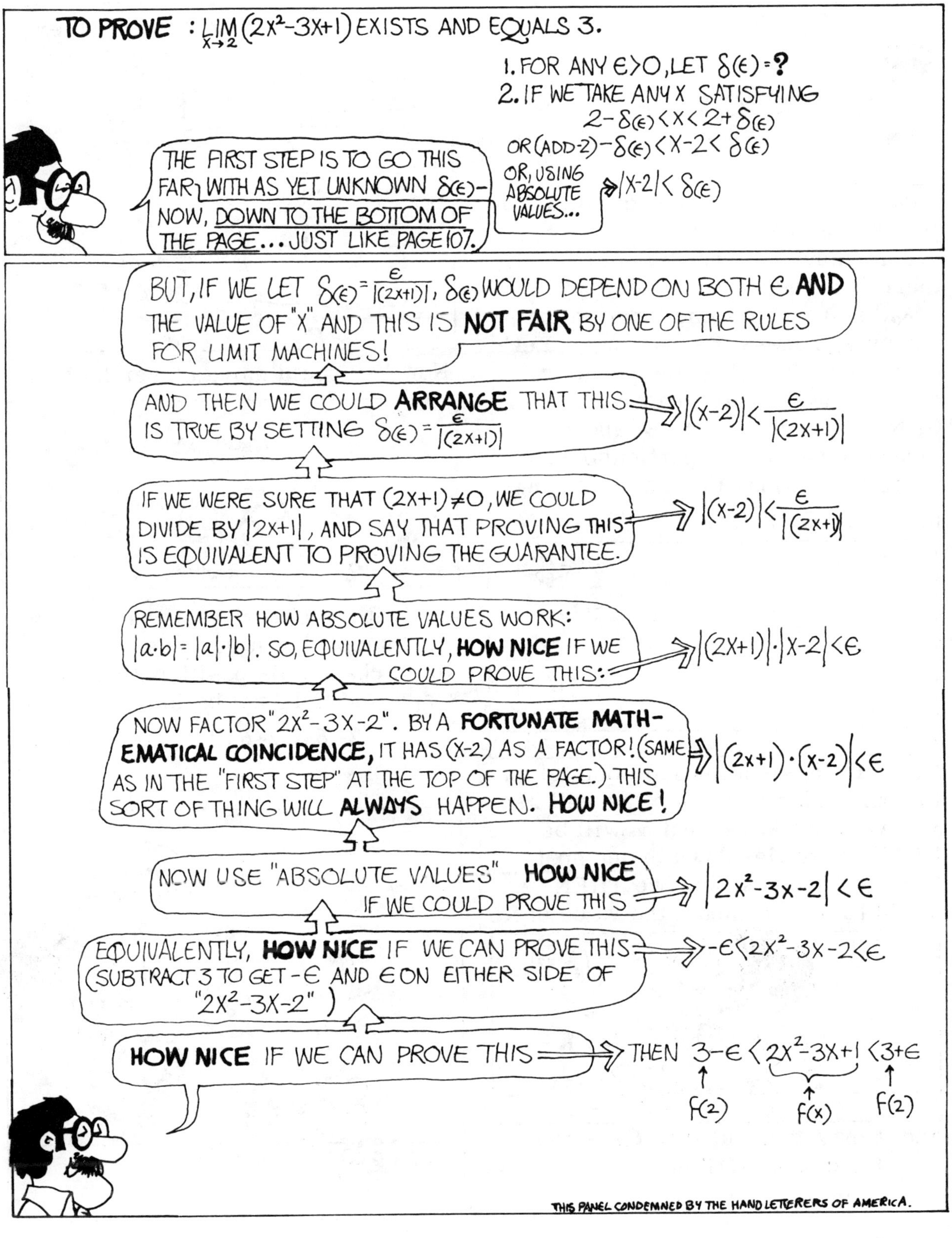

TO PROVE : $\lim_{x \to 2}(2x^2-3x+1)$ EXISTS AND EQUALS 3.
1. FOR ANY $\epsilon > 0$, LET $\delta(\epsilon) = ?$
2. IF WE TAKE ANY X SATISFYING $2-\delta(\epsilon) < x < 2+\delta(\epsilon)$
OR (ADD -2) $-\delta(\epsilon) < x-2 < \delta(\epsilon)$
OR, USING ABSOLUTE VALUES... $|x-2| < \delta(\epsilon)$
THE FIRST STEP IS TO GO THIS FAR WITH AS YET UNKNOWN $\delta(\epsilon)$ — NOW, DOWN TO THE BOTTOM OF THE PAGE... JUST LIKE PAGE 107.
HOW NICE IF WE CAN PROVE THIS ⟹ THEN $3-\epsilon < 2x^2-3x+1 < 3+\epsilon$
f(2) f(x) f(2)
EQUIVALENTLY, HOW NICE IF WE CAN PROVE THIS ⟹ $-\epsilon < 2x^2-3x-2 < \epsilon$ (SUBTRACT 3 TO GET $-\epsilon$ AND ϵ ON EITHER SIDE OF "$2x^2-3x-2$")
NOW USE "ABSOLUTE VALUES" HOW NICE IF WE COULD PROVE THIS ⟹ $|2x^2-3x-2| < \epsilon$
NOW FACTOR "$2x^2-3x-2$". BY A FORTUNATE MATHEMATICAL COINCIDENCE, IT HAS (X-2) AS A FACTOR! (SAME AS IN THE "FIRST STEP" AT THE TOP OF THE PAGE.) THIS SORT OF THING WILL ALWAYS HAPPEN. HOW NICE! ⟹ $|(2x+1)\cdot(x-2)| < \epsilon$
REMEMBER HOW ABSOLUTE VALUES WORK: $|a \cdot b| = |a| \cdot |b|$. SO, EQUIVALENTLY, HOW NICE IF WE COULD PROVE THIS: ⟹ $|(2x+1)| \cdot |x-2| < \epsilon$
IF WE WERE SURE THAT $(2x+1) \neq 0$, WE COULD DIVIDE BY $|2x+1|$, AND SAY THAT PROVING THIS IS EQUIVALENT TO PROVING THE GUARANTEE. ⟹ $|(x-2)| < \frac{\epsilon}{|(2x+1)|}$
AND THEN WE COULD ARRANGE THAT THIS IS TRUE BY SETTING $\delta(\epsilon) = \frac{\epsilon}{|(2x+1)|}$ ⟹ $|(x-2)| < \frac{\epsilon}{|(2x+1)|}$
BUT, IF WE LET $\delta(\epsilon) = \frac{\epsilon}{|(2x+1)|}$, $\delta(\epsilon)$ WOULD DEPEND ON BOTH ϵ AND THE VALUE OF "X" AND THIS IS NOT FAIR BY ONE OF THE RULES FOR LIMIT MACHINES!
THIS PANEL CONDEMNED BY THE HAND LETTERERS OF AMERICA.

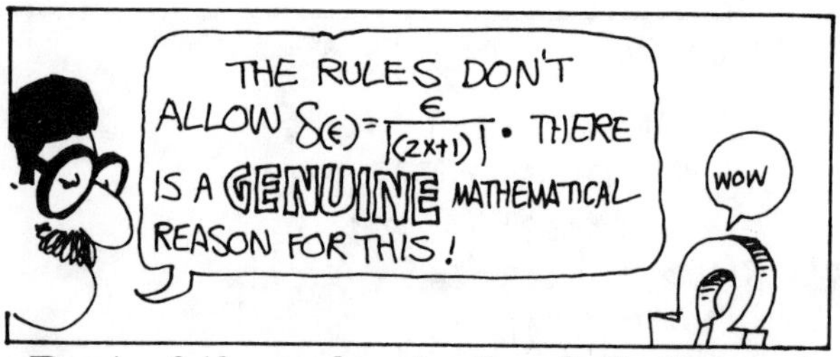

Part of the rules in the definition of limits goes like this:
"...for any challenge number $\epsilon > 0$, there is a response number $\delta_{(\epsilon)} > 0$...."
"$\delta_{(\epsilon)}$" stands for some function that can only process ϵ-challenges. Let's have a closer look at the peculiar cat-chasing-its-tail reason why $\delta_{(\,)}$ CAN'T depend on both ϵ and x .

Once x parks at a particular spot near 2 , HE may think he is sitting in $(2-\delta_{(\epsilon)}, 2+\delta_{(\epsilon)})$,

BUT, IF $\delta_{(\epsilon)}$ is allowed to depend on the particular value of x as well as ϵ , we won't know what $\delta_{(\epsilon)}$ will be UNTIL x settles down in one spot. Once x is in position, we THEN would have to compute the value of $\delta_{(\epsilon)}$

and it may turn out that $\delta_{(\epsilon)}$ is too small and x is NOT in

$(2-\delta_{(\epsilon)}, 2+\delta_{(\epsilon)})$ after all.

Now "x" could move closer to 2 . . .

but, with x changing his value, $\delta_{(\epsilon)}$ (still supposing it depends on x) would change his position also, and x again MIGHT not be in

$(2-\delta_{(\epsilon)}, 2+\delta_{(\epsilon)})$.

This could go on indefinitely, and

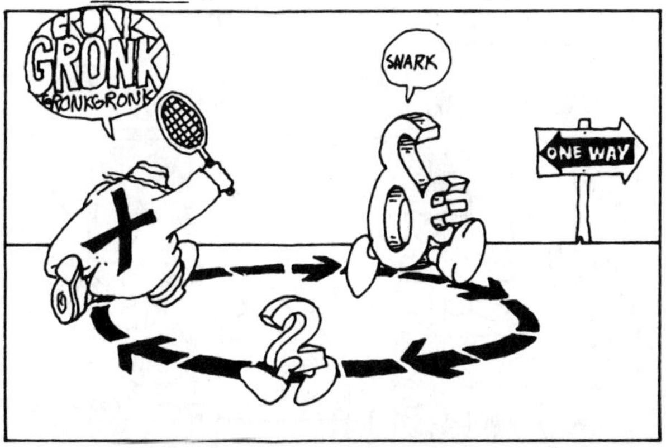

we would never get around to proving the guarantee, and that will never do! So $\delta_{(\epsilon)}$ CAN'T be allowed to depend on x as well as on ϵ .

So the secret computation, as far as we have taken it, still leaves us with the problem of defining $\delta_{(\epsilon)}$ in terms of ϵ ONLY, so that for all values of x satisfying $2-\delta_{(\epsilon)} < x < 2+\delta_{(\epsilon)}$ or equivalently, for all values of x such that $|x-2| < \delta_{(\epsilon)}$, the expression $|(x-2)|\cdot|(2x+1)| < \epsilon$ will hold true for all these values of x.

If we could pull this off, then we could run the reversible "How Nice..." argument in the secret computation backward and end up with a proof of the guarantee!

In order to do this, mathematicians first hunted for a way to control the troublesome factor "2x+1."

Consider for a moment the role of the $\delta_{(\epsilon)}$ that the limit-machine

"$\lim\limits_{x\to 2}(2x^2-3x+1)$ "

is supposed to produce. The $\delta_{(\epsilon)}$ is in the business of restricting the values of x under consideration to lie in a small interval in the vicinity of 2, and once x is required to be near 2, we will obviously restrict the range of values of "2x+1" as well.

For example, if we settle on a $\delta_{(\epsilon)}$ less than or equal to 1, so that our x's all remain within a distance 1 of the point 2 on the x-axis,

this could give us a key to controlling the value of "2x+1" and then we ought to be able to find a formula for $\delta_{(\epsilon)}$. So let's assume that we have made sure that any $\delta_{(\epsilon)}$ put out by the limit-machine is less than or equal to 1. And since that is our business—programming the limit-machine—there is no reason at all why we can't do it.

Now, if $|x-2| < \delta_{(\epsilon)}$ which in turn is less than or equal to 1: $|x-2| < \delta_{(\epsilon)} \leq 1$;
then $|x-2| < 1$.
This translates to $-1 < x-2 < 1$
or $1 < x < 3$
and such x's are truly within a distance 1 of 2.

Now let's see how this process of limiting $\delta_{(\epsilon)}$ to be less than or equal to 1 gives us some leverage on the size of the troublesome factor $|(2x+1)|$:

If x is safely in (1, 3), we can brew up (2x+1) from the inequality

$1 < x < 3$:

(times 2): $2 < 2x < 6$

(add 1): $3 < 2x+1 < 7$.

To get absolute values on 2x+1, note that $-7 < 3 < 2x+1 < 7$

or, for x's in $(1, 3)$, we have

$$-7 < 2x+1 < 7$$

or $$|(2x+1)| < 7 .$$

NOW, with 2x+1 under control, back to the secret computation!

We want to find some formula for $\delta_{(\epsilon)}$ so that we can show that for x's satisfying

$$|(x-2)| < \delta_{(\epsilon)},$$

we can guarantee that for these x's,

$$|(x-2)| \cdot |(2x+1)| < \epsilon .$$

We just showed that

$$|(2x+1)| < 7$$

if the x's are somewhere in $(1, 3)$. So, for x's in $(1, 3)$, we can multiply "$|(2x+1)| < 7$" by $|(x-2)|$ to get $|(x-2)| \cdot |(2x+1)| \leq |(x-2)| \cdot 7$.
(We need "$\leq$" instead of "$<$" since $|(x-2)|$ might be zero, and we have to take this into consideration.)
HOW NICE if $|(x-2)| \cdot 7 < \epsilon$,
for then $|(x-2)| \cdot |(2x+1)| \leq |(x-2)| \cdot 7 < \epsilon$,
which means that $|(x-2)| \cdot |(2x+1)| < \epsilon$,
and so the guarantee would go through!
But $|(x-2)| \cdot 7 < \epsilon$ is equivalent to $|(x-2)| < \frac{\epsilon}{7}$, and we can arrange to have this true! All we have to do is just make sure that $\delta_{(\epsilon)} \leq \frac{\epsilon}{7}$!

So it turns out that we need to customize a limit-machine

to make sure that, for any challenge $\epsilon > 0$, the $\delta_{(\epsilon)}$ that is produced satisfies

BOTH $\delta_{(\epsilon)} \leq 1$
AND $\delta_{(\epsilon)} \leq \frac{\epsilon}{7}$.

Then such a limit-machine should be guaranteeable.

What we do is make a limit-machine that takes any actual positive error-challenge number ϵ, divides it by 7, compares it with 1, and then puts out whichever of the following two expressions is smaller:

$$1 \text{ or } \left(\frac{\epsilon}{7}\right) .$$

This number is labeled

$$\min(1, \tfrac{\epsilon}{7})$$

("min" stands for minimum.)
So a formula for $\delta(\;)$ is

$$\delta(\;) = \min(1, \tfrac{(\;)}{7}) .$$

For example, if $\lim_{x \to 2}(2x^2 - 3x + 1)$

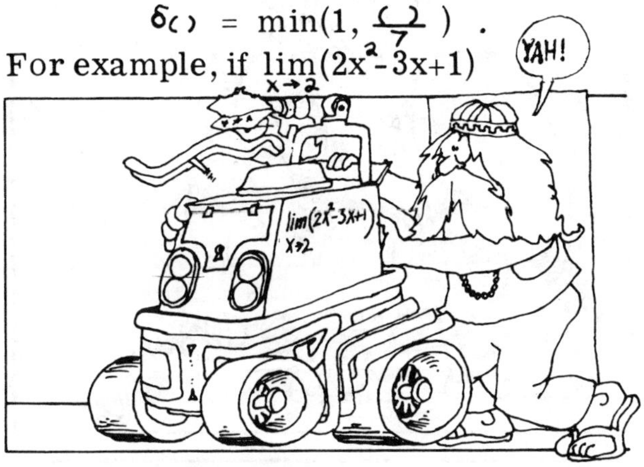

is challenged with an error-tolerance number LESS than 7 , like 5 , it computes $\frac{(\epsilon)}{7} = \frac{(5)}{7}$,

discovers that $\frac{5}{7} < 1$, and so produces $\frac{5}{7}$: $\delta(5) = \min(1, \frac{(5)}{7}) = \frac{5}{7}$.

If it is challenged with an error-tolerance number GREATER than 7, like 8 , it computes to get $\frac{8}{7} = 1\frac{1}{7}$, discovers that $1\frac{1}{7} > 1$ and it produces 1: $\delta(8) = \min(1, \frac{(8)}{7}) = 1$.

The graph of $\delta(\;) = \min(1, \frac{(\;)}{7})$ will look like

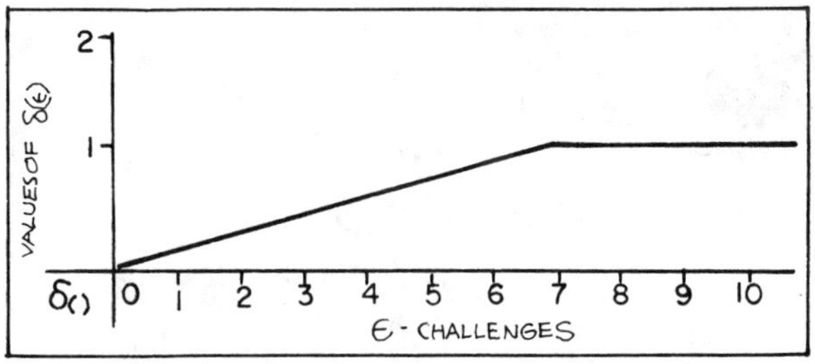

$$\delta(\epsilon) = \begin{cases} \frac{\epsilon}{7} & \text{if } 0 < \epsilon \le 7 \\ 1 & \text{if } \epsilon > 7 \end{cases}$$

In short, we design the limit-machine...

to put out $\delta_{(\epsilon)} = \frac{\epsilon}{7}$ for ϵ's satisfying $0 < \epsilon \leq 7$: The minute that $\frac{\epsilon}{7}$ itself is greater than 1, the machine changes over to putting out $\delta_{(\epsilon)} = 1$.

This makes sure that $\delta_{(\epsilon)} \leq 1$. We also have $\delta_{(\epsilon)}$ either equal to $\frac{\epsilon}{7}$ or less than $\frac{\epsilon}{7}$ (when $\frac{\epsilon}{7}$ gets bigger than 1) i.e. $\delta_{(\epsilon)} \leq \frac{\epsilon}{7}$ is true too.

So, with
$$\delta_{(\epsilon)} = \min(1, \tfrac{\epsilon}{7}),$$
if we consider x's satisfying
$$\left|(x - 2)\right| < \delta_{(\epsilon)},$$
both $\left|(x-2)\right| < \frac{\epsilon}{7}$ and $\left|(x-2)\right| < 1$ will be true AT THE SAME TIME, since $\delta_{(\epsilon)} \leq 1$ AND $\delta_{(\epsilon)} \leq \frac{\epsilon}{7}$.

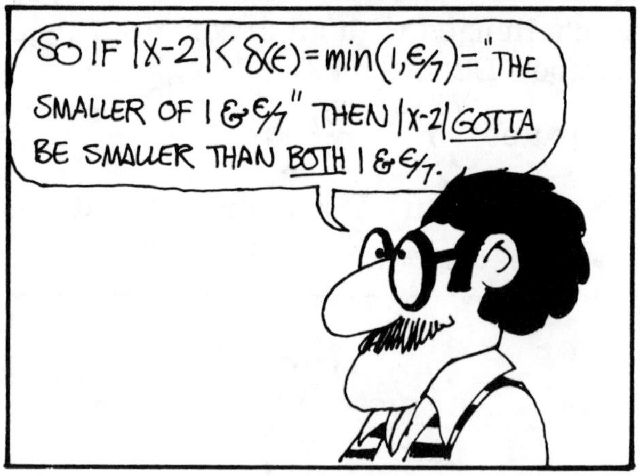

UNOFFICIAL COMPUTATION	OFFICIAL PROOF that
Follow the steps in order to see how to discover the official proof. The inequalities that go along with the steps are on the right.	$f(x) = 2x^2 - 3x + 1$ is continuous at $x = 2$. We must show that $\lim_{x \to 2}(2x^2 - 3x + 1)$ exists and equals $2(2)^2 - 3(2) + 1 = 3$.
STEP 1. Notice that $f(x) = 2x^2 - 3x + 1$ involves higher powers of x, like "$2x^2$". This means that you have to put in "$\delta_{(\epsilon)} = \min(1, \square)$." Leave $\square$ empty until step 4. Go as far as "$\lvert x - 2 \rvert < \delta_{(\epsilon)}$"; $(x - 2)$ will be one factor in step 2.	PROOF: 1. For any $\epsilon > 0$, let $\delta_{(\epsilon)} = \min(1, \boxed{\frac{\epsilon}{7}})$. 2. If $2 - \delta_{(\epsilon)} < x < 2 + \delta_{(\epsilon)}$ so $-\delta_{(\epsilon)} < x - 2 < \delta_{(\epsilon)}$ then $\lvert x - 2 \rvert < \delta_{(\epsilon)}$.
STEP 3. Use "$\lvert x - 2 \rvert < 1$" to see how big the "other factor" $(2x + 1)$ can get for these x's by brewing up $(2x + 1)$ from $1 < x < 3$.	Now, since $\delta_{(\epsilon)} = \min(1, \boxed{\frac{\epsilon}{7}})$, with $\lvert x - 2 \rvert < \delta_{(\epsilon)}$ we have $\lvert x - 2 \rvert < 1$ or $-1 < x - 2 < 1$ or $1 < x < 3$, so $2 < 2x < 6$ or $3 < 2x + 1 < 7$ or $-7 < 3 < 2x + 1 < 7$ or $\lvert 2x + 1 \rvert < 7$.
STEP 4. Since the size of the "other factor" is now controlled by "$\lvert 2x+1 \rvert < 7$," put $\frac{\epsilon}{7}$ into all $\boxed{\text{boxes}}$ above. STEP 5. See if you can finally get from "$\lvert x-2 \rvert < \frac{\epsilon}{7}$" to "$\lvert x-2 \rvert \cdot \lvert 2x+1 \rvert < \epsilon$" by putting in this line →	Since $\delta_{(\epsilon)} = \min(1, \boxed{\frac{\epsilon}{7}})$, $\lvert x-2 \rvert < \delta_{(\epsilon)}$ means $\lvert x-2 \rvert < \frac{\epsilon}{7}$ also, so $\lvert x-2 \rvert \cdot \lvert 2x+1 \rvert \le \lvert x-2 \rvert \cdot 7 < \frac{\epsilon}{7} \cdot 7 = \epsilon$,
STEP 2. Work backwards from "THEN" by a reversible "How Nice.." argument to "How nice if we can prove that $\lvert x-2 \rvert \cdot \lvert 2x+1 \rvert < \epsilon$." This identifies the crucial "other factor $(2x+1)$" as the factor NOT in step 1.	or $\lvert x-2 \rvert \cdot \lvert 2x+1 \rvert < \epsilon$ or $-\epsilon < (x-2)\cdot(2x+1) < \epsilon$ or $-\epsilon < 2x^2 - 3x - 2 < \epsilon$. THEN $3 - \epsilon < 2x^2 - 3x + 1 < 3 + \epsilon$.

Here is another example worked out. The unofficial way to discover the proof is shown by the five steps on the left.

	PROVE that $f(x) = 5-x-x^2$ is continuous at $x=-2$. Proof: We must show that $\lim_{x \to -2}(5-x-x^2)$ exists and equals $5-(-2)-(-2)^2 = 3$.
	1. For any $\epsilon > 0$, let $\delta_{(\epsilon)} = \min(1, \boxed{\frac{\epsilon}{4}})$.
STEP 1. Identifies factor $(x+2)$	2. IF $-2-\delta_{(\epsilon)} < x < -2+\delta_{(\epsilon)}$ so (add 2) $-\delta_{(\epsilon)} < x+2 < \delta_{(\epsilon)}$ or then $\lvert x+2 \rvert < \delta_{(\epsilon)}$
STEP 3. Determine size of "other factor" $(1-x)$	Also, since $\delta_{(\epsilon)} = \min(1, \boxed{\frac{\epsilon}{4}})$, with $\lvert x+2 \rvert < \delta_{(\epsilon)}$ we have $\lvert x+2 \rvert < 1$ or $-1 < x+2 < 1$ or $-3 < x < -1$. So (times "-1") $3 > -x > 1$ or (add 1) $4 > 1-x > 2 > -4$, which gives $\lvert 1-x \rvert < 4$.
STEP 4. $\frac{\epsilon}{4}$ in boxes.	Since $\delta_{(\epsilon)} = \min(1, \boxed{\frac{\epsilon}{4}})$, $\lvert x+2 \rvert < \delta_{(\epsilon)}$
STEP 5.	means $\lvert x+2 \rvert < \frac{\epsilon}{4}$ also, so $\lvert x+2 \rvert \cdot \lvert 1-x \rvert \leq \lvert x+2 \rvert \cdot 4 < \frac{\epsilon}{4} \cdot 4 = \epsilon$,
STEP 2. ↑ Identifies "other factor" $(1-x)$ (NOT the factor in step 1).	or $\lvert 2+x \rvert \cdot \lvert 1-x \rvert < \epsilon$, so $-\epsilon < (2+x) \cdot (1-x) < \epsilon$, and so $-\epsilon < 2-x-x^2 < \epsilon$, THEN $3-\epsilon < 5-x-x^2 < 3+\epsilon$.

Mathematical Note: We don't really need to use "1" in $\delta_{(\epsilon)} = \min(1, ?)$. For functions that are polynomials, we could use any positive number, like $\frac{1}{2}$ or 17, but "1" is nice and easy.

EXERCISES

coming up.

II.5.1 Prove that $3x^2 - 4x - 1$ is continuous at $x = 2$.

II.5.2 Prove that $3 - x - 2x^2$ is continuous at $x = -1$.

II.5.3 Prove that $2x^2 + x - 1$ is continuous at $x = \frac{1}{2}$.
(Hint: Here you need $(2x^2 + x - 1) = (x+1)\cdot(2x-1) = (2(x+1))\cdot(x - \frac{1}{2})$).

II.5.4 Prove that $f(x) = x^2$ is continuous at $x = a$ for ANY real number a.

II.5.5 Prove that $f(x) = x^3$ is continuous at $x = a$ for ANY real number a.

Proof: 1. For any $\epsilon > 0$, let $\delta(\epsilon) = \min(1, \frac{\epsilon}{3a^2+3|a|+1})$

2. If $a - \delta(\epsilon) < x < a + \delta(\epsilon)$, then $-\delta(\epsilon) < x - a < \delta(\epsilon)$ or $|x-a| < \delta(\epsilon)$.

Now $|x^2+ax+a^2| \leq |x^2| + |ax| + |a^2| = x^2 + |a|\cdot|x| + a^2$, so it will turn out to be enough to control the size of $(x^2 + |a|\cdot|x| + a^2)$. With $|x-a| < \delta(\epsilon) \leq 1$, $-1 < (x-a) < 1$ or $-(|a|+1) = -|a| - 1 \leq a-1 < x < a+1 \leq |a|+1$, and this gives $|x| < |a|+1$. Multiply this by $|a|$: $|a|\cdot|x| < (|a|^2 + |a|)$.

Square both sides of "$|x| < |a|+1$": $|x|^2 < (|a|+1)^2 = |a|^2 + 2|a| + 1$.

Add both of these: $|x|^2 + |a|\cdot|x| < 2|a|^2 + 3|a| + 1$.

Add $|a|^2$ to both sides: $|x|^2 + |a|\cdot|x| + |a|^2 < 3|a|^2 + 3|a| + 1$.

So $|x^2 + ax + a^2| \leq (x^2 + |a|\cdot|x| + a^2) < (3|a|^2 + 3|a| + 1) = (3a^2 + 3|a| + 1)$.

Since $|x-a| < \frac{\epsilon}{3a^2+3|a|+1}$ too,

$|x-a|\cdot|x^2+ax+a^2| \leq |x-a|\cdot(3a^2+3|a|+1) < \frac{\epsilon}{(3a^2+3|a|+1)}(3a^2+3|a|+1) = \epsilon$

or $|x-a|\cdot|x^2+ax+a^2| < \epsilon$

or $-\epsilon < (x-a)\cdot(x^2+ax+a^2) < \epsilon$

or $-\epsilon < x^3 - a^3 < \epsilon$,

THEN $a^3 - \epsilon < x^3 < a^3 + \epsilon$.

II.5.6 (Hard!) Prove that $f(x) = x^3 - 2x^2 + 3x - 4$ is continuous at $x = 2$.

II·6 FUNCTIONS THAT ARE NOT CONTINUOUS

In their search for a logical way to prove things about the **intuitive** notion that graphs of functions are smooth and squeezed together near a point, mathematicians came up with the ϵ, $\delta(\epsilon)$ limit-machine idea. They simply agreed that ONLY functions that DO have ϵ, $\delta(\epsilon)$ machines will be declared to be continuous.

Functions with straight lines or parabolas or cubics as their graphs certainly LOOK continuous, and it does turn out to be possible to construct suitable limit-machines for these functions.

An essential test of how well the limit-machine idea really fits our intuitive sense of "continuous" can be made by supposing that a function comes along whose graph is obviously NOT smooth and squeezed together near some point.

It had better NOT be possible to construct an ϵ, $\delta(\epsilon)$ limit-machine for such a function. Because if it were possible, mathematicians would find that they had legislated themselves into the awkward logical position of having to say that the function IS officially continuous, even though everyone could see that, from an intuitive point of view, it clearly is NOT continuous.

So mathematicians, having invented limit-machines in the first place, now set about to find a way to logically DEMOLISH any limit-machine that turns up CLAIMING to show that functions like this —

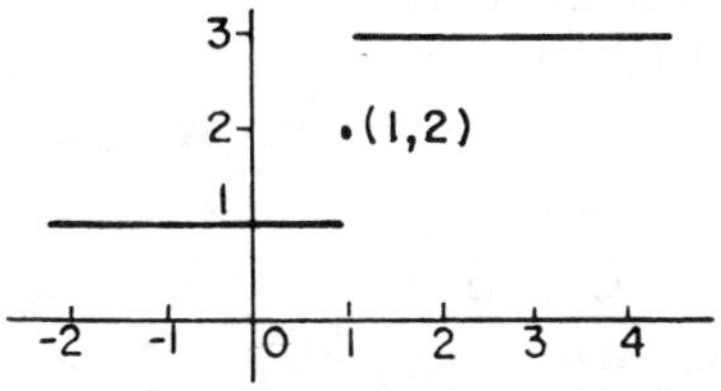

are smooth and squeezed together near points like (1, 2). They found a way to do this and that was enough to show that the function is not continuous.

The function with graph

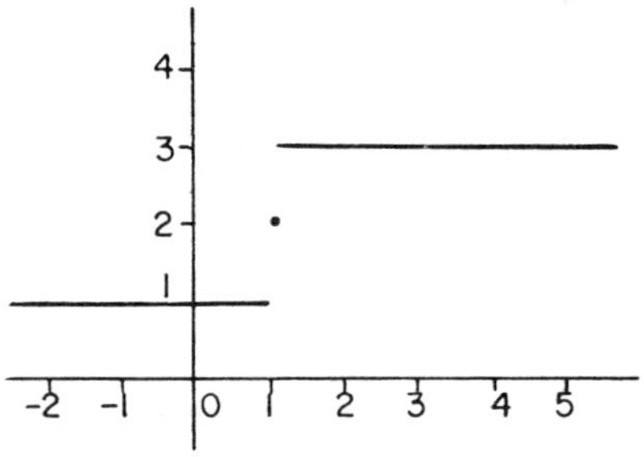

doesn't have a simple formula with parentheses to define it: The usual definition for f( ) is

$$f(x) = \begin{cases} 3 & \text{IF } x > 1 \\ 2 & \text{IF } x = 1 \\ 1 & \text{IF } x < 1 \end{cases} .$$

The demolition plan for showing that f( ) is not continuous goes like this:

Limit-machines are in the business of being challenged with error-tolerance ϵ -challenges and producing guaranteed $\delta_{(\epsilon)}$-responses.

Since any guaranteed limit-machine is supposed to be able to produce a guaranteed response $\delta_{(\epsilon)}$ for ANY $\epsilon > 0$, the only way to logically DEMOLISH the pretensions of a limit-machine

that comes in claiming to show that f( ) is continuous at x = 1 is to figure out some specific error-tolerance ϵ that clearly has NO guaranteeable response $\delta_{(\epsilon)}$. So suppose a limit-machine is busy claiming that

$\lim_{x\to 1} f(x)$ exists and equals $f(1) = 2$.

It has no trouble with ϵ's bigger than 1 —try $\epsilon = \frac{3}{2}$, for example:

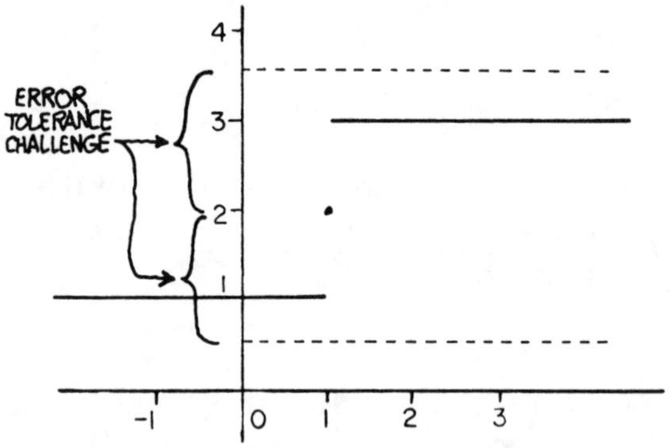

The machine can guarantee ANY positive number as response "$\delta(\frac{3}{2})$,"

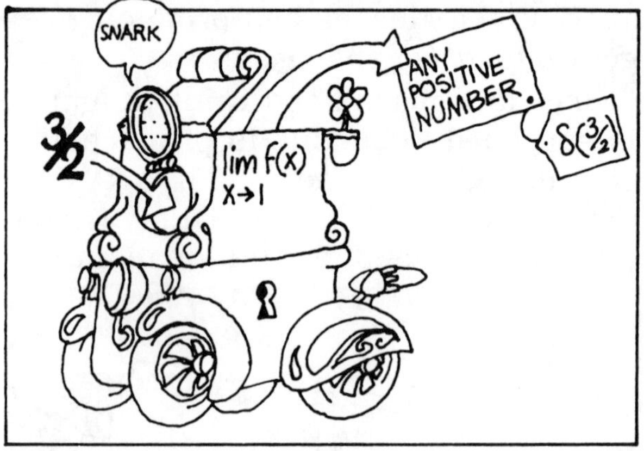

since all the guarantee has to show is

that $\underset{f(1)}{2} - \frac{3}{2} < f(x) < \underset{f(1)}{2} + \frac{3}{2}$

or $\frac{1}{2} < f(x) < 3\frac{1}{2}$.

But this is ALWAYS true no matter what value x takes, since f(x) only takes values 1 , 2 or 3 , and there is no trouble at all with inequality

$$\frac{1}{2} < \left\{ \begin{matrix} 1 \\ 2 \\ \text{or } 3 \end{matrix} \right\} < 3\frac{1}{2}$$

being true.

On the other hand, if we try an ϵ-challenge less than 1 , like $\epsilon = \frac{1}{2}$,

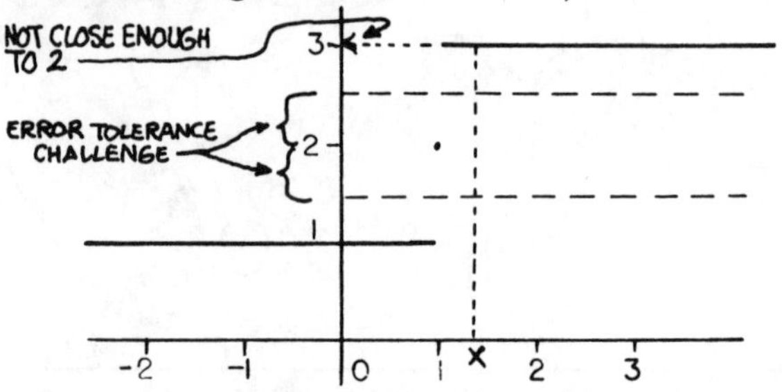

it looks like there is NO WAY to find an interval $(1-\delta(\frac{1}{2}), 1+\delta(\frac{1}{2}))$ around 1 so that <u>all</u> x's in the interval end up with f(x)'s between $2-\frac{1}{2}$ and $2+\frac{1}{2}$.

But "looks like. . ." is not enough to definitely demolish the guarantee on any limit-machine's responses. What we need is a general procedure for LOGICALLY demolishing the guarantee on any $\delta(\frac{1}{2})>0$, and here it is:

No matter what $\delta(\frac{1}{2})>0$ is produced, the point $x = (1+\frac{2}{3}\cdot\delta(\frac{1}{2}))$, for example, is automatically in the response interval $(1-\delta(\frac{1}{2}), 1+\delta(\frac{1}{2}))$,

since $1-\delta(\frac{1}{2}) < 1+\frac{2}{3}\cdot\delta(\frac{1}{2}) < 1+\delta(\frac{1}{2})$

is certainly true. (Instead of "$\frac{2}{3}$", we could use any number "c" with $0 < c < 1$, just as long as $1 < 1+c\cdot\delta(\frac{1}{2}) < 1+\delta(\frac{1}{2})$ is true.)

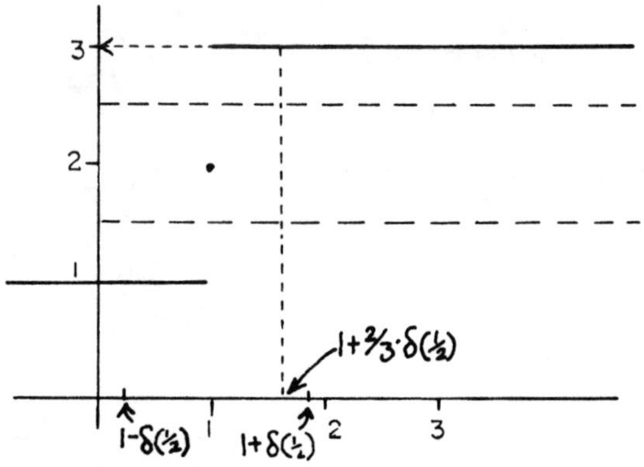

Yet $(1+\frac{2}{3}\delta(\frac{1}{2})) > 1$, since $\delta(\frac{1}{2})$ must be greater than zero, so

$$f(1+\tfrac{2}{3}\delta(\tfrac{1}{2})) = 3$$

and the right hand side of

$$\underset{f(1)}{2} - \tfrac{1}{2} < \underbrace{f(1+\tfrac{2}{3}\delta(\tfrac{1}{2}))}_{3} < \underset{f(1)}{2} + \tfrac{1}{2}$$

WRONG HERE

is clearly false; 3 is NOT $< 2+\frac{1}{2}$.

So NO $\delta(\frac{1}{2})>0$ is guaranteeable — we can always show that there is at least one x in $(1-\delta(\frac{1}{2}), 1+\delta(\frac{1}{2}))$ ($x = (1+\frac{2}{3}\delta(\frac{1}{2}))$ in particular) for which

$$f(1) - \tfrac{1}{2} < f(x) < f(1) + \tfrac{1}{2}$$

is FALSE. If no $\delta(\frac{1}{2})> 0$ is guaranteeable, no guaranteed limit-machine exists, and so both intuitively and officially by the mathematical definition,

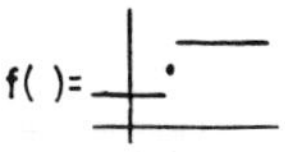

is NOT continuous at $x = 1$.

OFFICIAL PROOF that

$$f(x) = \begin{cases} 3 & \text{if } x > 1 \\ 2 & \text{if } x = 1 \\ 1 & \text{if } x < 1 \end{cases}$$

is NOT continuous at $x = 1$.

Proof: We must show that NO guaranteeable [lim F(x), x→1] exists and equals $f(1) = 2$.

Let $\epsilon = \frac{1}{2}$. If any $\delta(\frac{1}{2})$ exists, it has to be greater than zero.

So point $x = (1+\frac{2}{3}\delta(\frac{1}{2})) > 1$ and we have $f(1+\frac{2}{3}\delta(\frac{1}{2})) = 3$.

Then $x = (1+\frac{2}{3}\delta(\frac{1}{2}))$ satisfies

$$1 - \delta(\tfrac{1}{2}) < 1+\tfrac{2}{3}\delta(\tfrac{1}{2}) < 1 + \delta(\tfrac{1}{2}),$$

yet $\underset{f(1)}{2} - \frac{1}{2} < \underbrace{f(1+\frac{2}{3}\delta(\frac{1}{2}))}_{3} < \underset{f(1)}{2} + \frac{1}{2}$

WRONG HERE

is FALSE. No matter what positive value $\delta(\frac{1}{2})$ takes, we can always make this argument to show that the guarantee doesn't hold for at least one x in $(1-\delta(\frac{1}{2}), 1+\delta(\frac{1}{2}))$. So NO $\delta(\frac{1}{2})>0$ can EVER be guaranteed. This means that no limit-machine can successfully process ϵ-challenge "$\frac{1}{2}$." But a limit-machine is supposed to be able to produce a guaranteed response $\delta(\epsilon)$ for any $\epsilon > 0$: This means that no [lim F(x), x→1] can exist, since any candidate will fail to meet challenge "$\frac{1}{2}$."

AS FAR AS MATHEMATICIANS ARE CONCERNED, THE OFFICIAL PROOF SHOWS THAT f(x)= {3 IF X>1, 2 IF X=1, 1 IF X<1} IS OFFICIALLY NOT CONTINUOUS AT X=1
CRAZY GAMES MATHEMATICIANS PLAY
AROUND THE RUGGED ROCK THE RAGGED RASCALS RAN...

EXERCISES

II.6.1 If

$$f(x) = \begin{cases} 2 & \text{if } x > 1 \\ 3 & \text{if } x = 1 \\ 4 & \text{if } x < 1 \end{cases}$$

prove that f( ) is not continuous at $x = 1$.

II.6.2 If

$$f(x) = \begin{cases} 3 & \text{if } x > 1 \\ 2 & \text{if } x \leq 1 \end{cases}$$

prove that f( ) is not continuous at $x = 1$.

II.6.3 If

$$f(x) = \begin{cases} 3 & \text{if } x \geq 1 \\ 2 & \text{if } x < 1 \end{cases}$$

prove that f( ) is not continuous at $x = 1$.

II.6.4 If

$$f(x) = \begin{cases} 1.02 & \text{if } x > 2 \\ 1.00 & \text{if } x \leq 2 \end{cases}$$

prove that f( ) is not continuous at $x = 2$.

II·7 ON BEING CONTINUOUS IN AN INTERVAL

We have legislated that the graph of a function f( ) is smooth and "squeezed together" near (a, f(a)) if "$\lim_{x \to a} f(x)$ exists and equals f(a)." This definition, complicated as it is, still talks about only the ONE point of the graph — (a, f(a)). However, the "smooth curves" that come up in REAL LIFE

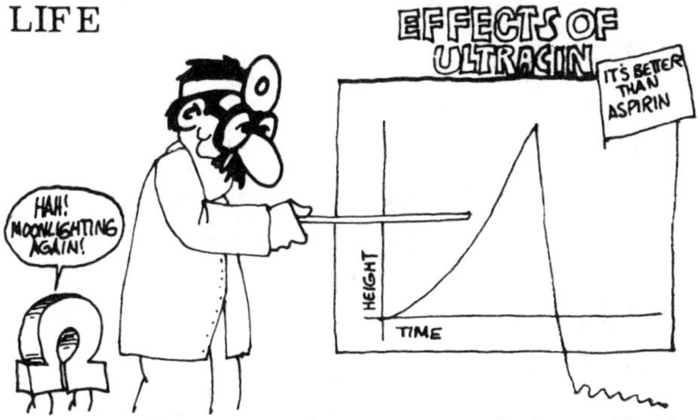

are continuous at every point on their graphs—or at least at every point above or below a whole chunk of the x - axis. This means cooking up a new definition:

DEFINITION: f( ) is continuous on interval (a, b) if f( ) is continuous at every point c between a and b, that is, if at every c : $a < c < b$,

$$\lim_{x \to c} f(x) \text{ exists and equals } f(c).$$

Unfortunately, to put this definition into action—which we must, if we want to show that the graph of a function is smooth above an interval (a, b) —we would have to go through the ritual of constructing a guaranteed limit-machine for every number c between a and b . This would mean constructing an infinite number of limit-machines and there is no way anybody can construct that many limit-machines.

Mathematicians found a way around this problem for many functions. They discovered that, given a specific function, they could construct one limit-machine for that function that would work at ALL c in intervals like $(-1, 4)$.

For example, functions having straight lines for their graphs are easy to deal with. Take a function like $f(\ ) = 2(\ ) - 1$.

We can construct a limit-machine that works at every point "c." Check this out:

Prove: $f(\ ) = 2(\ ) - 1$ is continuous at every number c .

Proof: We need to show that $\lim_{x \to c}(2x-1)$ exists and equals $f(c) = 2c - 1$.

1. For any $\epsilon > 0$, let $\delta(\epsilon) = \frac{\epsilon}{2}$.
2. If $c - \delta(\epsilon) < x < c + \delta(\epsilon)$,
   so $c - \frac{\epsilon}{2} < x < c + \frac{\epsilon}{2}$ SECRET
   then $2c - \epsilon < 2x < 2c + \epsilon$,
   so $(2c-1) - \epsilon < 2x-1 < (2c-1) + \epsilon$,
   THEN $f(c) - \epsilon < f(x) < f(c) + \epsilon$.

So we can arrange that this proof will always go through by constructing

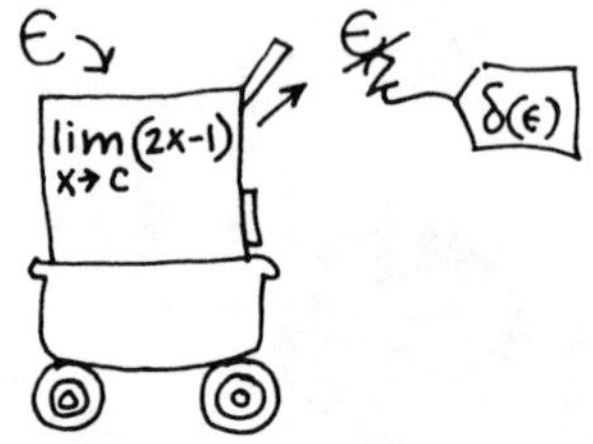

which will work no matter what c is, and we can guarantee it! As far as a mathematician is concerned, this is enough to show that

$$f(x) = 2x - 1$$

is a genuine smooth curve, squeezed together (continuous) at every point of its graph.

It is even possible to find ONE limit-machine that works for more complicated functions like

$$f(x) = 3x^2 + 2x - 1$$

for all "c" in intervals like $(-1, 4)$! See how the proof is put together here:

Prove: $f(\ ) = 3(\ )^2 + 2(\ ) - 1$ is continuous at every c ; $-1 < c < 4$.

Proof: We need to show that

$$\lim_{x \to c}(3x^2 + 2x - 1)$$

exists and equals $f(c) = 3c^2 + 2c - 1$ (if $-1 < c < 4$).

1. For any $\epsilon > 0$, let $\delta(\epsilon) = \frac{\epsilon}{26}$
2. IF $c - \delta(\epsilon) < x < c + \delta(\epsilon)$
   then $|x - c| < \delta(\epsilon)$.
   Now, $-1 < c < 4$
   and we only want to consider x's satisfying $-1 < x < 4$ as well.*
   So (both times 3) $-3 < 3x < 12$
   and $-3 < 3c < 12$.
   Add: $-6 < 3x + 3c < 24$.
   And add 2 :
   $(-26 <)\ -4 < 3x + 3c + 2 < 26$
   or $|3x + 3c + 2| < 26$
   Since $|x - c| < \frac{\epsilon}{26}$,
   $|x-c| \cdot |3x+3c+2| \leq |x-c| \cdot 26 < \frac{\epsilon}{26} 26 = \epsilon$.
   or $|x-c| \cdot |3x+3c+2| < \epsilon$.
   So $-\epsilon < (x-c) \cdot (3 \cdot (x+c) + 2) < \epsilon$
   or $-\epsilon < 3 \cdot (x-c) \cdot (x+c) + 2(x-c) < \epsilon$
   or $-\epsilon < 3 \cdot (x^2 - c^2) + 2x - 2c < \epsilon$
   or $-\epsilon < (3x^2 + 2x - 1) - (3c^2 + 2c - 1) < \epsilon$,

THEN

$(3c^2 + 2c - 1) - \epsilon < (3x^2 + 2x - 1) < (3c^2 + 2c - 1) + \epsilon$.

*Mathematical Note! To make sure x stays in $(-1, 4)$, we really should let $\delta(\epsilon) = \min(\frac{\epsilon}{26}, c-(-1), 4-c)$. That'll do it!

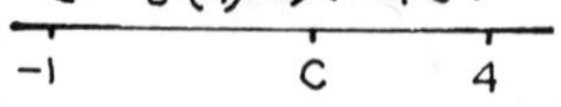

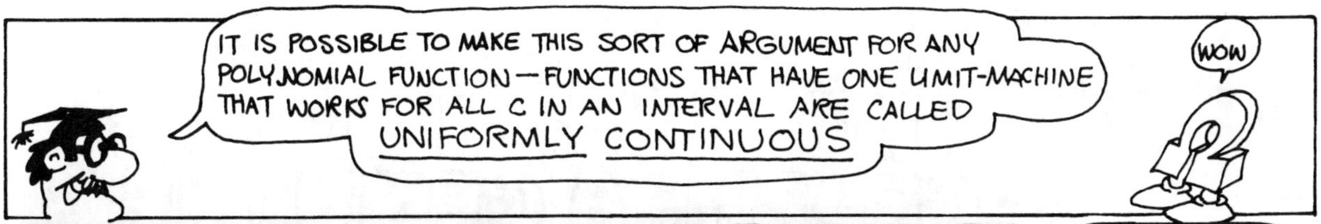

IT IS POSSIBLE TO MAKE THIS SORT OF ARGUMENT FOR ANY POLYNOMIAL FUNCTION—FUNCTIONS THAT HAVE ONE LIMIT-MACHINE THAT WORKS FOR ALL C IN AN INTERVAL ARE CALLED UNIFORMLY CONTINUOUS
WOW

EXERCISES

II.7.1 Prove that $f(x) = 2x^2$ is continuous at every c ; $0 < c < 4$.

II.7.2 Prove that $f(x) = x^2+3x-7$ is continuous at every c ; $1 < c < 5$.

II·8 WHAT HAPPENS WHEN A POINT IS MISSING FROM A GRAPH: A GENUINE THEOREM IS PROVED!

In the next chapter we will encounter functions whose domains contain all points near a particular point but, unfortunately, these functions can't decide how to process the point itself. Suppose for the moment that some function h( ) wants to process its missing point "a" so that it can be

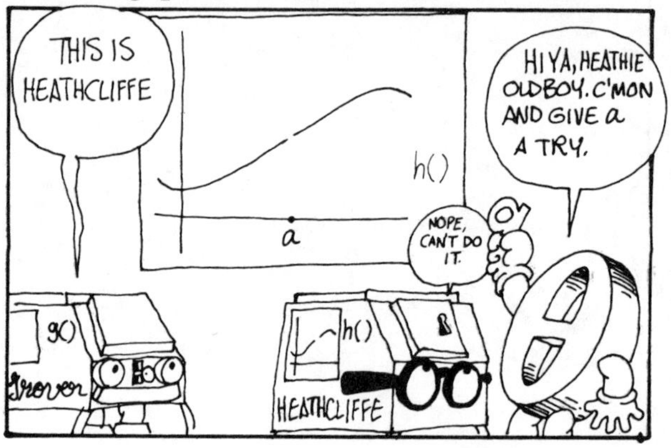

continuous at "a." Then "a" will have to be added to h( )'s domain and we will also need to guess what number value h( ) should take at "a" in order to be continuous there. This is the same as trying to "extend" h( ) to a new function $\bar{h}$( ) (translated "h-bar") that agrees with h( ) for all x's in the domain of h( ) and can also process "a," and then asking if it is continuous at "a." It is easy to "extend" h( ) — it is the same as adding one more point to the graph of h( ) vertically above "a" — but we want to add just the right point so that $\bar{h}$( ) is continuous at "a."

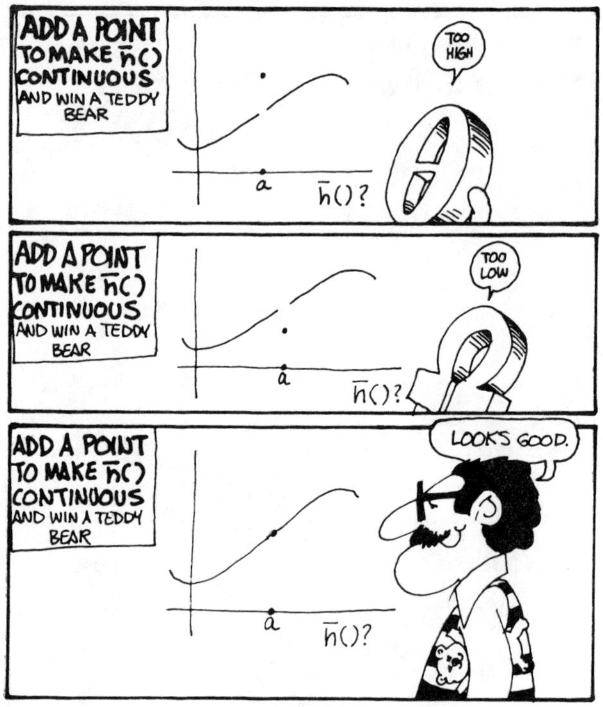

It may not be possible to do this at all. For example, if the graph of h( ) takes a hop above a ,

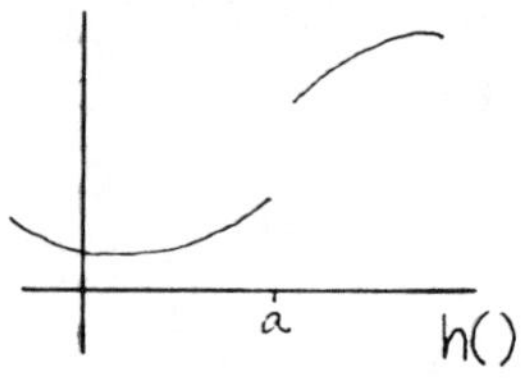

it doesn't look like there is any suitable point to put in the graph to make it continuous. You could try making the midpoint between the two pieces of the graph the value that $\overline{h}(\ )$ should take above a ,

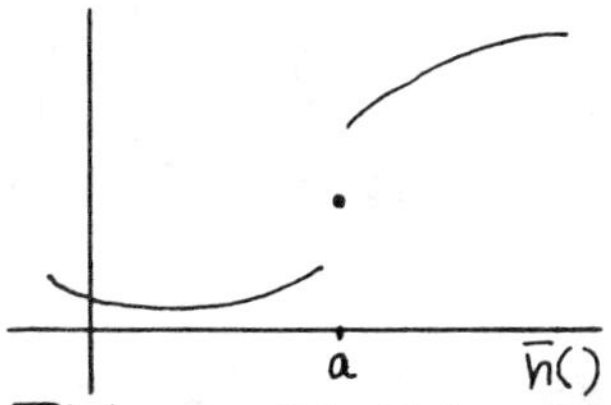

but this $\overline{h}(\ )$ wouldn't be continuous at x = a. (Remember what happened to any limit-machine trying to handle this function:

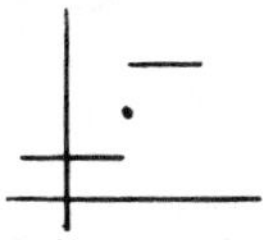

IF it is possible to find a new function $\overline{h}(\ )$ that

1. agrees with h( ) on h( )'s domain and
2. the new $\overline{h}(\ )$ is continuous at "a,"

it appears from the picture that there can be only one possible value for $\overline{h}(a)$; i.e., there is only one point (a, L) that will "fill in" the graph of h( ) to make it a smooth curve, and height "L" should be the value of $\overline{h}(a)$, which will also be the limiting number value for $\overline{h}(x)$ as x approaches "a."

This turns out to be extremely important in the next chapter and, since pictures can be tricky, we will show that our legislated definitions of "limit" and "continuous" are good enough to PROVE that there is only one such point. The proof needs a "Theorem" to get the job done.

This proving that there is only one point is done by supposing—just for now—that there are two possibly different functions f( ) and g( ), both continuous at x=a, that extend h( ).

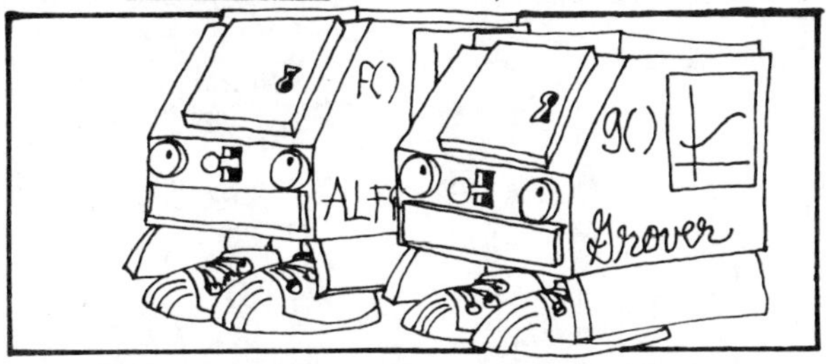

In other words, we suppose that

1. f( ) and g( ) agree with h( ) on h( )'s domain, i.e., for all x except possibly x=a,

$$f(x) = g(x) .$$

We don't know whether or not f(a) = g(a).

And we suppose that

2. f( ) and g( ) CAN process "a" and are CONTINUOUS at x=a.

F() BEING CONTINUOUS AT a
YOU CAN TELL WHICH δ(ε)'S COME OUT OF THIS MACHINE-THEY GOT LITTLE "F'S" ON 'EM!
f()
ALFRED
lim F(x) x→a
g() BEING CONTINUOUS AT a
YOU CAN TELL WHICH δ(ε)'S COME OUT OF THIS MACHINE BY THE LITTLE g'S ON THEM.
g()
Grover
lim g(x) x→a
THEN, SUPPOSING THESE TWO THINGS ABOUT F() AND g(), THE THEOREM WILL SHOW THAT F() AND g() HAVE TO AGREE AT "a", TOO. SO F(x) = g(x) WILL BE TRUE FOR ALL X'S INCLUDING "a".

AS A RESULT OF THE THEOREM, THERE WILL BE AT MOST ONE WAY TO EXTEND h() TO h̄() SO THAT h̄() CAN PROCESS "a" AND BE CONTINUOUS AT "a" — THE FUNCTIONS F() AND g() BOTH DO IT AND THEY TURN OUT TO BE EQUAL EVERYWHERE, INCLUDING "a."
WOW
IT'S THE TRUTH, FOLKS! CLEAN LIVING TRIUMPHS AGAIN!
SO, ONCE THE THEOREM IS SHOWN, THERE WILL BE AT MOST ONE WAY TO FILL IN THE HOLE IN THIS GRAPH OF h() SO THAT THE NEW FUNCTION h̄() IS CONTINUOUS AT X = a ...
AND THAT IS BY PUTTING IN POINT (a, F(a)) = (a, g(a))!
(a,F(a))=(a,g(a))
F(a) = g(a)
a
SO, IF h() WANTS TO ADD "a" TO ITS DOMAIN, THERE IS AT MOST ONE VALUE THAT h() COULD TAKE AS IT'S LIMITING NUMBER AS X GOES TO "a" IN ORDER TO BE CONTINUOUS THERE.
BUT WE STILL HAVE TO PROVE THE THEOREM...

THEOREM II-1

Suppose f( ) and g( ) are functions such that
1. $f(x) = g(x)$ if $x \neq a$ and
2. f( ) and g( ) are continuous at $x = a$.

Conclusion: Then $f(a) = g(a)$.

IT'S SHOW TIME
F()
lim f(x) x→a
GOOD SHIP LOLLYPOP
f()
ALFRED
WHAT'RE YOU GUYS DOIN' HERE?
g()
NOTHIN' LIKE SHOW BIZ
lim g(x) x→a
g()
Grover
LRW
ACT 1
NOW, TO START THE THEOREM, WE FIRST GET ALFIE AND GROVER TO COMPUTE f(a) AND g(a)...
f(a)
g(a)
a
a
f()
g()
Alfred
Grover
WHEN WE COMPARE f(a) AND g(a), IT MAY TURN OUT THAT f(a)=g(a), AND THATS WHAT WE HOPE WILL HAPPEN.
f(a)
g(a)

BUT, IF f(a) DOESN'T EQUAL g(a), THE ONLY OTHER POSSIBILITIES ARE THAT EITHER f(a) IS BIGGER THAN g(a); I.E., f(a)>g(a)...
WE GOT SHORT CHANGED
f(a)
g(a)
...OR g(a) IS BIGGER THAN f(a); I.E., g(a)>f(a).
HEY MAKE UP YOUR MIND
f(a)
g(a)
THE PLOT OF THE THEOREM OUTLINES LIKE THIS: MATHEMATICIANS COULDN'T FIGURE OUT A WAY TO PROVE DIRECTLY THAT f(a)=g(a)... SO...
THEY CONCENTRATED INSTEAD ON THE OTHER TWO POSSIBILITIES. LOOK AT IT THIS WAY— CONTESTANT MELVIN, FOR TWO GLORIOUS WEEKS AT BIG AL'S SCHOOL FOR JUNIOR HOODLUMS WHICH OF THESE WOULD YOU SAY IS TRUE?
McSQUARED'S QUIZ SHOW
EXACTLY ONE OF THESE IS TRUE. GUESS WHICH ONE!
BROUGHT TO YOU BY WM KAUFMANN INC.
A
f(a)
g(a)
B
f(a)
g(a)
C
f(a)
g(a)
G.V.
M.M.

QUIZ SH
A! I BET IT'S C.
A B C
f(a) g(a)
G.V. M.M.
A!
NOPE! IN ACT II WE WILL SEE HOW MATHEMATICIANS FOUND A WAY TO LOGICALLY **DEMOLISH** THIS POSSIBILITY THAT f(a)>g(a), ASSUMING THE THEOREM'S SUPPOSITIONS ABOUT f( ) AND g( ).
SO Ⓐ IS OUT. MM DOESN'T GET ANY POINTS THIS ROUND
GRRR
!
LOGIC
NOW, WILL B OR C BE CORRECT?
B MAYBE?
HEH, HEH, NOPE! WRONG AGAIN. IN THE EXERCISES WE WILL LOGICALLY DEMOLISH THE POSSIBILITY THAT g(a)>f(a) TOO. YESSIR, IT CAN BE DONE!
FINK CLOD RAT I WUZ FRAMED IDIOT DUNDERHEAD TURNCOAT IT WUZ STOOLIE RIGGED QUISLING!
SO Ⓒ IS OUT...
AND SINCE EXACTLY ONE OF THE THREE IS TRUE, THAT LEAVES ONLY f(a)=g(a)!
ARRGH! I WANNA GET Ⓑ TOO!!
DON'T HIT IT! IT'S THE ONLY ONE LEFT! IT'S GOTTA BE TRUE!
YOWP
Arrrrr

THAT'S THE GENERAL OUTLINE OF THE PROOF. MATHEMATICIANS TRY TO USE THIS TRICKY INDIRECT METHOD OF LOGICALLY DEMOLISHING PROPOSITIONS THEY HOPE ARE FALSE WHENEVER THEY CAN'T PROVE SOMETHING DIRECTLY. NOW, WE START ACT II DIRECTLY BELOW.
ACT II
SO, TO WRAP UP THE PROOF WE HAVE TO FIND A WAY TO LOGICALLY DEMOLISH THE POSSIBILITY THAT f(a) > g(a)....
f(a)
g(a)
LOGIC
NOW, f(x) = g(x) FOR ALL X'S EXCEPT X = a. SO, f() AND g() WILL HAVE THE SAME GRAPH EXCEPT ABOVE X = a.
f(a)
a f()
g(a)
a g()
CAN f(a)>g(a) BE TRUE?
TUNE IN NEXT PANEL FOR ANOTHER EXCITING CHAPTER
LOGIC
NOW WE GOTTA FIGURE OUT WHAT'S WRONG WITH f(a) > g(a)... HMMM...
OH, YAH... LOOKIT THIS–BOTH f() AND g() ARE CONTINUOUS AT X = a. THAT MEANS THAT THE GRAPH WOULD HAVE TO GO UP TO BE SQUEEZED TOGETHER NEAR (a, f(a)) AND ALSO GO DOWN TO BE SQUEEZED TOGETHER NEAR (a, g(a)) IF f(a) > g(a). LOOKS LIKE THAT'S IMPOSSIBLE.
YUP, ALFIE'S LIMIT-MACHINE SAYS THE GRAPH GOTTA BE UP HERE
(a, f(a))
NO WAY! GROVER'S LIMIT-MACHINE SAYS IT GOTTA BE DOWN HERE
(a, g(a))
HEY! NO FAIR CHANGING MY GRAPH!
f()

BUT "LOOKS LIKE" ISN'T VERY LOGICAL. SO WE GOTTA WHEEL OUT OUR LOGIC EQUIPMENT AND REALLY SHOW THAT THERE IS NO WAY THAT f(a)>g(a) IF f() AND g() ARE CONTINUOUS AT X=a.
HA! THIS WILL SHOW THAT THE GRAPH GOTTA BE UP NEAR (a,f(a)) IF X IS CLOSE TO a. SO THERE IS NO WAY THAT g(x) DOWN BELOW CAN BE CONTINUOUS AT X=a!
HA! THIS WILL SHOW THAT THE GRAPH GOTTA BE DOWN NEAR (a,g(a)) IF X IS CLOSE TO a. SO THERE IS NO WAY THAT f(x) UP ABOVE CAN BE CONTINUOUS AT X=a!
lim x→a f(x)
lim x→a g(x)
GROWR
I'LL SHOW THAT OMEGA! I'M GONNA FENCE IN THE GRAPH NEAR (a,f(a)) WITH OL' lim x→a f(x) SO THERE'S NO WAY THAT IT CAN GET DOWN NEAR (a,g(a))! THAT'LL SHOW THAT g() ISN'T CONTINUOUS... HMMM... TRY ε=½(f(a)-g(a))... THAT'S HALF THE DISTANCE BETWEEN f(a) AND g(a)
FENCE!
(a,f(a))
f(a)
f(a)-g(a)
g(a)
(a,g(a))
a
lim x→a f(x)
½(f(a)-g(a))
POOF
δ^f(½(f(a)-g(a)))
YETCH BLEAGH
lim x→a f(x)
OUT COMES A GUARANTEED δ^f(½(f(a)-g(a))) >0.
ε
NOW, FOR ALL X'S SATISFYING
a-δ^f(½(f(a)-g(a))) < x < a+δ^f(½(f(a)-g(a)))
THE GUARANTEE SAYS WE GOTTA HAVE
f(a)-½(f(a)-g(a)) < f(x) < f(a)+½(f(a)-g(a))
½(f(a)+g(a))

THIS "½(f(a)+g(a))<f(x)" MEANS THAT THE GRAPH GOTTA BE ABOVE THIS LINE SO IT CAN'T GET DOWN NEAR (a,g(a)) AND g() CAN'T BE CONTINUOUS. HA!
f(a)
½(f(a)+g(a))
g(a)
(a,f(a))
(a,g(a))
a
δ^f(½(f(a)-g(a)))
THIS TIME THE GUARANTEE SAYS FOR X'S SATISFYING a-δ^g(½(f(a)-g(a)))<x<a+δ^g(½(f(a-g(a))) WE GOTTA HAVE g(a)-½(f(a)-g(a))<g(x)<g(a)+½(f(a)-g(a))
½(f(a)+g(a))
AND THAT MEANS THE GRAPH HAS GOTTA BE DOWN HERE SO IT CAN'T GET UP NEAR (a,f(a)) AND SO f() CAN'T BE CONTINUOUS AT a! HAHA!
f(a)
½(f(a)+g(a))
g(a)
(a,f(a))
(a,g(a))
a
δ^g(½(f(a)-g(a)))
OH, YAH? MY LIMIT-MACHINE IS JUST AS GUARANTEED AS YOURS, AND I CAN FENCE IN THE GRAPH TOO-AND WITH THE SAME ε=½(f(a)-g(a))!
½(f(a)-g(a))
δ^g(½(f(a)-g(a)))
lim g(x) x→a
BLORK
NOW WAITAMINUTE! f(x)=g(x) WHEN x≠a, SO THE GRAPHS ARE THE SAME WHEN x≠a AND MY LIMIT-MACHINE SAYS THE GRAPH GOTTA BE UP HERE!
SO YOU BETTER WATCH IT!
NOWAY! lim g(x) x→a SAYS THE GRAPH GOTTA BE DOWN HERE!
?
f(a)
½(f(a)+g(a))
g(a)
(a,f(a))
(a,g(a))
?
a
SO CUT OUT THIS "UP HERE" JAZZ
THAT'S WHAT I SAY!
lim g(x) x→a

THE GRAPH'S GOTTA BE UP!
OUCH
WRONG! THE GRAPH'S GOTTA BE DOWN!
YOWP!
LOGIC
IT'S FENCED UP! YOU LOGICAL LOSER
DOWN! YOU REASONING RUFFIAN
TO THE RESCUE
FIGHT! OBOY
lim f(x) x→a
lim g(x) x→a
LRW
GSL
ALFRED
Good Vibes
NO! WAIT! WAIT! EVERYBODY'S RIGHT! EVERYTHING IS GUARANTEED!
IT'S THE ORIGINAL SUPPOSITION THAT "f(a) > g(a)" THAT'S GOTTA BE WRONG! STOP!
OH, MEL...
UNHAND ME
WOW
OOPS
WOMEN AND CHILDREN FIRST!
LOGIC

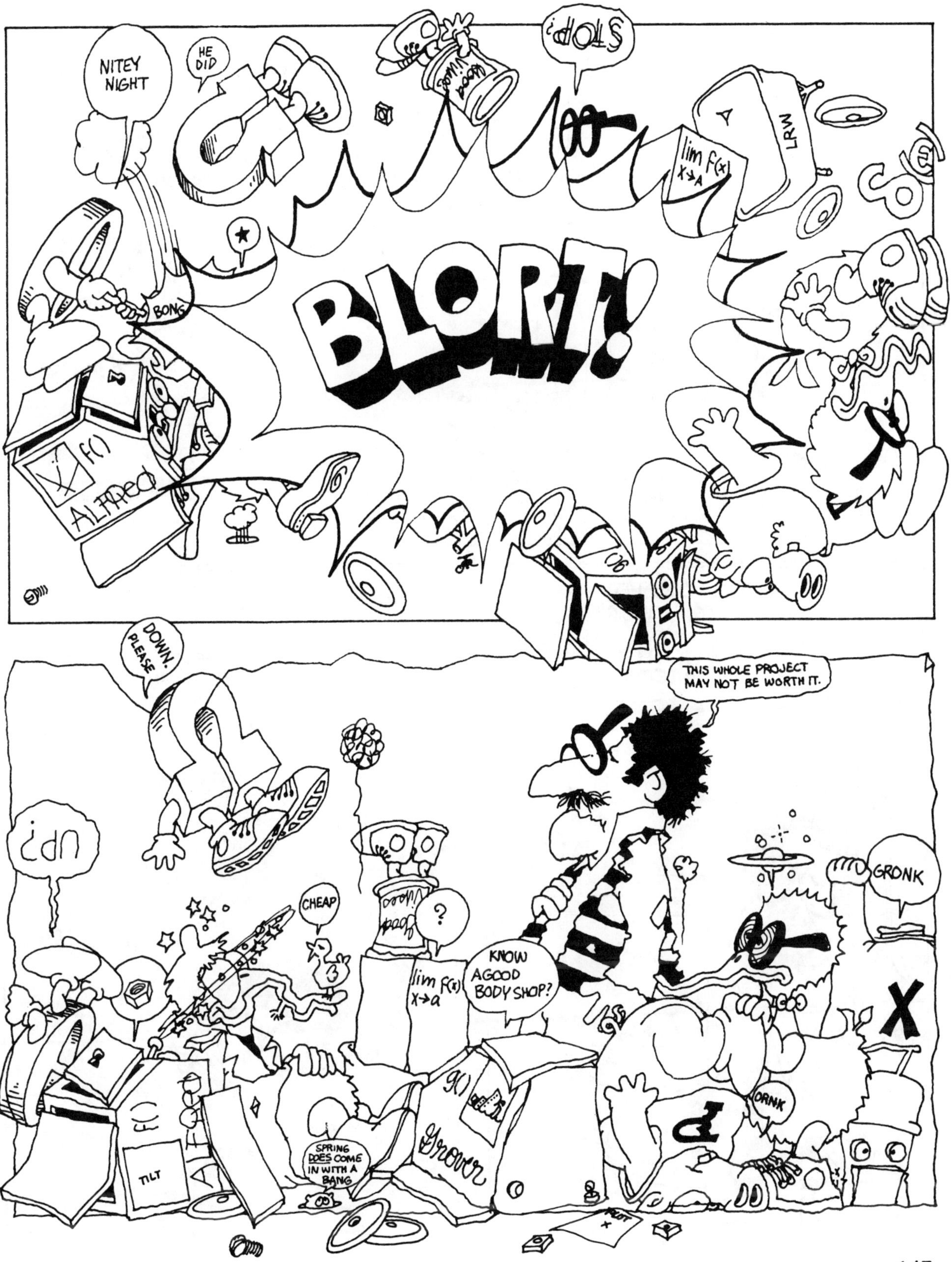
NITEY NIGHT
HE DID
STOP?
BONG
BLORT!
lim f(x) x→A
A
LRW
f()
ALFRED
DOWN. PLEASE
THIS WHOLE PROJECT MAY NOT BE WORTH IT.
UP?
CHEAP
?
lim f(x) x→a
KNOW A GOOD BODY SHOP?
GRONK
X
ORNK
g()
Grover
TILT
SPRING DOES COME IN WITH A BANG

C'MON, GANG-LET'S CLEAN UP THE ACT. WE'RE TRYING TO SHOW $f(a)>g(a)$ IS **IMPOSSIBLE** IF BOTH FUNCTIONS ARE CONTINUOUS AT $x=a$. I THINK WE'VE GOT A WAY TO DO IT, SINCE BOTH PARTIES IN THIS ARGUMENT ARE RIGHT IF THE GUARANTEES HOLD.
SOMETHING'S GOTTA GIVE, AND I BET IT'S THE IDEA THAT "$f(a)>g(a)$."
GLUB...
f()
TAKE THIS POINT HERE--- IT'S $c=a+\frac{1}{2}(\min(\delta^f(\frac{1}{2}(f(a)-g(a))), \delta^g(\frac{1}{2}(f(a)-g(a))))$ THIS "C" IS BIGGER THAN a, SO $f(c)=g(c)$, AND "C" IS IN **BOTH** GUARANTEED INTERVALS.
?
$\delta^g(\frac{1}{2}(f(a)-g(a)))$
$\delta^f(\frac{1}{2}(f(a)-g(a)))$
a
c
SO THE GUARANTEE ON $\delta^f(\frac{1}{2}(f(a)-g(a)))$ SAYS THAT $f(a)-\frac{1}{2}(f(a)-g(a))<f(c)$ [$=\frac{1}{2}(f(a)+g(a))$] GOTTA BE TRUE AND THE GUARANTEE ON $\delta^g(\frac{1}{2}(f(a)-g(a)))$ SAYS THAT $g(c)<g(a)+\frac{1}{2}(f(a)-g(a))$ [$=\frac{1}{2}(f(a)+g(a))$] GOTTA BE TRUE, TOO. PUT THESE TOGETHER, AND REMEMBER THAT $f(c)=g(c)$: WE GET $g(c)<\frac{1}{2}(f(a)+g(a))<f(c)$ [$=g(c)$]...
H.T.RYNM.
HOW TO REPAIR YOUR NEW MATH BY PENNY ANTE

THIS IS A GENUINE FOOLPROOF LOGICAL IMPOSSIBILITY— THERE'S NO WAY THAT A NUMBER LIKE $g(c)$ CAN BE STRICTLY LESS THAN ITSELF! AYUP, UHHUH...
SO, THERE IS NO WAY TO HAVE ① $f(x)=g(x)$ IF $x \neq a$ ② BOTH LIMIT-MACHINES GUARANTEED AT $x=a$ AND ③ $f(a) > g(a)$ ALL TRUE AT THE SAME TIME— WE GET THE GENUINE LOGICAL CONTRADICTION "$g(c) < g(c)$" IF THEY ARE.
CAUTION ALL WET
THE STATEMENT OF THE THEOREM SAYS THAT WE ARE ASSUMING THAT ① AND ② ARE TRUE: SO ③ HAS GOTTA GO— THE POSSIBILITY THAT $f(a) > g(a)$ IS DEMOLISHED!
ONCE YOU GUYS DEMOLISH THE POSSIBILITY THAT $g(a) > f(a)$ IN THE EXERCISES, THE ONLY POSSIBILITY LEFT IS "$f(a) = g(a)$" AND SO THIS HAS TO BE TRUE. THE THEOREM IS PROVED!

Here is an official mathematical proof of Theorem II-1:

THEOREM II-1.

Suppose f( ) and g( ) are functions such that

1. $f(x) = g(x)$ if $x \neq a$ and
2. f( ) and g( ) are continuous at $x = a$.

Then $f(a) = g(a)$.

Proof: If $f(a) > g(a)$, let $\epsilon = \frac{1}{2}(f(a)-g(a))$.

Since f( ) is continuous at a, there is a $\delta^f(\frac{1}{2}(f(a)-g(a)))>0$ such that, for x's satisfying

(i) $a-\delta^f(\frac{1}{2}(f(a)-g(a)))<x<a+\delta^f(\frac{1}{2}(f(a)-g(a)))$,

$f(a)-\frac{1}{2}(f(a)-g(a))<f(x)<f(a)+\frac{1}{2}(f(a)-g(a))$

so $\frac{1}{2}(f(a)+g(a))<f(x)$.

Since g( ) is continuous at a, there is a $\delta^g(\frac{1}{2}(f(a)-g(a)))$ such that, for x's satisfying

(ii) $a-\delta^g(\frac{1}{2}(f(a)-g(a)))<x<a+\delta^g(\frac{1}{2}(f(a)-g(a)))$,

$g(a)-\frac{1}{2}(f(a)-g(a))<g(x)<g(a)+\frac{1}{2}(f(a)-g(a))$

so $g(x)<\frac{1}{2}(f(a)+g(a))$.

Now $\min(\delta^f(\frac{1}{2}(f(a)-g(a))), \delta^g(\frac{1}{2}(f(a)-g(a))))>0$,

so $c=a+\frac{1}{2}\min(\delta^f(\frac{1}{2}(f(a)-g(a))), \delta^g(\frac{1}{2}(f(a)-g(a)))) > a$.

Therefore $f(c)=g(c)$

and furthermore, c is in the guaranteed intervals given by (i) and (ii). As a result, $\frac{1}{2}(f(a)+g(a))<f(c)$

and $g(c)<\frac{1}{2}(f(a)+g(a))$.

Put all these together:

$g(c) < \frac{1}{2}(f(a)+g(a)) < f(c)=g(c)$,

a contradiction.

If $f(a) < g(a)$, you obtain a similar contradiction.

Therefore, we must have $f(a)=g(a)$.

EXERCISE

II.8.1 Suppose f( ) and g( ) are functions such that

1. $f(x)=g(x)$ if $x\neq a$ and
2. f( ) and g( ) are continuous at $x=a$.

Prove that $g(a) > f(a)$ is impossible.

II·9 DEFINITION OF A GENERAL LIMIT POINT

The results of the last section summarize like this:

If we have a function h( ) whose domain contains all numbers EXCEPT some number a ,

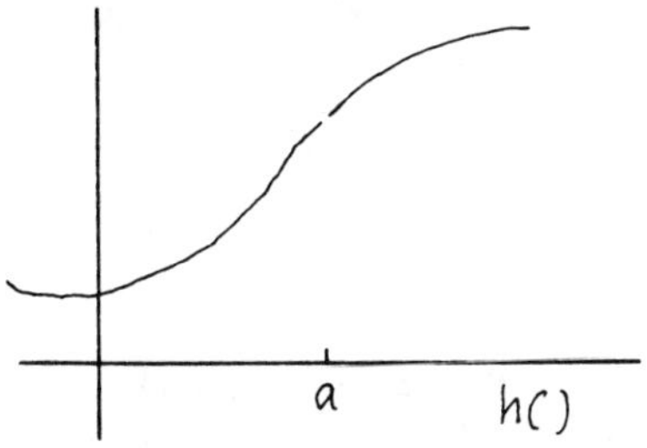

and we suppose for the moment that h( ) wants to add "a" to its domain, then there is at most one value "L" that h( ) can take as its limit number as x goes to "a" in order to be continuous there.

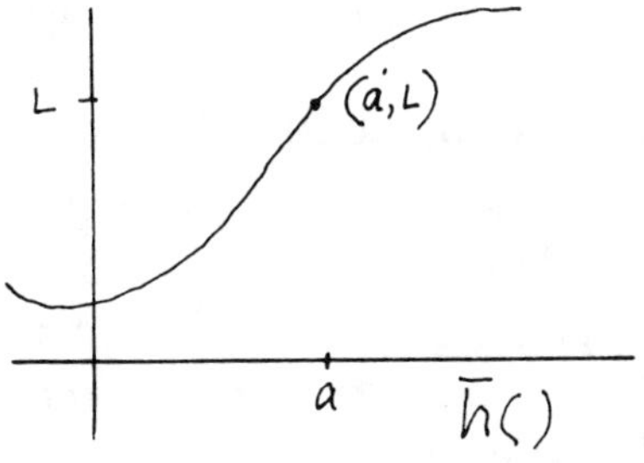

But the theorem in the last section doesn't give us any clue as to how to actually find this limit number value "L" and we will need such limit numbers in Chapter III.

Let's have a look at an actual function that has a point missing from its graph, like

$$h(x) = \frac{2x^2-3x-2}{x-2} .$$

We have to exclude x=2 from h( )'s domain, since the bottom of the fraction is "x-2," and this is zero when x=2 and dividing by zero is NOT FAIR. It turns out that we can factor the top of the fraction:

$$2x^2-3x-2 = (2x+1)(x-2) .$$

So $h(x)= \dfrac{2x^2-3x-2}{x-2} = \dfrac{(2x+1)(x-2)}{(x-2)}$.

When $x \neq 2$, we can cancel the "(x-2)'s" to get a new formula for h( ): $h(x)=\begin{cases} 2x+1 & \text{if } x \neq 2 \\ \text{undefined} & \text{if } x=2 \end{cases}$.

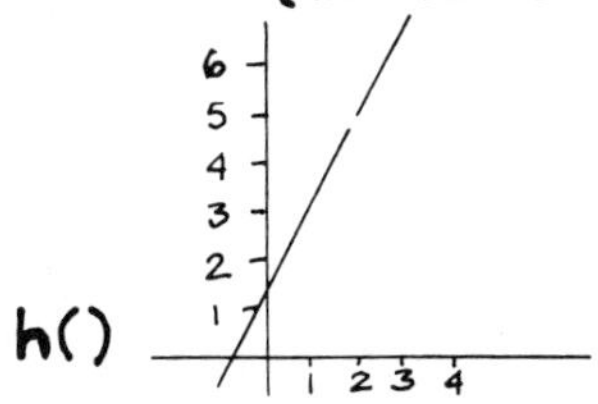

Now from the graph, it looks like the best guess as to the limit value that h( ) wants to take at x = 2 to be continuous there is 5. This is because the graph of h( ) shows where the correspondence-arrows for function h( ) make their right-angled turn, coming from x and going to h(x). And as x goes to 2, it certainly looks like h(x) wants to go to 5.

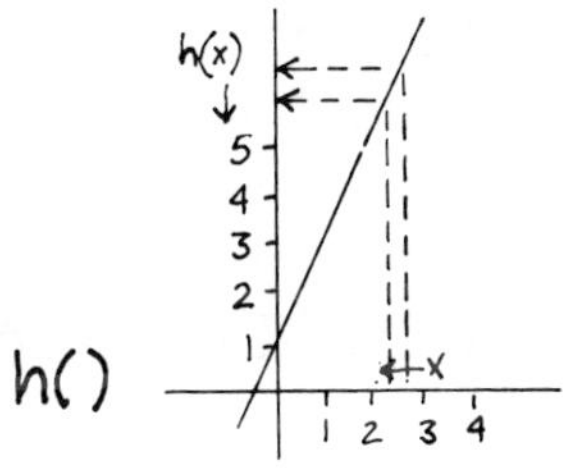

The same guess "5" also comes from a look at the new formula for h( ): $h(x) = \begin{cases} 2x+1 & \text{if } x \neq 2 \\ \text{undefined} & \text{if } x=2 \end{cases}$

Make the guess by cheating a little and putting x=2 into the formula for h( ) even though it is officially unfair: $h(2) = 2\cdot 2 + 1 = 5$! This time, the reason that 5 should be the correct guess is because we can define function $\bar{h}(x) = 2x+1$ for all "x" INCLUDING "2" and now it is obvious that $\bar{h}(x)$

1. agrees with h( ) on h( )'s domain, and
2. $\bar{h}(x) = 2x+1$ looks continuous everywhere, including x=2, and
3. $\bar{h}(2) = 2\cdot 2+1 = 5$.

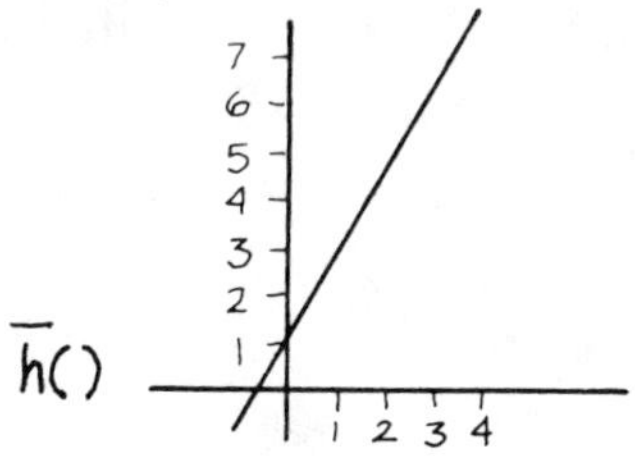

But to be really sure, we need an ϵ, $\delta_{(\epsilon)}$ argument: the genuine theorem from the last section proved that if the ϵ, $\delta_{(\epsilon)}$ argument works with value "5" assigned to $\bar{h}(2)$, 5 <u>must</u> be the right guess. For the theorem showed that there is <u>at most</u> one value that h( ) can take at 2 where you can make an ϵ, $\delta_{(\epsilon)}$ argument showing continuity. If the ϵ, $\delta_{(\epsilon)}$ argument doesn't work, "5" is the wrong guess, and we must guess again until either

1. we find a number instead of "5" that results in a correct ϵ, $\delta_{(\epsilon)}$ argument or
2. there is no "correct" guess and so h( ) can't be extended to be continuous at 2.

But 5 does work! We can prove it! With $h(x) = \dfrac{2x^2-3x-2}{x-2}$, here is an argument that shows that 5 is the correct value for h( ) to take at x=2 to be continuous there: We first construct a function $\bar{h}(x)$ that agrees with h(x) when $x \neq 2$ and takes value "5" at x = 2.

$$\text{LET } \bar{h}(x) = \begin{cases} h(x) & \text{IF } x \neq 2 \\ 5 & \text{IF } x = 2 \end{cases}$$

$$= \begin{cases} \dfrac{2x^2-3x-2}{x-2} & \text{IF } x \neq 2 \\ 5 & \text{IF } x = 2 \end{cases} = \begin{cases} \dfrac{(2x+1)(x-2)}{(x-2)} & \text{IF } x \neq 2 \\ 5 & \text{IF } x = 2 \end{cases}$$

$$\overset{\text{(CANCEL)}}{=} \begin{cases} 2x+1 & \text{IF } x \neq 2 \\ 5 & \text{IF } x = 2 \end{cases}$$

$$= 2x+1 \text{ FOR ALL X INCLUDING 2.}$$
$$(\text{O.K. SINCE } 2\cdot 2+1 = 5)$$

Now we prove that $\bar{h}(x) = 2x+1$ is continuous at x=2:

1. For any $\epsilon > 0$, let $\delta_{(\epsilon)} = \frac{1}{2}\epsilon$.
2. If $2 - \frac{1}{2}\epsilon < x < 2 + \frac{1}{2}\epsilon$
   so $4 - \epsilon < 2x < 4 + \epsilon$
   Then $5 - \epsilon < 2x+1 < 5 + \epsilon$
   or $5 - \epsilon < \bar{h}(x) < 5 + \epsilon$.

So this finally proves that 5 is the correct value for h( ) to take at x=2 to be continuous there.

The procedure we just followed consisted of

1. <u>Guessing</u> what number value h( ) wants to take at x=a in order to be continuous at "a"

and then

2. Proving that the guess is the correct one by an ϵ, $\delta_{(\epsilon)}$ argument.

If this procedure can be done, <u>mathematicians</u> <u>still</u> <u>say</u>

"$\lim_{x \to a} h(x)$ exists"

<u>even</u> <u>though</u> "a" <u>may</u> <u>NOT</u> <u>be</u> <u>in</u> <u>the</u> <u>domain</u> <u>of</u> <u>h</u>( ).

Here is the official mathematical definition of "limit." It doesn't exactly describe this procedure, but we will see in a minute how it fits in with what we have just done.

> DEFINITION: "$\lim_{x \to a} h(x)$ exists and equals L" if there is a number "L" such that, for every challenge number $\epsilon > 0$, there is a response number $\delta_{(\epsilon)} > 0$ with the following guarantee:
>
> If $x \neq a$ and x satisfies $a - \delta_{(\epsilon)} < x < a + \delta_{(\epsilon)}$, | or, equivalently, if $0 < |x-a| < \delta_{(\epsilon)}$,
>
> Then, for these values of x,
>
> $L - \epsilon < h(x) < L + \epsilon$ | or, equivalently, $|h(x) - L| < \epsilon$.

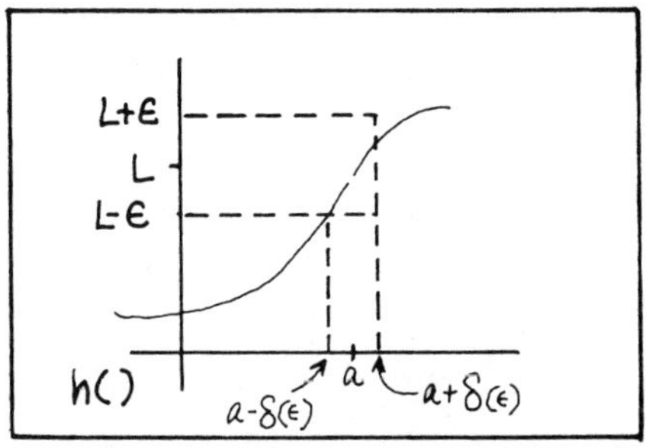

NUMBER "L" IS THE VALUE WE GUESS THAT h() WANTS TO TAKE AT X = a TO BE CONTINUOUS AT "a"

THIS MAKES SURE THAT X≠a. h() MAY NOT BE DEFINED AT "a," SO WE MUST EXCLUDE THE POSSIBILITY "x = a" FROM THE "THEN"... PART OF THE DEFINITION.

Now, how does this new definition fit in with first guessing what value h( ) wants to take at "a" to be continuous there and then proving that our guess is correct? Assume that h( ) satisfies the new definition. From the graph of h( ) it certainly looks like h( ) wants to take value "L" at "a" in order to be continuous there. So try "L" as our guess and define $\bar{h}(x) = \begin{cases} h(x) & \text{if } x \neq a \\ L & \text{if } x = a \end{cases}$

Now, according to Theorem II-1, if we can find an ϵ, $\delta_{(\epsilon)}$ argument that proves that $\bar{h}(\ )$ is continuous, then "L" MUST be the correct guess. Since we are assuming that the new definition holds, try using the guaranteed $\delta_{(\epsilon)}$ that is the response to any challenge ϵ in the new definition. The guarantee goes like this:

If $x \neq a$ and x satisfies

$$a - \delta_{(\epsilon)} < x < a + \delta_{(\epsilon)},$$

then by the guarantee on $\delta_{(\epsilon)}$,

$$L - \epsilon < h(x) < L + \epsilon.$$

Since $\bar{h}(a) = L$ and $\bar{h}(x) = h(x)$ when $x \neq a$, this last inequality translates to "Then $\underset{\uparrow L}{\bar{h}(a)} - \epsilon < \bar{h}(x) < \underset{\uparrow L}{\bar{h}(a)} + \epsilon$."

And when $x = a$, then

$a - \delta_{(\epsilon)} < a < a + \delta_{(\epsilon)}$ is certainly true, and so is

$$\bar{h}(a) - \epsilon < \bar{h}(a) < \bar{h}(a) + \epsilon.$$

These two arguments show that the $\delta_{(\epsilon)}$ that comes with the new definition of "$\lim_{x \to a} h(x)$ exists" also serves to prove that $\bar{h}(x)$ (taking value "L" at "a") is continuous at "a." And thus performing the ritual described in the new definition will also guarantee that there is a function "$\bar{h}(\ )$" that extends h( ) and is continuous at x=a.

Conversely, performing the ritual of
1. finding a function $\overline{h}(\ )$ that agrees with $h(\ )$ except possibly at "a" and then
2. proving that $\overline{h}(\ )$ is continuous at "a"
is a perfectly good way of showing that "$\lim_{x \to a} h(x)$ exists" according to the new definition. The appropriate "guess" will be "Let $L = \overline{h}(a)$," and while proving that $\overline{h}(\ )$ is continuous at "a" we assume that we will have found a guaranteed $\delta_{(\epsilon)}$ for any challenge ϵ such that

If $a - \delta_{(\epsilon)} < x < a + \delta_{(\epsilon)}$
Then $\overline{h}(a) - \epsilon < \overline{h}(x) < \overline{h}(a) + \epsilon$.

It follows certainly, that

If $a - \delta_{(\epsilon)} < x < a + \delta_{(\epsilon)}$ AND $x \neq a$
Then $\overline{h}(a) - \epsilon < h(x) < \overline{h}(a) + \epsilon$

DESIGNATED VALUE FOR "L" — O.K. TO PUT THIS HERE, SINCE $x \neq a$ NOW, AND $\overline{h}(x) = h(x)$ IF $x \neq a$. — "L"

and this is precisely the guarantee that is needed in the new definition.

Here's how to prove that $\lim_{x \to 2} h(x)$ exists using the ritual in the new definition, where $h(x) = \dfrac{2x^2-3x-2}{x-2}$. (This time we have to figure out both L and $\delta_{(\epsilon)}$. To figure out L , first write the "then. . . " part of the proof without knowing L; factor the top of the fraction; secretly cancel "(x-2)," leaving just "2x+1" (O.K. since the "If. . . " part of the proof says $x \neq 2$, so $(x-2) \neq 0$); finally unofficially substitute 2 in 2x+1 to guess $L=5=2 \cdot 2+1$ as the value that $h(\ )$ wants to take at x=2 to be continuous there. Next, to find $\delta_{(\epsilon)}$, just proceed as you would if you were proving that "2x+1" is continuous at x=2. Remember that the proof must read logically from top to bottom —that is why the "uncanceling" sentence must be put in.)

Proof: Let $L = 5$. For any $\epsilon > 0$,
let $\delta_{(\epsilon)} = \frac{1}{2}\epsilon$.
If x satisfies
$2 - \frac{1}{2}\epsilon < x < 2 + \frac{1}{2}\epsilon$ AND $x \neq 2$,
then (times 2) $4 - \epsilon < 2x < 4 + \epsilon$
so (add 1) $5 - \epsilon < 2x+1 < 5 + \epsilon$.
Now, since $x \neq 2$, we can "uncancel" to get $\dfrac{x-2}{x-2} = 1$, so we can multiply $2x+1$ by $\dfrac{x-2}{x-2}$ to get

$$L - \epsilon < \frac{(2x+1)(x-2)}{x-2} < L + \epsilon,$$

UNOFFICIAL— CANCEL (x-2) AND GUESS THAT L=5 =2·(2)+1.

SECRET

$$\text{or } L - \epsilon < \frac{2x^2-3x-2}{x-2} < L + \epsilon$$

then $L - \epsilon < h(x) < L + \epsilon$

The distinction between a "limit existing" and a function being "continuous" is sometimes a little confusing.

When talking only about "$\lim_{x \to a} h(x)$ existing" in the new definition (and not mentioning continuity), it doesn't matter whether or not "a" is in the domain of $h(\ )$ — and even if "a" <u>is</u> in the domain of $h(\ )$, it doesn't matter whether or not $h(a) = L$, the limit number. The question just never comes up, since x is <u>not allowed</u> to equal "a" while proving the guarantee in the new definition. The number "L" is simply the value that the function wants to take (but may not actually do so) at x=a in order to be continuous at "a."

If $h(\ )$ is <u>continuous</u> at x=a, the value of h(a) is important. The old definition of continuity went like this:

Definition: h( ) is continuous at x=a if $\lim_{x \to a} h(x)$ exists and equals h(a).

$\lim_{x \to a} h(x)$ exists and equals h(a) if, for any challenge number $\epsilon > 0$, there is a response number $\delta_{(\epsilon)} > 0$ guaranteed as follows:

If x satisfies the condition

$$a - \delta_{(\epsilon)} < x < a + \delta_{(\epsilon)} \quad ,$$

then, for these values of x ,

$$h(a) - \epsilon < h(x) < h(a) + \epsilon \quad .$$

Compare this with the new definition of "$\lim_{x \to a} h(x)$ exists." Note that for a function h( ) to be "continuous", we require both that "$\lim_{x \to a} h(x)$ exists" according to the new definition and, in addition, we must have

1. "a" in the domain of h( ) (since "h(a)" must "make sense") and
2. h(a) = L, the limit number in the new definition.

This business of excluding consideration of the value (if any) that the function may take at "a" in the new definition of "limits existing" sometimes leads to somewhat peculiar logical situations like the following: If we consider the function

$$g(x) = \begin{cases} 2x+1 & \text{if } x \neq 2 \\ 3 & \text{if } x = 2 \end{cases} \quad ,$$

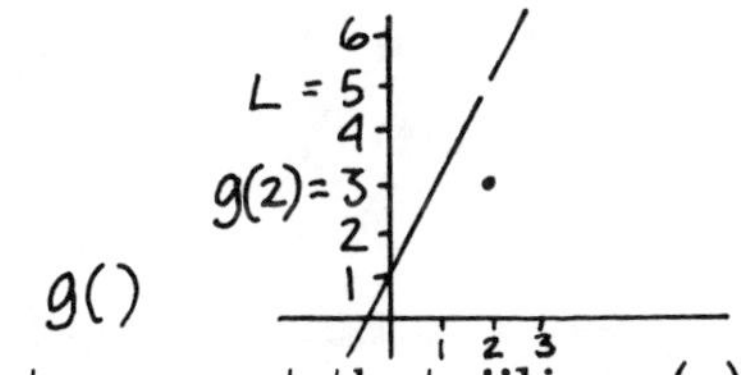

it turns out that "$\lim_{x \to 2} g(x)$ exists and equals 5" — the official proof is similar to the one in the example— since x will not be allowed to equal 2 in the proof. But g(x) is NOT continuous at 2, since g(2) = 3 and the limit-number L=5, so g(2) ≠ L, and g( ) fails to meet the further extra requirements for continuity.

On the other hand, if we follow the old ritual from section II.4 and prove that some function f( ) is continuous at "a" using the old definition,

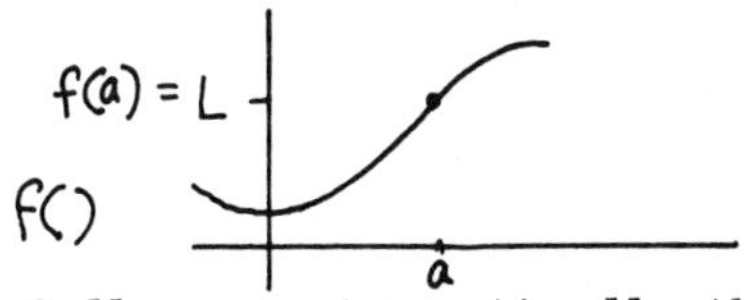

it follows automatically that

1. "a" must be in the domain of f( ) and
2. "$\lim_{x \to a} f(x)$ exists" according to the new definition— the appropriate "L" IS f(a) and the same guaranteed ϵ , $\delta_{(\epsilon)}$ argument that goes along with proving f( ) is "continuous" also works to show that "$\lim_{x \to a} f(x)$ exists" in the new definition.

Mathematicians often say "the limit of h(x) as x goes to 'a' IS L" or write

"$\lim_{x \to a} h(x) = L$ " instead of

"$\lim_{x \to a} h(x)$ exists and equals L ."

This gets us back to the intuitive idea that, as x moves toward "a" from the right or the left, h(x) approaches L as its limit number.

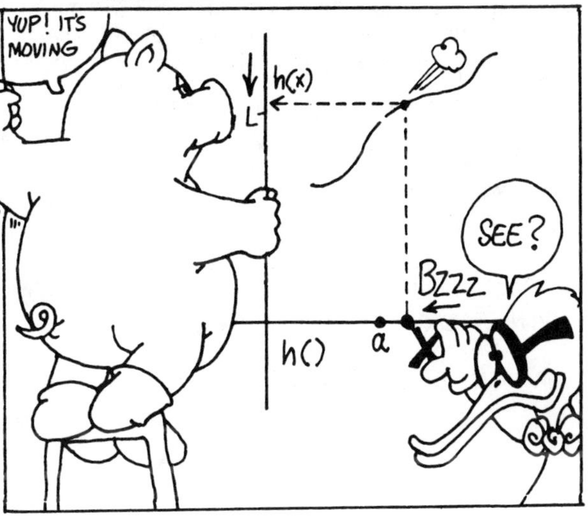

But the precise logical meaning of "limit" is not only the limit number "L" (which we can only guess at) but also it assumes an implied ϵ, $\delta(\epsilon)$ process proving that the guess "L" is the correct one to make the extended function $\overline{h}(\ )$ (taking value L at "a") continuous at "a."

So, to sum things up:

1. A "limit exists" for a function h( ) at a point "a" if we can guess what number value "L" h( ) wants to take at x=a (but may not actually do so) in order to be continuous at "a" and then prove that the guess is the correct one by a suitable ϵ, $\delta(\epsilon)$ argument. Theorem II-1 proves that there is only one possible value "L" for which this ritual can be done. The value (if any) that the function may take at "a" is ignored completely in the proof. If "$\lim_{x \to a} h(x)$ exists and equals L," then, as x moves toward "a" from the right or the left, value "h(x)" approaches "L" as its limit number.

2. A function h( ) is "continuous" at a point "a" if the limit exists as in 1. above AND this time the point "a" must be in the domain of h( ) and h(a) must equal the limit number L. Proving that a function is continuous at "a" automatically shows that "$\lim_{x \to a} h(x)$ exists": the limit number "L" IS "h(a)."

3. If you can find a function $\overline{h}(\ )$ that agrees with h( ) everywhere except possibly at "a" and prove that $\overline{h}(\ )$ is continuous at "a," then $\lim_{x \to a} h(x)$ exists as in 1. above; the correct guess for "L" is $\overline{h}(a)$.

EXERCISES

Sample:
Prove that $\lim_{x \to -2} \frac{3x^2+7x+2}{x+2}$ exists.

(TO FIND "L", START AT "THEN" AS USUAL; FACTOR THE TOP OF THE FRACTION; CANCEL; THEN "GUESS" WHAT "L" SHOULD BE BY SETTING $x=-2$: $L \stackrel{?}{=} 3(-2)+1=-5$!)

PROOF:

LET $L=-5$. FOR ANY $\epsilon>0$, LET $\delta(\epsilon)=\epsilon/3$

IF $-2-\epsilon/3 < x < -2+\epsilon/3$ AND $x \neq -2$

SO $-6-\epsilon < 3x < -6+\epsilon$

AND $-5-\epsilon < 3x+1 < -5+\epsilon$

SINCE $x \neq -2$, $\frac{x+2}{x+2}=1$, SO

THEN $-5-\epsilon < \frac{(3x+1)(x+2)}{x+2} < -5+\epsilon$

OR THEN $-5-\epsilon < \frac{3x^2+7x+2}{x+2} < -5+\epsilon$

II.9.1
Prove that $\lim_{x \to 3} \frac{x^2-x-6}{x-3}$ exists.

II.9.2

Prove that $\lim_{x \to -4} \frac{-2x^2-5x+12}{x+4}$ exists.

II.9.3

Prove that $\lim_{x \to 2} \frac{x^2-4x+4}{x-2}$ exists.

II.9.4

Prove that $\lim_{x \to 0} \frac{0}{x}$ exists.

II.9.5

Prove that $\lim_{h \to 0} \frac{2h^3+3h^2+h}{h}$ exists.

(Just treat "h" like "x.")

II.9.6

Prove that $\lim_{x \to 2} \frac{x^3-4x^2+7x-6}{x-2}$ exists.

II·10 TWO LIMIT THEOREMS

Constructing limit-machines and proving guarantees every time we need them is time-consuming if not downright boring. They are of no use in computing anything scientific — their use is a logical one: mathematicians need to know that they can construct them to make sure that limits exist and guessed limit numbers are correct or that functions are continuous, thus keeping differential calculus on a sound logical footing.

But knowing that one could without fail construct a limit-machine is different from actually constructing a limit-machine and showing that it is guaranteed. The needs of logic only require that we be sure that we could construct a limit-machine if we had to. So mathematicians set out to find an argument that would show that they had a foolproof method for constructing guaranteed limit-machines for lots of functions. The argument had to be so convincing that everyone would agree that, for lots of functions, if a limit had to be shown to exist, an appropriate infallible limit-machine definitely could be constructed, so no one would ever have to bother actually constructing one.

To show all this, mathematicians built an argument around a series of "limit theorems." These theorems prove that if guaranteed limit-machines exist for some functions, then automatically there will be guaranteed limit-machines for certain more complicated functions made out of the original functions, and tell us correct guesses for new limits.

Mathematicians then proved that limits exist for a few simple functions by giving definite ϵ, $\delta_{(\epsilon)}$ arguments. Then, when they brewed up all kinds of complicated functions from the simple functions, the limit theorems automatically guaranteed that suitable ϵ, $\delta_{(\epsilon)}$ proofs could be made showing that limits exist for the complicated functions. So no one need bother actually writing down the proofs.

One easy way to make new functions out of old functions is to multiply the whole function by a constant number. For example, if

$$f(x) = x^2 + 2x,$$

then $\quad$ "$2 \cdot f(x)$" $= 2(x^2 + 2x) = 2x^2 + 4x$.

The graph of "$2 \cdot f(x)$" is just a stretched-along-the-y-axis version of f( ). The y-distance of a point $(x, 2 \cdot f(x))$ on the graph of "$2 \cdot f(\ )$" is just 2 times as far from the x-axis as the corresponding point $(x, f(x))$ on the graph of f( ):

Another way to make new functions out of old ones is to "add" two functions. For example, we can look at

$$h(x) = x^2 + \; x$$

as the "sum" of functions

$$f(x) = x^2 \quad \text{and} \quad g(x) = \; x$$

and this is what writing

$$\text{"}h(x) = f(x) + g(x)\text{"}$$

means. A point on the graph of "f(x) + g(x)" is (x, f(x)+g(x)): it can be located by just adding the y-distances of points (x, f(x)) and (x, g(x)) above the same "x" point on the x-axis.

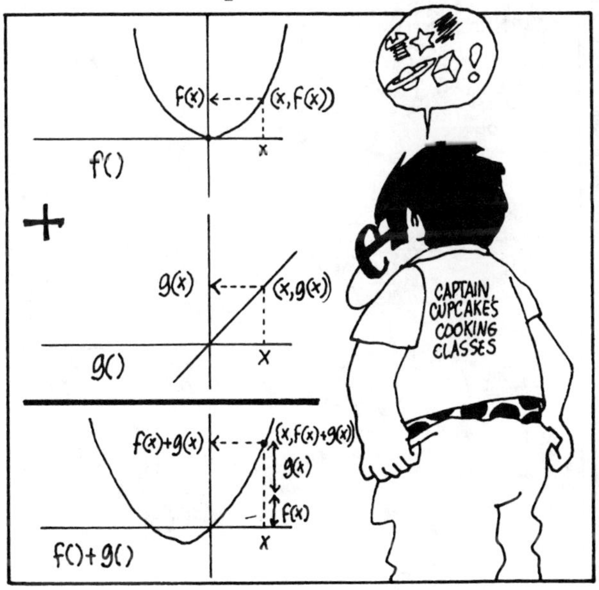

We will see in a minute how these two methods of making complex functions from simpler ones will produce the entire family of polynomials! But first, let's have a look at some limit theorems.

The limit theorems prove that each time we use one of these methods of constructing new functions from old ones, the new functions <u>inherit</u> any limit properties that the old functions may have had. For these limit theorems to be of any use, they must start by assuming that the old functions have limits. This leads to the somewhat peculiar practice of starting the statement of a theorem with an expression like "If $\lim_{x \to a} f(x)$ exists..." without giving any formula for f( ). Then the course of argument in the theorem itself will compound the problem by asserting that there must be $\delta^f_{(\epsilon)}$'s guaranteed as usual to satisfy the statement

"If $a - \delta^f_{(\epsilon)} < x < a + \delta^f_{(\epsilon)}$ and $x \neq a$
Then $L - \epsilon < f(x) < L + \epsilon$."

This statement is supposed to hold true even though there is no formula for $\delta^f_{(\epsilon)}$ either. There can't be, as there isn't any formula for f( ) in the first place. And, obviously, with no formula for $\delta^f_{(\epsilon)}$, there can be no algebraic demonstration to show that the "If... Then" guarantee is true. But the theorems don't want to be any more specific. They simply wait for an actual function to come along that has a formula and guaranteed $\delta^f_{(\epsilon)}$ before they become operational. So, even though we don't see any proof of it, it is all right for the theorem to go ahead and assume that $\delta^f_{(\epsilon)}$ is guaranteed—it will be when the theorem is actually used.

The first limit theorem shows that having limits is inherited when we multiply a function by a constant and tells us what the new limit should be.

THEOREM II-2 (The "K·f theorem") (Official statement)	(Unofficial Translation)
If $\lim_{x\to a} f(x)$ exists and equals L,	If f( ) wants to take value "L" at x = a in order to be continuous there and has a limit-machine to show that "L" is the right number.
then $\lim_{x\to a} K\cdot f(x)$ exists and equals K·L, for any constant number K.	then K·f( ) wants to take value "K·L" at x = a in order to be continuous there and we can construct a limit-machine to show that "K·L" is the right number.
Proof:	

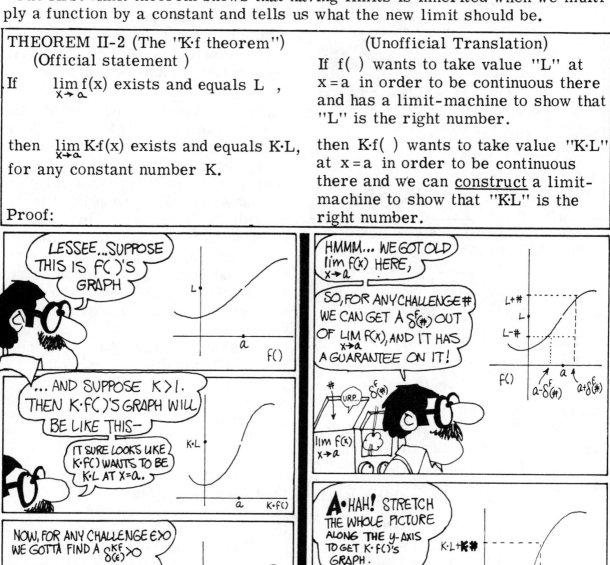

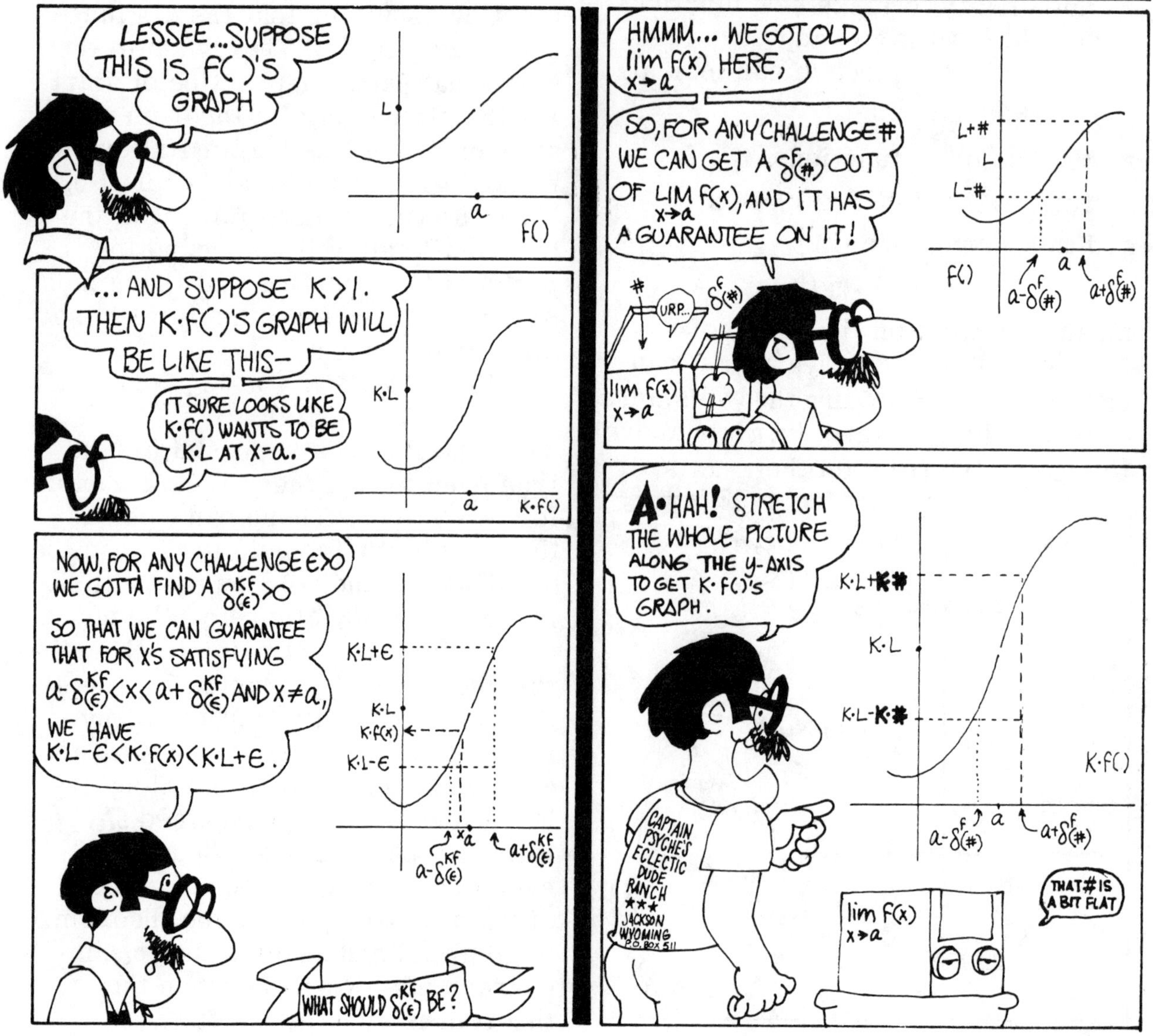

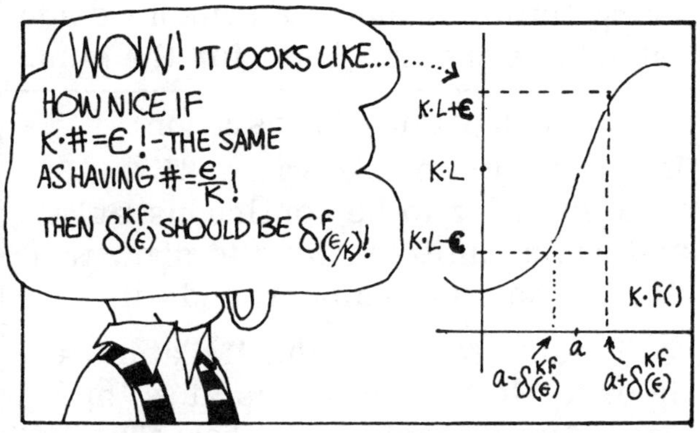

GUARANTEE

If x satisfies

$a - \delta^{f}_{(\epsilon/K)} < x < a + \delta^{f}_{(\epsilon/K)}$ and $x \neq a$,

then, for these x's,

$L - \frac{\epsilon}{K} < f(x) < L + \frac{\epsilon}{K}$.

NOW we can multiply by K to get

$$KL - \epsilon < Kf(x) < KL + \epsilon$$

and that is exactly what we want for "K·f( )."

Here is the official proof. We use

$$\# = \frac{\epsilon}{|K|}$$

to make sure that # is positive. (If K = 0, there is nothing to prove, since then K·f( ) = 0·f( ) = 0 for ALL x, and the function that takes value 0 all the time is certainly continuous everywhere.)

THEOREM II-2.

If $\lim_{x \to a} f(x)$ exists and equals L, then $\lim_{x \to a} K{\cdot}f(x)$ exists and equals K·L for any constant number K.

Proof: If K = 0, there is nothing to prove. So assume $K \neq 0$. For any $\epsilon > 0$, let $\delta^{Kf}_{(\epsilon)} = \delta^{f}_{(\epsilon/|K|)}$

If $0 < |x - a| < \delta^{f}_{(\epsilon/|K|)}$

then, for these x's , the guarantee on $\delta^{f}_{(\epsilon/|K|)}$ says that $|f(x) - L| < \frac{\epsilon}{|K|}$

or (times |K|),

$|K| \cdot |f(x) - L| < \frac{\epsilon}{|K|}|K| = \epsilon$.

So,

$$|K \cdot (f(x) - L)| < \epsilon$$

or $|K{\cdot}f(x) - KL| < \epsilon$.

The guarantee is O.K. !

The second limit theorem shows that having limits is inherited when we add two functions together and tells us what should be the new limit for the sum.

THEOREM II-3 (The "f + g" theorem)	(Unofficial Translation)
IF 1. $\lim_{x\to a} f(x)$ exists and equals L and	If f( ) wants to take value "L" at x = a in order to be continuous at "a" and has a limit-machine to show that "L" is the right number and if
2. $\lim_{x\to a} g(x)$ exists and equals M	g( ) wants to take value "M" at x = a in order to be continuous at "a" and has a limit-machine to show that "M" is the right number,
THEN $\lim_{x\to a}(f(x)+g(x))$ exists and equals L+M.	then f( ) + g( ) wants to take value "L + M" at x = a in order to be continuous at "a" and we can <u>construct</u> a limit-machine to show that "L + M" is the right number.

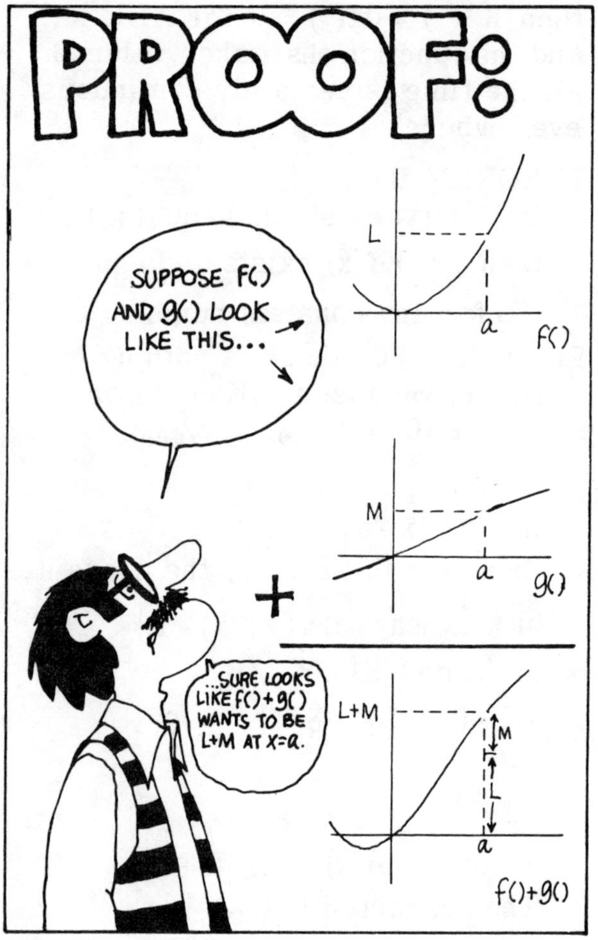

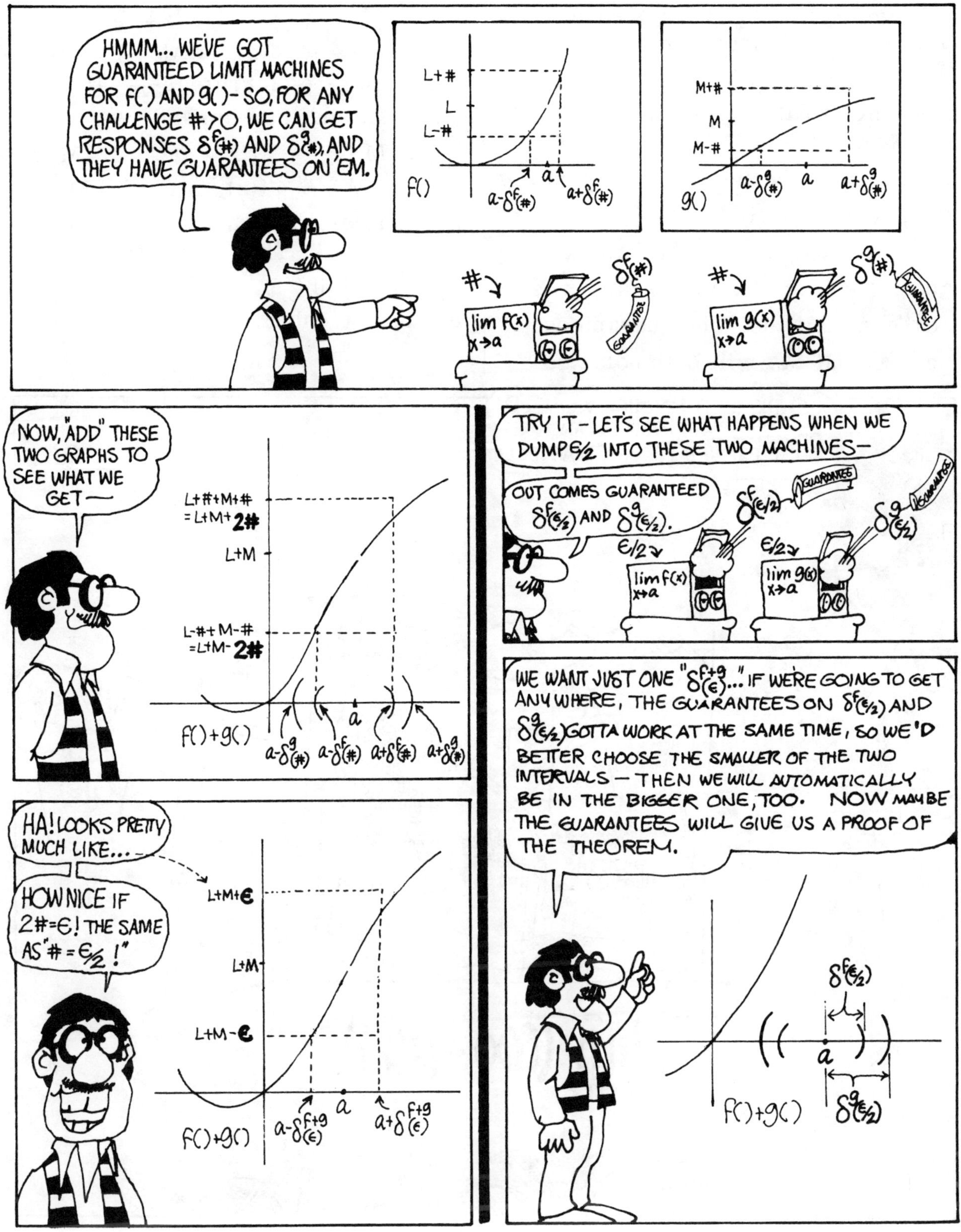
HMMM... WE'VE GOT GUARANTEED LIMIT MACHINES FOR f( ) AND g( ) - SO, FOR ANY CHALLENGE # > 0, WE CAN GET RESPONSES $\delta^f(\#)$ AND $\delta^g(\#)$, AND THEY HAVE GUARANTEES ON 'EM.
L+#
L
L-#
f( )
a
a-$\delta^f(\#)$
a+$\delta^f(\#)$
M+#
M
M-#
g( )
a-$\delta^g(\#)$
a+$\delta^g(\#)$
#
lim f(x) x→a
GUARANTEE
lim g(x) x→a
NOW, "ADD" THESE TWO GRAPHS TO SEE WHAT WE GET—
L+#+M+# = L+M+2#
L+M
L-#+M-# = L+M-2#
f( )+g( )
a-$\delta^g(\#)$
a-$\delta^f(\#)$
a+$\delta^f(\#)$
a+$\delta^g(\#)$
TRY IT - LET'S SEE WHAT HAPPENS WHEN WE DUMP ε/2 INTO THESE TWO MACHINES—
OUT COMES GUARANTEED $\delta^f(\epsilon/2)$ AND $\delta^g(\epsilon/2)$.
ε/2
lim f(x) x→a
lim g(x) x→a
GUARANTEE
HA! LOOKS PRETTY MUCH LIKE...
HOW NICE IF 2# = ε! THE SAME AS "# = ε/2 !"
L+M+ε
L+M
L+M-ε
f( )+g( )
a-$\delta^{f+g}(\epsilon)$
a+$\delta^{f+g}(\epsilon)$
WE WANT JUST ONE "$\delta^{f+g}(\epsilon)$..." IF WE'RE GOING TO GET ANYWHERE, THE GUARANTEES ON $\delta^f(\epsilon/2)$ AND $\delta^g(\epsilon/2)$ GOTTA WORK AT THE SAME TIME, SO WE'D BETTER CHOOSE THE SMALLER OF THE TWO INTERVALS - THEN WE WILL AUTOMATICALLY BE IN THE BIGGER ONE, TOO. NOW MAYBE THE GUARANTEES WILL GIVE US A PROOF OF THE THEOREM.
$\delta^f(\epsilon/2)$
a
f( )+g( )
$\delta^g(\epsilon/2)$

To get x in both guaranteed intervals means we have to try

$$\delta^{f+g}_{(\epsilon)} = \min(\delta^{f}_{(\epsilon/2)}, \delta^{g}_{(\epsilon/2)}) \quad .$$

Let's see what happens with this $\delta^{f+g}_{(\epsilon)}$:

If x satisfies

$0<|x-a|<\delta^{f+g}_{(\epsilon)}$ = the smaller of $\delta^{f}_{(\epsilon/2)}$ and $\delta^{g}_{(\epsilon/2)}$,

then such x's will satisfy

both $a - \delta^{f}_{(\epsilon/2)} < x < a + \delta^{f}_{(\epsilon/2)}$ (and $x \neq a$)

and $a - \delta^{g}_{(\epsilon/2)} < x < a + \delta^{g}_{(\epsilon/2)}$ (and $x \neq a$)

at the same time! The guarantees on $\delta^{f}_{(\epsilon/2)}$ and $\delta^{g}_{(\epsilon/2)}$ will both hold, so

$$L - \frac{\epsilon}{2} < f(x) < L + \frac{\epsilon}{2}$$

and $\quad M - \frac{\epsilon}{2} < g(x) < M + \frac{\epsilon}{2}$

will both be true at the same time! Now just add them up and we get

$$L + M - \frac{\epsilon}{2} - \frac{\epsilon}{2} < f(x) + g(x) < L + M + \frac{\epsilon}{2} + \frac{\epsilon}{2}$$

or $\quad L + M - \epsilon < f(x) + g(x) < L + M + \epsilon$,

which is exactly what we want for "f( ) + g( )!" This shows that

$$\delta^{f+g}_{(\epsilon)} = \min(\delta^{f}_{(\epsilon/2)}, \delta^{g}_{(\epsilon/2)})$$

is guaranteeable!

Here is the official proof:

THEOREM II-3

IF 1. $\lim_{x\to a} f(x)$ exists and equals L

and 2. $\lim_{x\to a} g(x)$ exists and equals M

THEN $\lim_{x\to a}(f(x)+g(x))$ exists and equals L+M.

Proof: For any $\epsilon > 0$, let

$\delta^{f+g}_{(\epsilon)} = \min(\delta^{f}_{(\epsilon/2)}, \delta^{g}_{(\epsilon/2)})$.

Now, if x satisfies

$$0 < |x-a| < \delta^{f+g}_{(\epsilon)},$$

such values of x will also satisfy

both $a-\delta^{f}_{(\epsilon/2)} < x < a+\delta^{f}_{(\epsilon/2)}$ (and $x \neq a$)

and $a-\delta^{g}_{(\epsilon/2)} < x < a+\delta^{g}_{(\epsilon/2)}$ (and $x \neq a$).

So $L-\frac{\epsilon}{2} < f(x) < L+\frac{\epsilon}{2}$

and $M-\frac{\epsilon}{2} < g(x) < M+\frac{\epsilon}{2}$.

Adding these two we get

$$L+M-\frac{\epsilon}{2}-\frac{\epsilon}{2} < f(x)+g(x) < L+M+\frac{\epsilon}{2}+\frac{\epsilon}{2}$$

or $L+M-\epsilon < f(x)+g(x) < L+M+\epsilon$.

Now we start with a few simple functions that already have limit-machines, and so have ϵ, $\delta_{(\epsilon)}$ arguments showing that they are continuous. Then we use the "K·f( )" method and the "f( )+g( )" method to brew up any cubic polynomial. The limit theorems prove that these cubics will then inherit the property of having limits and being continuous.

On page 113 an exercise showed that f(x)=x has a limit-machine

that shows that f(x) = x is continuous for all "a."

On page 126, an exercise showed that $f(x) = x^2$ is continuous at any "a", with machine

On page 127 we showed that $f(x) = x^3$ is continuous at any "a" with machine

Finally, the function f(x) = 1 certainly looks continuous,

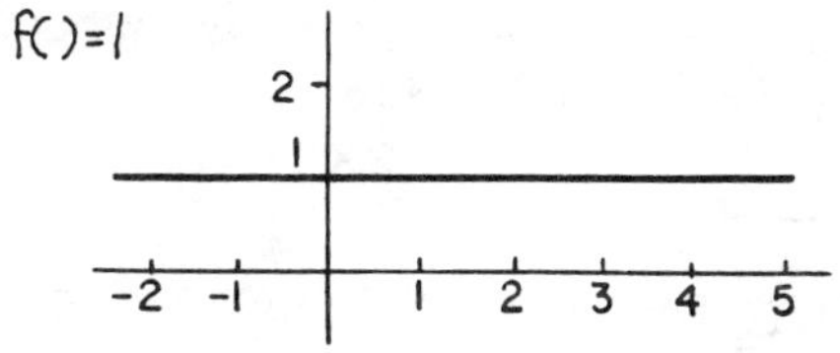

although proving it is continuous is somewhat peculiar. Here is the proof:

Show that $\lim_{x\to a} 1$ exists and equals 1.

Proof:

For any $\epsilon > 0$, let $\delta_{(\epsilon)}$ = anything positive! (IT DOESN'T MATTER WHAT

If $a-\delta_{(\epsilon)} < x < a+\delta_{(\epsilon)}$ ← THIS SAYS, SINCE

Then $1-\epsilon < 1 < 1+\epsilon$. ← THIS IS ALWAYS TRUE)

So functions

"1," "x," "x^2" and "x^3"

are continuous at any "a." Since proving that a function is continuous at a point automatically proves that the appropriate limit exists, we can say that "limits exist" for these functions too.

Any polynomial no more complicated than a cubic can be produced from these functions by using the "K·f( )" method and the "f( )+g( )" method over and over again, building up a "family tree" that eventually produces the polynomial. The limit theorems prove that each time these methods are used, the new functions so produced will inherit the property of having limits and being continuous. A typical argument for, say, function $f(x) = 3x^3 - \frac{14}{3}x^2 + 2x - 5$ shows the "family tree" and goes like this:

x^3 x^2 x 1

THESE HAVE LIMITS FOR ANY "a"

$K=3$ $K=(-\frac{14}{3})$ $K=2$ $K=(-5)$

$3x^3$ $(-\frac{14}{3})\cdot x^2$ $2\cdot x$ $(-5)\cdot 1$

THEN THE "K·F" THEOREM SAYS THAT THESE HAVE GUARANTEED LIMITS

$3x^3 - \frac{14}{3}x^2$ $2x - 5$

THEN THE "F+G" THEOREM SAYS THAT THESE HAVE GUARANTEED LIMITS

$(3x^3 - \frac{14}{3}x^2) + (2x-5)$

FINALLY, THE "F+G" THEOREM USED AGAIN SAYS THAT THIS HAS A GUARANTEED LIMIT.

The whole function will be continuous as well, since any "a" is in the function's domain, and all the limit numbers along the way are the component function's values at "a."

Just to check out everything, let's try the construction techniques in the "K·f" theorem and the "f+g" theorem to construct a guaranteed limit-machine to show that function

$$f(x) = x^2 + 5x$$

is continuous at a=2: $f(2)=(2)^2+5\cdot(2)=14$.

We start with these two machines:

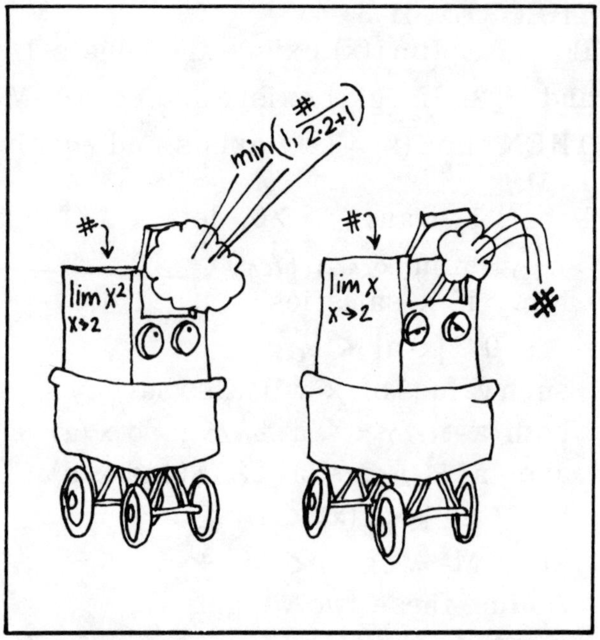

and we know that they are guaranteed. So now brew up "$x^2 + 5x$":

For "5x," the "Kf" theorem gives limit-machine

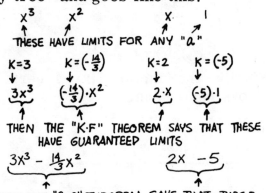

So $\delta_{(\epsilon)}$ turns out to be

$$\delta_{(\epsilon)} = \min(1, \tfrac{\epsilon}{10}).$$

Check the guarantee:

If $\quad |x-2| < \delta_{(\epsilon)} = \min(1, \frac{\epsilon}{10})$,

then $\quad |x-2| < 1$,

so $\quad -1 < x-2 < 1$,

or $\quad 1 < x < 3$

or $\quad -5 < 3 < x+2 < 5$

or $\quad |x+2| < 5$.

Now we make the following calculation:

$$|x^2+5x-14| = |(x^2-4)+5\cdot(x-2)| =$$

$$|(x+2)\cdot(x-2)+5\cdot(x-2)| \leq$$

$$|(x+2)||(x-2)|+5|x-2| \leq$$

$$5\cdot|x-2| + 5\cdot|x-2| < 5\tfrac{\epsilon}{10} + 5\tfrac{\epsilon}{10} = \epsilon,$$

since $|x-2| < \frac{\epsilon}{10}$ also.

So $\quad |x^2+5x-14| < \epsilon$,

or $\quad -\epsilon < x^2+5x-14 < \epsilon$

or $\quad 14 - \epsilon < x^2+5x < 14 + \epsilon$.

So $\quad \delta_{(\epsilon)} = \min(1, \frac{\epsilon}{10})$ actually works!

The two limit theorems we already have could be used to show that ALL polynomials are continuous if we had a proof that

$$x^4,\ x^5,\ x^6,\ \ldots$$

are continuous. A bootstrap sort of argument can be made to establish this by regarding x^5, say, as the "product" of the two functions x^3 and x^2 and then proving a further limit theorem about the product of functions. Here is the theorem; the proof is in the extra exercises and excursions at the end of the chapter.

THEOREM:

If $\lim_{x \to a} f(x)$ exists and equals L and if $\lim_{x \to a} g(x)$ exists and equals M, then $\lim_{x \to a}(f(x) \cdot g(x))$ exists and equals LM.

Then, since we already have specific ϵ, $\delta_{(\epsilon)}$ arguments to show that x^3 and x^2 are continuous (and thus have limits), the product limit theorem says that x^5 will automatically have a limit, so we need not supply any further ϵ, $\delta_{(\epsilon)}$ proof.

The bootstrap argument then continues by regarding x^6 as $x^5 \cdot x$ and using the product limit theorem once again; etc.

Other functions that come up in engineering, physics, chemistry and economics, like "$\sin(x)$," "$\cos(x)$," "e^x," "$\log(x)$," etc. have been shown to be continuous with the help of many more limit theorems. But although every actual function that represents behavior that is intuitively "continuous" can also be shown to satisfy the official ϵ, $\delta_{(\epsilon)}$ definition of continuity, the subject continues to intrigue mathematicians. The ϵ, $\delta_{(\epsilon)}$ formulation of "continuous" certainly appears to be a concrete and logically satisfactory way to deal with the mysterious and somewhat mystical ideas of "infinity" and "infinitesimally small," but one is never quite sure...

EASY READER
IRVINGTON DOGTROT
IRVINGTON DOGTROT
IRVINGTON DOGTROT
MG2
2%
GOOD VIBES
RENO NEVADA

EXTRA EXERCISES & EXCURSIONS

Extra exercises are given for many sections in chapter 2, and they are numbered, for example, like EII.4.1 to indicate that they belong to section II.4. Selected answers and solutions can be found at the end of the book.

For Section II.4.

> Exercise EII.4.1.
> For any constants K and C, prove that f( ) = K( )+C (f(x)=Kx+C) is continuous at x=a for any a.

For Section II.5.

EXCURSION!

In the discussion of limits in section II.5 we argued that we can't allow $\delta(\epsilon)$ to depend on x as well as ϵ. Some readers of the previous edition of McSquared's Calculus Primer have asked for an example showing how we can go wrong if we let $\delta(\epsilon)$ depend on x as well as ϵ. Here is such an example:

$$\text{Define } f(x) = \begin{cases} \frac{1}{|x|} & \text{if } x \neq 0 \\ 0 & \text{if } x = 0 \end{cases}$$

f( )'s graph is sketched in the following picture:

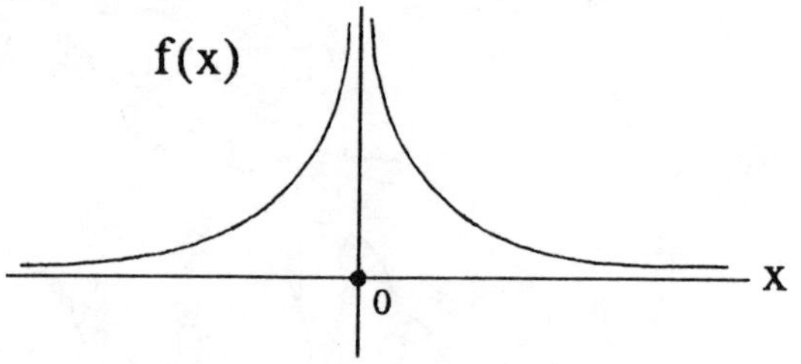

Intuitively, the graph of f(x) is certainly NOT smooth and squeezed together near (0,f(0)) = (0,0).

IF we allow $\delta(\epsilon)$ to depend on x, we could let $\delta(\epsilon) = \epsilon|x|^2$ and make the following argument that looks like a 'proof' that f(x) is continuous at x=0:

If $0 < |x-0| < \epsilon|x|^2$, we can divide both sides of this inequality by positive $|x|^2$ to get

$$\frac{1}{|x|} = \frac{|x-0|}{|x|^2} < \epsilon\frac{|x|^2}{|x|^2} = \epsilon.$$

So $\frac{1}{|x|} = \left|\frac{1}{|x|} - 0\right| = |f(x)-f(0)| < \epsilon.$

If x=0, then $|f(0)-f(0)| = 0 < \epsilon$ also. So we appear to have a 'proof' that f(x) is continuous at x=0.

But the requirement that

$$|x| = |x-0| < \epsilon|x|^2$$

does NOT make $|x|$ close to zero for small ϵ, which is what '$|x-0| < \delta(\epsilon)$' is supposed to do. On the contrary, if we just divide the inequality $|x| < \epsilon|x|^2$ by $|x|\epsilon$, which is positive (when $x \neq 0$), we get

$$\frac{1}{\epsilon} = \frac{|x|}{\epsilon|x|} < \frac{\epsilon|x|^2}{\epsilon|x|} = |x|,$$

so $|x|$ is bounded away from 0. For example, if $\epsilon=1/10$, we get

$$10 = \frac{1}{1/10} < |x| \;.$$

So the alleged 'proof' above does not allow the fence building near $(0,0) = (0,f(0))$ that motivates the definition of limit. By insisting that $\delta(\epsilon)$ depend only on ϵ, we avoid such false 'proofs', which is fortunate, for intuitively,

$$f(x) = \begin{cases} \frac{1}{|x|} & \text{if } x \neq 0 \\ 0 & \text{if } x = 0 \end{cases}$$

is not continuous at $x=0$.

Exercise EII.5.1.

Prove that $4x^2+2x+2$ is continuous at $x = -1/2$.

Exercise EII.5.2. Sample.

Prove that $f(x) = x^3-3x^2+2x+2$ is continuous at $x=1$.

Solution: Here we use the results discussed in the extra exercises for chapter I, section 3.

We must show that

$\lim_{x\to 1}(x^3-3x^2+2x+2)$ exists and equals $1^3-3\cdot 1^2+2\cdot 1+2 = 2$.

1. For any $\epsilon>0$,

let $\delta(\epsilon) = \min(1, \boxed{\frac{\epsilon}{4}})$.

2. If $1-\delta(\epsilon) < x < 1+\delta(\epsilon)$
so $-\delta(\epsilon) < x-1 < \delta(\epsilon)$
or $|x-1| < \delta(\epsilon)$.

Also, since $\delta(\epsilon) = \min(1, \boxed{\frac{\epsilon}{4}})$, with $|x-1| < \delta(\epsilon)$, we have $|x-1| < 1$ or $-1 < x-1 < 1$ or $0 < x < 2$. Multiply this inequality by -2; we get $0 > -2x > -4$ or $-4 < -2x < 0$.

Also, with $0 < x < 2$, $0 < x^2 < 4$. Adding these two inequalities, we get

$$\begin{array}{rcl} -4 < & -2x & < 0 \\ 0 < & x^2 & < 4 \\ \hline -4 < & x^2-2x & < 4, \end{array}$$

so $|x^2-2x| < 4$.

Since $\delta(\epsilon) = \min(1, \boxed{\frac{\epsilon}{4}})$,

$|x-1| < \delta(\epsilon)$ means that $|x-1| < \frac{\epsilon}{4}$ also, so

$|x-1||x^2-2x| \leq |x-1|\cdot 4 < \frac{\epsilon}{4}\cdot 4 = \epsilon$,

or $|x-1||x^2-2x| < \epsilon$.

So $-\epsilon < (x-1)(x^2-2x) < \epsilon$

or $-\epsilon < x^3-3x^2+2x < \epsilon$

or $2-\epsilon < x^3-3x^2+2x+2 < 2+\epsilon$.

Exercise EII.5.3.

Prove that $2x^3-4x^2-3x+7$ is continuous at $x=2$.

Exercise EII.5.4.

Prove that $-x^3+2x^2-6$ is continuous at $x=3$.

Exercise EII.5.5.

Prove that x^4 is continuous at $x=1$.

For Section II.7.

Exercise EII.7.1.

Prove that $f(x) = -2x^2+4x-5$ is continuous at every c; $1 < c < 4$.

Exercise EII.7.2.

Prove that $g(x) = x^3-2x^2+3x-4$ is continuous at every c; $1<c<3$. (You will need the results shown in the extra exercises for section I.3.)

For Section II.9.

Remember, if a polynomial function p(x) (like a quadratic or cubic function) takes value 0 at some point x=a, then (x-a) will be a factor of p(x).

Exercise EII.9.1.

Prove that $\lim_{x\to -3}\frac{2x^2+2x-12}{x+3}$ exists.

Exercise EII.9.2.

Prove that $\lim_{x\to 1}\frac{2x^3-3x^2+3x-2}{x-1}$ exists.

For Section II.10.

Exercise EII.10.1. Sample.
If $f(x) = 3x+2$, prove that $\lim_{x\to 2} f(x) = 8$ and implement the methods of theorem II-2 (the 'KF' theorem) to prove that

$$\lim_{x\to 2} 15x+10 = \lim_{x\to 2} 5(3x+2) = \lim_{x\to 2} 5f(x) = 5\cdot 8 = 5\cdot\lim_{x\to 2} f(x).$$

Solution: From section II·4, defining $\delta^f(\epsilon)$ to be $\epsilon/3$ should work to prove that $\lim_{x\to 2} f(x) = 8$. Let's check to see if this is true.

We should be able to prove that

IF $0<|x-2|<\epsilon/3$

THEN $|(3x+2)-8|<\epsilon$.

Proof: If $0<|x-2|<\epsilon/3$, then $|(3x+2)-8|=|3x-6|=3\cdot|x-2|<3(\epsilon/3)= \epsilon$.

So $\lim_{x\to 2} f(x) = 8$ is true.

If we implement the 'Kf' theorem with K=5, the instructions are to use $\delta^{5f}(\epsilon) = \delta^f(\epsilon/5) = \frac{\epsilon/5}{3} = \epsilon/15$.

Let's see if this δ works: It should be true that

if $0<|x-2|<\epsilon/15$, then $|(15x+10)-40|=|15x-30|=15|x-2|<\epsilon$.

Since $|x-2|<\epsilon/15$ by assumption, $15|x-2| < 15(\epsilon/15) = \epsilon$, and we have verified that the 'Kf' theorem works in this case.

Exercise EII.10.2.
If $f(x) = 2x-3$, prove that $\lim_{x\to 2} f(x) = 1$ and implement the methods of theorem II-2 (the 'KF' theorem) to prove that

$$\lim_{x\to 2} -8x+12 = \lim_{x\to 2} -4(2x-3) = \lim_{x\to 2} -4f(x) = -4\cdot 1 = -4\lim_{x\to 2} f(x).$$

Exercise EII.10.3.
If $f(x) = 4-5x$, prove that $\lim_{x\to -1} f(x) = 9$ and implement the methods of theorem II-2 (the 'KF' theorem) to prove that

$$\lim_{x\to -1} -12+15x = \lim_{x\to -1} -3(4-5x) = \lim_{x\to -1} -3f(x) = -3\cdot 9 = -3\cdot \lim_{x\to -1} f(x).$$

Exercise EII.10.4. Sample.
If $f(x) = 3x+2$ and $g(x) = 4-5x$, prove that $\lim_{x\to 2} f(x) = 8$ and $\lim_{x\to 2} g(x) = -6$. Implement the methods of theorem II-3 (the 'f+g' theorem) to prove that

$$\lim_{x\to 2} (3x+2)+(4-5x) = 8+(-6) = \lim_{x\to 2} (3x+2) + \lim_{x\to 2} (4-5x).$$

Solution: From exercise EII.10.1, defining $\delta^f(\epsilon)$ to be $\epsilon/3$ works to prove that $\lim_{x\to 2} f(x) = 8$.

From section II-4, defining $\delta^g(\epsilon)$ to be $\epsilon/5$ should work to prove that $\lim_{x\to 2} g(x) = -6$. Let's check to see if this is true.

We should be able to prove that

IF $0<|x-2|<\epsilon/5$

THEN $|(4-5x)-(-6)|<\epsilon$.

Proof: If $0<|x-2|<\epsilon/5$, then $|(4-5x)-(-6)|=|10-5x|=5\cdot|2-x|= 5|x-2| < 5(\epsilon/5)=\epsilon$.

So $\lim_{x\to 2} g(x) = -6$ is true.

If we implement the program described in theorem II-4,

$$\delta(\epsilon) = \min\{\delta^f(\epsilon/2),\delta^g(\epsilon/2)\} = \min\{\frac{\epsilon/2}{3}, \frac{\epsilon/2}{5}\} = \min\{\epsilon/6,\epsilon/10\}$$

$= \epsilon/10$ should be a good $\delta(\epsilon)$ to prove that if $0<|x-2|<\delta(\epsilon)=\epsilon/10$, then $|(3x+2)+(4-5x)-[8+(-6)]| < \epsilon$. Now $|(3x+2)+(4-5x)-[8+(-6)]|=$

$$|(3x+2)-8 + (4-5x)-(-6)| \le$$

$$|(3x+2)-8| + |(4-5x)-(-6)| =$$

$|3x-6|+|10-5x| = 3|x-2|+5|2-x| = 3|x-2|+5|x-2| = 8|x-2| < 8(\epsilon/10) < \epsilon$, since we are assuming that $0<|x-2|<\delta(\epsilon)= \epsilon/10$. We have verified that the theorem works in this case.

Exercise EII.10.5.
If $f(x) = 5x-6$ and $g(x) = (x/2)-3$, prove that $\lim_{x\to 2} f(x) = 4$ and $\lim_{x\to 2} g(x) = -2$. Implement the methods of theorem II-3 (the 'f+g' theorem) to prove that

$$\lim_{x\to 2} (5x-6)+((x/2)-3) = 4+(-2) = \lim_{x\to 2} (5x-6) + \lim_{x\to 2} ((x/2)-3).$$

Exercise EII.10.6.
If $f(x) = 3x+2$ and $g(x) = 4-5x$, prove that $\lim_{x\to -1} f(x) = -1$ and $\lim_{x\to -1} g(x) = 9$. Implement the methods of theorem II-3 (the 'f+g' theorem) to prove that

$$\lim_{x\to -1} (3x+2)+(4-5x) = -1+9 = \lim_{x\to -1} (3x+2) + \lim_{x\to -1} (4-5x).$$

EXCURSION!

Here we show two more limit theorems; a product limit theorem and a quotient limit theorem.

The product of two functions like

$f(x) = x^2-3x+1$ and $g(x) = 2x^2+6$

is the new function

$h(x) = f(x) \cdot g(x) = (x^2-3x+1) \cdot (2x^2+6)$.

From section II-10, the product limit theorem goes like this:

THEOREM II-4.

(The product limit theorem.)

If $\lim_{x \to a} f(x)$ exists and equals L and $\lim_{x \to a} g(x)$ exists and equals M, then $\lim_{x \to a} f(x) \cdot g(x)$ exists and equals $L \cdot M = \lim_{x \to a} f(x) \cdot \lim_{x \to a} g(x)$.

The theorem shows that the property of having limits is inherited by the new function formed by multiplying the two functions.

The theorem means that

IF (1) we guess L to be the limit for f(x) as $x \to a$ and have an $\epsilon, \delta^f(\epsilon)$ limit machine proving that L is correct

AND (2) we guess M to be the limit for g(x) as $x \to a$ and have an $\epsilon, \delta^g(\epsilon)$ limit machine proving that M is correct,

THEN $L \cdot M$ is the correct guess for the limit of the product $f(x) \cdot g(x)$ as $x \to a$ and we can prove it correct by giving an $\epsilon, \delta(\epsilon)$ argument using the limit machines that come with

$$\lim_{x \to a} f(x) \text{ and } \lim_{x \to a} g(x).$$

Proof: Somehow, for any $\epsilon > 0$, we have to find some $\delta(\epsilon) > 0$ so that

IF $0 < |x-a| < \delta(\epsilon)$,

THEN $|f(x) \cdot g(x) - L \cdot M| < \epsilon$.

Since we have $\lim_{x \to a} f(x) = L$ and $\lim_{x \to a} g(x) = M$, for any $\# > 0$, we can make $|f(x)-L| < \#$ and $|g(x)-M| < \#$.

We can write

$$|f(x)g(x) - LM| =$$
$$|f(x)g(x) - f(x)M + f(x)M - LM| =$$
$$|f(x)(g(x)-M) + M(f(x)-L)| \leq$$
$$|f(x)||g(x)-M| + |M||f(x)-L|.$$

If we could arrange that both of these are less than $\epsilon/2$, then we would have a proof.

Making $|M||f(x)-L| < \epsilon/2$ is easy. If M=0, we are done. Otherwise, if x satisfies

$$0 < |x-a| < \delta^f(\epsilon/(2|M|)),$$

then the guarantee on $\delta^f(\epsilon/(2|M|))$ makes $|f(x)-L| < \epsilon/(2|M|)$, or

$$|M||f(x)-L| < \epsilon/2.$$

The first term is harder to deal with. We can't use the same sort of argument and require that

$$0 < |x-a| < \delta^g(\epsilon/(2|f(x)|)),$$

because, as in section II-5, we can't let δ^g depend on x. But section II-5 also suggests a way around this. We can get $|f(x)|$ under control by first computing $\delta^f(1)$. If $0 < |x-a| < \delta^f(1)$, then $L-1 < f(x) < L+1$. Since $L+1 \leq |L|+1$ and $-|L|-1 \leq L-1$, we get

$$-(|L|+1) = -|L|-1 \leq L-1 < f(x) < L+1 \leq |L|+1,$$

or $|f(x)| < |L|+1$. So, multiplying this inequality by $|g(x)-M|$,

if $0 < |x-a| < \delta^f(1)$, then

$$|f(x)||g(x)-M| \leq (|L|+1)|g(x)-M|.$$

We CAN arrange that this is less than $\epsilon/2$. Just make

$$0 < |x-a| < \delta^g(\epsilon/(2(|L|+1)));$$

Now, if x satisfies

(1) $0 < |x-a| < \delta^f(1)$ and

(2) $0 < |x-a| < \delta^g(\epsilon/(2(|L|+1)))$

then

$$|f(x)||g(x)-M| \leq (|L|+1)|g(x)-M| < (|L|+1)\frac{\epsilon}{2(|L|+1)} = \frac{\epsilon}{2}.$$

If (3) $0 < |x-a| < \delta^f(\epsilon/(2|M|))$ also, then

$$|f(x)g(x) - LM| = |f(x)g(x) - f(x)M + f(x)M - LM| = |f(x)(g(x)-M) + M(f(x)-L)| \leq |f(x)||g(x)-M| + |M||f(x)-L| < \frac{\epsilon}{2} + \frac{\epsilon}{2} = \epsilon.$$

Thus we need to make (1), (2) and (3) all hold at the same time; we can define $\delta(\epsilon) = \min\{\delta^f(1),\delta^g(\epsilon/(2(|L|+1))),\delta^f(\epsilon/(2|M|))\}$, and this is the $\delta(\epsilon)$ we need to prove the theorem.

Exercise EII.10.7. Sample.
In exercise EII.10.4 we proved that if $f(x) = 3x+2$ and $g(x) = 4-5x$, then $\lim_{x\to 2} f(x) = 8$ and $\lim_{x\to 2} g(x) = -6$. Implement the methods of theorem II-3 (the product limit theorem) to prove that

$$\lim_{x\to 2} (3x+2)\cdot(4-5x) = 8\cdot(-6) = \lim_{x\to 2} (3x+2)\cdot\lim_{x\to 2} (4-5x).$$

Solution. From exercise EII.10.4, we know that $\delta^f(\#) = \#/3$ and $\delta^g(\#) = \#/5$ are good δ's for proving that the limits for f(x) and g(x) are correct.

The instructions in the product limit theorem II-3 tell us that setting $\delta(\epsilon) = \min\{\delta^f(1),\delta^g(\epsilon/(2(|L|+1))),\delta^f(\epsilon/(2|M|))\}$ will produce a δ that we can use to prove that $\lim_{x\to 2} f(x)\cdot g(x)$ exists and equals $L\cdot M$. Here, $L = \lim_{x\to 2} f(x) = 8$; $M = \lim_{x\to 2} g(x) = -6$, $\delta^f(\ ) = (\ )/3$ and $\delta^g(\ ) = (\ )/5$, so the δ prescribed by the theorem should be

$$\min\{\delta^f(1),\delta^g(\epsilon/(2(|8|+1))),\delta^f(\epsilon/(2|-6|))\}$$
$$=\min\ \{\delta^f(1),\delta^g(\epsilon/18),\delta^f(\epsilon/12)\}$$
$$= \min\ \{\tfrac{1}{3}, \frac{\epsilon/18}{5}, \frac{\epsilon/12}{3}\}$$
$$= \min\ \{\tfrac{1}{3}, \frac{\epsilon}{90}, \frac{\epsilon}{36}\} = \min\ \{\tfrac{1}{3}, \frac{\epsilon}{90}\}.$$

This $\delta(\epsilon)$ is supposed to work to show that if $0<|x-2|<\delta(\epsilon)$, then $|(3x+2)(4-5x)-8\cdot(-6)| = |f(x)\cdot g(x) - L\cdot M| < \epsilon$.

We follow the program prescribed by the theorem to see if this is the case.

First, this should ensure that $|M||f(x)-L|<\epsilon/2$. So, if $0<|x-2|<\delta(\epsilon) = \min\ \{\tfrac{1}{3}, \frac{\epsilon}{90}\}$, we should have $6|(3x+2)-8| = 6|3x-6|=6\cdot 3|x-2|<\epsilon/2$. Since we are assuming that $|x-2|<\epsilon/90$, this means that we will have $18|x-2|<18(\epsilon/90) =\epsilon/5$, which is less than $\epsilon/2$ as required.

The theorem also assures us that if $|x-2|<\delta^f(1)=1/3$, then we should have $|f(x)|<|L|+1 = 8+1=9$. To check this, we need to show that $|3x+2|<9$.

Now $|x-2|<1/3$ means that $-1/3<x-2<1/3$. We use the methods of section I.3 to brew up an inequality for $f(x) = 3x+2$. Multiplying by 3 we get $-1 = 3(-1/3)<3x-6<3(1/3) = 1$; add 8 and we get $-7=8-1<3x+2<8+1=9$, so $|3x+2|<9$ as required.

Then $|f(x)||g(x)-M| =$

$$|3x+2||(4-5x)-(-6)| =$$
$$|3x+2||10-5x| < 9|10-5x|$$

$= 9{\cdot}5|2-x| = 45|x-2|$. This should be less than $\epsilon/2$ if $|x-2|<\delta(\epsilon)\leq \epsilon/90$; again the assertion is true, for $45(\epsilon/90) = \epsilon/2$.

If we put these all together and follow the program in the theorem, we get

$$|(3x+2)(4-5x)-8\cdot(-6)| =$$
$$|(3x+2)(4-5x)-(3x+2)(-6)$$
$$+ (3x+2)(-6)-8\cdot(-6)| =$$
$$|[3x+2][(4-5x)-(-6)]+$$
$$(-6)[(3x+2)-8]| \leq$$
$$|3x+2||10-5x|+|-6||3x-6|\leq$$
$$9\cdot 5|2-x| + 6\cdot 3|x-2| = 63|x-2|.$$

Since we are assuming that $|x-2|<\epsilon/90$, $63|x-2|<63(\epsilon/90) = (7/10)\epsilon < \epsilon$; we have verified that the theorem works in this case.

Exercise EII.10.8.

In exercise EII.10.5 we proved that if $f(x) = 5x-6$ and $g(x) = (x/2)-3$, then $\lim_{x\to 2} f(x) = 4$ and $\lim_{x\to 2} g(x) = -2$. Implement the methods of theorem II-4 to prove that

$$\lim_{x\to 2} (5x-6)\cdot((x/2)-3) = 4\cdot(-2) =$$
$$\lim_{x\to 2} (5x-6) \cdot \lim_{x\to 2} ((x/2)-3).$$

Exercise EII.10.9.

In exercise EII.10.6 we proved that if $f(x) = 3x+2$ and $g(x) = 4-5x$, then $\lim_{x\to -1} f(x) = -1$ and $\lim_{x\to -1} g(x) = 9$. Implement the methods of theorem II-4 to prove that

$$\lim_{x\to -1} (3x+2)\cdot(4-5x) = (-1)\cdot 9 =$$
$$\lim_{x\to -1} (3x+2) \cdot \lim_{x\to -1} (4-5x).$$

These exercises are fairly difficult and they make us appreciate the power of the product limit theorem. The theorem tells us that we never have to make such computations; the products of functions with limits automatically <u>inherit</u> the property of having <u>guaranteed</u> limits; no such demonstrations are necessary.

Another way to form new functions is to divide one function by another. For example, if $f(x) = 5-3x$ and $h(x) = 2x+4$, then the <u>quotient</u> of $f(x)$ and $h(x)$ is the function $\frac{5-3x}{2x+4}$. This new function will make sense for any x such that the bottom function $2x+4 \neq 0$.

We wish to prove that if the two functions have limits, then the quotient of the two functions will inherit this property.

First, let's have a look at a special case.

THEOREM II-5.

If $\lim_{x \to a} h(x)$ exists and equals M, then $\lim_{x \to a} \frac{1}{h(x)}$ exists and equals $\frac{1}{M}$.

We assume that $M \neq 0$ so that this will make sense.

Proof:

For any $\epsilon>0$, we must find some $\delta(\epsilon)>0$ so that

if $0<|x-a|<\delta(\epsilon)$,

then $\left|\frac{1}{h(x)} - \frac{1}{M}\right| < \epsilon$.

If we do a little algebra, we get

$$\left|\frac{1}{h(x)} - \frac{1}{M}\right| = \left|\frac{M}{h(x)M} - \frac{h(x)}{Mh(x)}\right| =$$

$$\frac{|M - h(x)|}{|M||h(x)|} = \frac{1}{|M||h(x)|}|M-h(x)|.$$

Now $|M-h(x)| = |(-1)(h(x)-M)| = |-1||h(x)-M| = 1 \cdot |h(x)-M| = |h(x)-M|$. We can make this term as small as we please, since $\lim_{x \to a} h(x) = M$.

We must get $\frac{1}{|M||h(x)|}$ under control. M is some constant $\neq 0$, so there is no problem there. To deal with $1/|h(x)|$, this time we will find some constant $K>0$ so that $|h(x)|>K$ for x close to a.

Suppose that $M>0$.

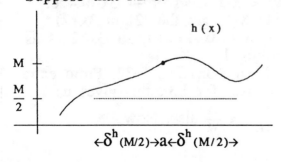

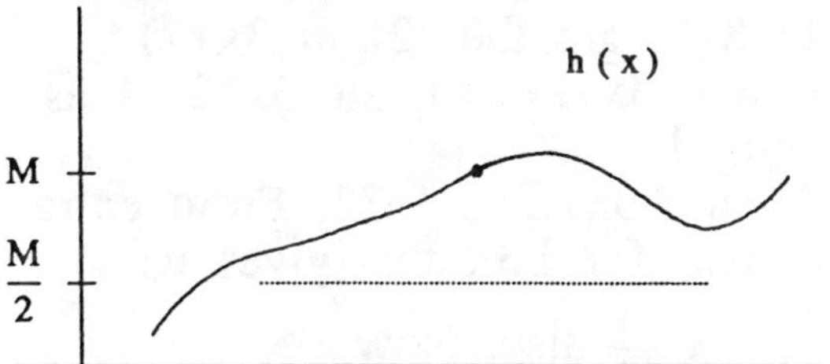

With $M>0$, $M/2 = |M|/2 > 0$ too, so we can use the $\delta^h(\cdot)$ that comes with $\lim_{x \to a} h(x)$ to compute $\delta^h(|M|/2)$; the guarantee on $\delta^h(|M|/2)$ tells us that

if $0<|x-a|<\delta^h(|M|/2)$, then

$$|M|/2 = M - M/2 < h(x) < M + M/2.$$

So $0< M/2 < h(x) = |h(x)|$; h(x) will be positive for these x's, and bounded away from 0 by $M/2 = |M|/2$.

If $M<0$, the algebra is a little harder. $|M| = -M$ will be > 0, and so $-M/2 = |M|/2$ is > 0 too.

Again, we compute $\delta^h(|M|/2) = \delta^h(-M/2)$; the guarantee on $\delta^h(-M/2)$ tells us that

if $0<|x-a|<\delta^h(-M/2)$, then

$$M-(-M/2) < h(x) < M+(-M/2) = M/2.$$

Now $M/2 < 0$, so, for these x's, $h(x) < M/2 < 0$; if we multiply this inequality by (-1), it will 'flip' the inequality signs, and we will get

$$|h(x)| = -h(x) > -M/2 = |M|/2.$$

Thus, again, if

$0<|x-a|<\delta^h(|M|/2)$,

then $|h(x)| > |M|/2 > 0$.

We have found a way to control x to make $|h(x)| > |M|/2 > 0$; so $|M||h(x)| > |M||M|/2$ also.

From the results in the Extra Exercises for I.3, this means that

$$\frac{1}{|M||h(x)|} < \frac{1}{|M||M|/2} = \frac{2}{|M||M|}.$$

Return to our original problem. We need to arrange that

$$\left|\frac{1}{h(x)} - \frac{1}{M}\right| < \epsilon.$$

We have shown that

$$\left|\frac{1}{h(x)} - \frac{1}{M}\right| = \frac{1}{|M||h(x)|}|h(x)-M|.$$

If $0<|x-a|<\delta^h(|M|/2)$, then we know that

$$\frac{1}{|M||h(x)|} < \frac{2}{|M||M|}.$$

We can multiply this by $|h(x)-M|$ to get

$$\frac{1}{|M||h(x)|}|h(x)-M| \leq \frac{2}{|M||M|}|h(x)-M|.$$

We can make this less than ϵ if we make $0 < |x-a| < \delta^h(|M|^2\epsilon/2)$

Finally, we put this all together. Define $\delta(\epsilon) =$ $\min\{\delta^h(|M|/2),\delta^h(|M|^2\epsilon/2)\}$.

If $0<|x-a|< \delta(\epsilon)$,

the guarantee on $\delta^h(|M|/2)$ will make

$$\frac{1}{|M||h(x)|}|h(x)-M| \leq \frac{2}{|M||M|}|h(x)-M|.$$

The guarantee on $\delta^h(|M|^2\epsilon/2)$ will make

$$\frac{2}{|M||M|}|h(x)-M| < \frac{2}{|M||M|}(|M|^2\epsilon/2)$$

$= \epsilon$.

Then

$$\left|\frac{1}{h(x)} - \frac{1}{M}\right| = \frac{1}{|M||h(x)|}|h(x)-M| <$$

ϵ;

we have a proof of the theorem.

Exercise EII.10.10. Sample.
In exercise EII.10.1 we proved that if $h(x) = 3x+2$ then

$$\lim_{x\to 2} h(x) = 8.$$

Implement the methods of theorem II-5 to prove that

$$\lim_{x\to 2} \frac{1}{3x+2} \text{ exists and equals } \frac{1}{8}.$$

Solution. From exercise EII.10.4, we know that $\delta^h(\#) = \#/3$ is a good δ for proving that the limit for h(x) is correct.

The instructions in theorem II-4 tell us that setting $\delta(\epsilon) =$

$$\min\{\delta^h(|M|/2),\delta^h(|M|^2\epsilon/2)\}$$

will produce a δ that we can use to prove that $\lim_{x\to 2} \frac{1}{h(x)}$ exists and equals 1/M. Here, $M = \lim_{x\to 2} h(x) =$ 8 and $\delta^h(\ ) = (\ )/3$, so the δ prescribed by the theorem should be $\min\{\delta^h(|M|/2),\delta^h(|M|^2\epsilon/2)\} =$

$$\min\{\delta^h(8/2),\delta^h(8^2\epsilon/2)\} =$$

$\min\{\delta^h(4),\delta^h(32\epsilon)\} = \min\{4/3,32\epsilon/3\}$.

The theorem tells us that if $0<|x-2|<4/3$, then we should have $|3x+2|=|h(x)|>|M|/2 = 8/2 = 4$. Let's see if this is true.

With $|x-2|<4/3$, we will have $-4/3 < x-2 < 4/3$ or $-4<3x-6<4$ or (add 8 to get the '2' in 3x+2) $4 = 8-4<3x+2<8+4$, so $|3x+2|>4$ as promised.

Then $8|3x+2|>8\cdot 4=32$. From extra exercises for I.3., this gives us $\frac{1}{8|3x+2|} < \frac{1}{32}$ also. Now

$$\left|\frac{1}{h(x)} - \frac{1}{M}\right| = \left|\frac{1}{3x+2} - \frac{1}{8}\right| =$$

$$\left|\frac{8}{(3x+2)8} - \frac{3x+2}{8(3x+2)}\right| =$$

$$\left|\frac{8-(3x+2)}{8(3x+2)}\right| = \left|\frac{6-3x}{8(3x+2)}\right| =$$

$$\frac{1}{8|3x+2|}|6-3x|.$$

Since, with $|x-2|<4/3$, we have shown that $\frac{1}{8|3x+2|} < \frac{1}{32}$,

$$\frac{1}{8|3x+2|}|6-3x| \le \frac{1}{32}|6-3x| =$$

$$\frac{3}{32}|2-x| = \frac{3}{32}|x-2|.$$

This is less than ϵ if $|x-2| < 32\epsilon/3$; we have verified that the theorem works in this case.

Exercise EII.10.11.
In exercise EII.10.4 we proved that if $h(x) = 4-5x$ then

$$\lim_{x\to 2} h(x) = -6.$$

Implement the methods of theorem II-5 to prove that

$$\lim_{x\to 2} \frac{1}{4-5x} \text{ exists and equals } -\frac{1}{6}.$$

Exercise EII.10.12.
In exercise EII.10.3 we proved that if $h(x) = 4-5x$ then

$$\lim_{x\to -1} h(x) = 9.$$

Implement the methods of theorem II-5 to prove that

$$\lim_{x\to -1} \frac{1}{4-5x} \text{ exists and equals } \frac{1}{9}.$$

Theorem II-5 was just a special case of the quotient limit theorem; the full theorem we wish to prove is the following:

THEOREM II-6.

(The quotient limit theorem.)
If $\lim_{x\to a} f(x) = L$ and $\lim_{x\to a} h(x) = M$ and $M\neq 0$,

then $\lim_{x\to a} \frac{f(x)}{h(x)} = \frac{L}{M} = \frac{\lim_{x\to a} f(x)}{\lim_{x\to a} h(x)}$.

Proof: Theorem II.5 shows that $\lim_{x\to a} \frac{1}{h(x)} = \frac{1}{M}$. Let $g(x) = \frac{1}{h(x)}$, so we know that $\lim_{x\to a} g(x) = \frac{1}{M}$. Then we can just use the product limit theorem II.4 on $f(x)$ and $g(x)$ to get $\lim_{x\to a} \frac{f(x)}{h(x)} =$

$$\lim_{x\to a} f(x)\frac{1}{h(x)} = \lim_{x\to a} f(x)\cdot g(x)$$

exists and equals

$\lim_{x\to a} f(x) \cdot \lim_{x\to a} g(x) = L \frac{1}{M} = \frac{L}{M}$, and that is all we need for the proof.

HEY, KIDS, IT'S RECESS!
AAAH... TANGENT TOURS? MAJOR JAEGER FOR THE FIELD TRIP?
RIGHTO. WHERE'S EVERYBODY'S BLOOMIN' GEAR?
OK, GANG- EVERYONE GET YOUR GEAR TOGETHER!
WHERE'S THE JACUZZI?
f()
ALFRED
FRIED EPOXY
Brown's USED GUM COLLECTION
FREEZE DRIED ANCHOVY BURRITOS
BAD JOKES FOR CHAPTE 3
STEAMBOAT SPRINGS
CHAGRIN FALLS
KPIG
EXOTIC IOWA
MOTO OIL
1 GROSS ZIP GUNS

C'MON, HUP HUP LET'S MOVE IT!
DID THAT LAST CHAPTER MAKE SENSE?
BUS STOP ON ALTERNATE THURSDAYS WHEN THE SITUATION WARRANTS IT OR OTHERWISE, NO.
NO EWE TURN
YUP...
AN ARMCHAIR GENERAL
PEORIA
OSHKOSH
Las Vegas
WATCH YOUR MIND
ACNE BUS LINES
PETROL
Good Vibes
SIOUX CITY
DUBUQUE
WATCH YOUR STEP
WHAT'S TO EAT?
I AM FILLED WITH KNOWLEDGE!
DON'T MAKE ANY SUDDEN MOVES!
I'M GOING TO DO THE RIEMANN ZETA FUNCTION!
g( )
Grover
WHAT AM I GONNA DO WHEN WE GET THROUGH ALL THIS?
YOU CAN ALWAYS GET A JOB AS A BOOKEND!
ACNE BUS LINES
f( )
CHICAGO
ALFRED

YOU BETTER STOP MAKING FUN OF ME OR I'LL TELL McSQUARED AND GROVER AND YOU'LL GET YOURS!
HAH!
f()
ALFRED
C'MON, KNOBNOSE- CAN'T YA TAKE A JOKE?
WATCH IT...
f()
ALFRED
WOWP?
PHONK
f()
ALFRED
WHAT'S GOING ON?
GRR...
WHADJA DO THAT FOR?
?
f()
ALFRED
I'M TIRED OF BEING THE DUMMY IN THIS BOOK!
(I WANNA BE A HERO
YOU OK?
WHAT HAPPENED?
WHAZZA MATTER? SHORTENED SNIFFER?
HE BENT MY SCHNOZ
f()
ALFRED
I WANNA BE A NUCLEER FIZZICIST!
YOU DON'T KNOW PHYSICS FROM ATOM!
THAT'S WHAT I'M DOING - GOING FISSION!
f()
ALFRED
BAIT
I THINK HE MEANS IT.

WE'RE JUST ABOUT SET!
JOLLY GOOD
LOOKIT THAT CRAZY ALFRED FREAKING OUT! I GOTTA HELP OL' R.D.
CALCULUS PROPS
g()
Grover
WONK
WONK
f()
IF ONLY TOM MIX COULD SEE ME NOW!
CALCULUS PROPS
g()
Grover
f()
G.V. TO THE RESCUE!
SOME FRIENDS...
CENSORED
R.D.! I SAVED YA!
GLUK
GOTCHA!
f()
GRAB

?
YOU GRABBED ALFIE!
YOU SWINE
f()
ALFRED
g()
Grover
Good Vibes
OOG! ITS A LONG BOOK!
HERE, ALFIE-HAVE A DRINK
MOTOR OIL
COME ON, LADS-THE BUS IS ALL PACKED AND WE'RE READY TO LEAVE IN 5 MINUTES
SOMEONE UNPACKED THE BUS
WHERE'S MELVIN?
SOME BLIGHTER UNPACKED THE BLASTED TYRES, TOO
DUM DE DUM...
ACME BUS LINES

FAST EDDY WILL GIMME 5 BUCKS EACH FOR THESE!
L.R.W.
MUMF YOU
GIMME BACK MY
WAGON
YOU B
AHAH!
IT'S THAT RUDDY DUCK
I DIDN'T DO IT!
LRW
I THINK IT'S TIME TO SPLIT
LRW
HAH!
RATS

PEORIA
OSHKOSH
YALE
BRITANNIA WAIVES THE RULES
WATCH YOUR MIND
CHAPTER 3 NEXT PAGE →
OZ
ANIMAL FARM
NIRVANA
ACNE BUS LINES
SICK GLORIOUS TRANSIT
f()
ALFRED
lim f(x) x→2
PETROL
CALIFORNIA JITNEY
IRVINGTON DOGTROT

CHAPTER III

DERIVATIVES...

In which we see how mathematicians found an answer to the third question:

The answer to this question also gives a way to compute velocities and thus get the whole scientific trip off the ground.

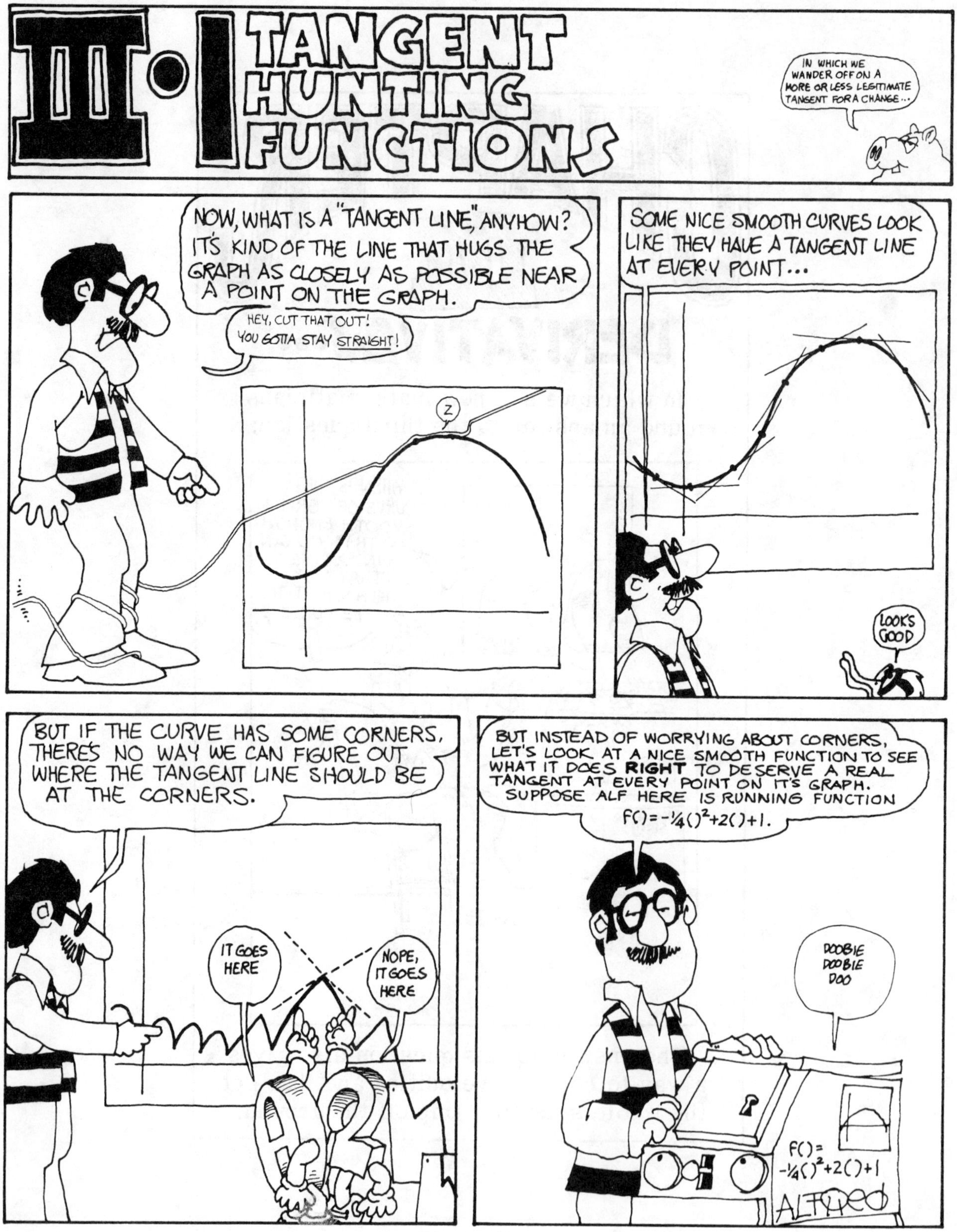
III·1 TANGENT HUNTING FUNCTIONS
IN WHICH WE WANDER OFF ON A MORE OR LESS LEGITIMATE TANGENT FOR A CHANGE...
NOW, WHAT IS A "TANGENT LINE," ANYHOW? IT'S KIND OF THE LINE THAT HUGS THE GRAPH AS CLOSELY AS POSSIBLE NEAR A POINT ON THE GRAPH.
HEY, CUT THAT OUT! YOU GOTTA STAY STRAIGHT!
Z
SOME NICE SMOOTH CURVES LOOK LIKE THEY HAVE A TANGENT LINE AT EVERY POINT...
LOOKS GOOD
BUT IF THE CURVE HAS SOME CORNERS, THERE'S NO WAY WE CAN FIGURE OUT WHERE THE TANGENT LINE SHOULD BE AT THE CORNERS.
IT GOES HERE
NOPE, IT GOES HERE
BUT INSTEAD OF WORRYING ABOUT CORNERS, LET'S LOOK AT A NICE SMOOTH FUNCTION TO SEE WHAT IT DOES **RIGHT** TO DESERVE A REAL TANGENT AT EVERY POINT ON IT'S GRAPH. SUPPOSE ALF HERE IS RUNNING FUNCTION
F( ) = -1/4( )^2+2( )+1.
DOOBIE DOOBIE DOO
F( ) = -1/4( )^2+2( )+1
ALFRED

This function's graph is a parabola, and it looks smooth enough to have a tangent line at any point on the graph. In particular, try the point above x=2:

$f(2) = -\frac{1}{4}(2)^2 + 2\cdot(2) + 1 = 4,$

so the point is $(2, f(2)) = (2, 4)$.

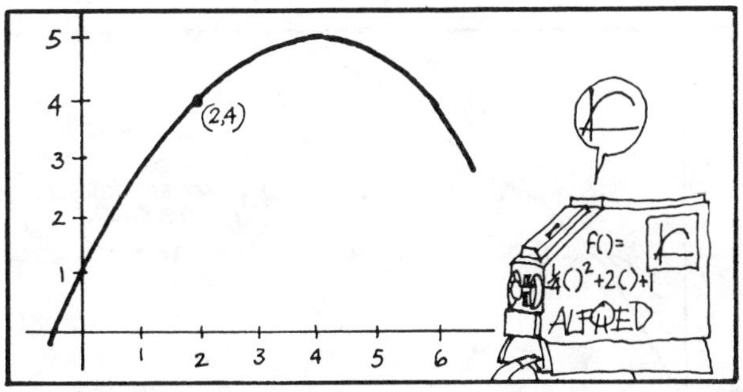

If there really is a tangent line to the graph at (2, 4), we should be able to figure out the line's formula. From Chapter I, formulas for straight lines need both a point and a slope. We know that the line should go through point (2, 4), but we don't know what direction the line should take to be tangent. Somehow we need to invent a way to determine the proper slope for the line — we need to find the best triangular wedge to prop up the line so that it does the job of hugging the graph of f( ) as closely as possible near (2, 4) and actually touching the curve only at (2, 4).

As yet we don't have any way to solve the problem of finding the precise number that gives the slope for the tangent line in one easy step. Instead, to help discover what the slope should be, we construct a new "tangent-hunting" function that reports various approximations to what might be a good slope for the tangent line.

The only way we can get some actual numbers to approximate what intuitively should be the slope of the tangent line is to start with lines that go through (2, 4) and some other point on the graph near (2, 4). The point on the graph above x=4 looks pretty close to (2, 4).

$$f(4) = -\tfrac{1}{4}(4)^2 + 2\cdot(4) + 1 = 5 ,$$

so the point is (4, f(4)) = (4, 5) . Call the line through (4, 5) and (2, 4) the "first official approximate tangent-line candidate."

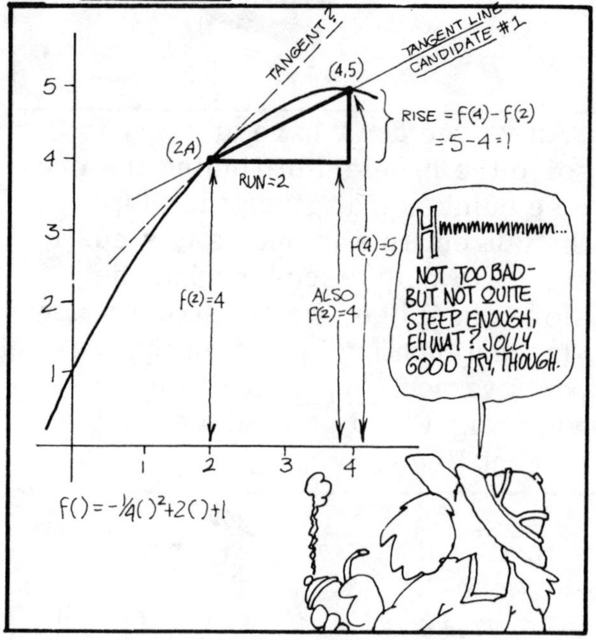

The slope of this line is easy to figure out: just use the triangular wedge with one vertex at (2, 4) and top vertex at (4, 5) .

The "run" is 4 - 2 = 2 ;

The "rise" is f(4) - f(2) = 5-4=1;

so the slope is

$$\text{Slope} = \frac{\text{rise}}{\text{run}} = \frac{f(4)-f(2)}{4-2} = \frac{5-4}{2} = \frac{1}{2}.$$

The line doesn't look steep enough to be declared the official tangent line—besides, it goes through two points—so its slope "$\frac{1}{2}$" is probably too small. It looks like a better approximation to what should be the tangent line could be found by taking the line through (2, 4) and a second point like (3, f(3)) = (3, $4\frac{3}{4}$) that is a little closer to (2, 4) than (4, 5) is.

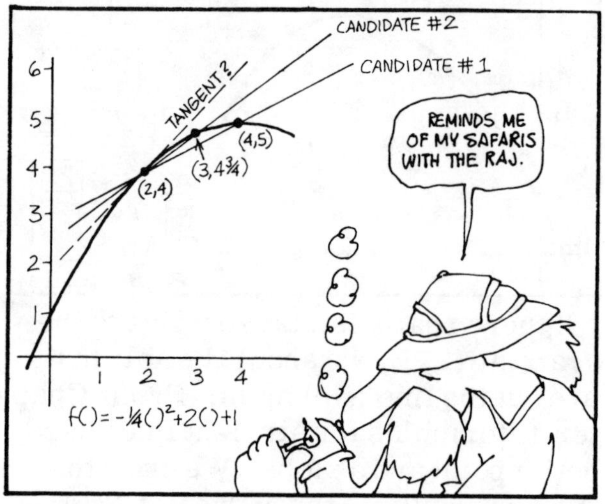

Tangent-line candidate #2 still isn't steep enough, but it does indeed look closer to the tangent line than candidate #1. In fact, the closer the second point is to (2, 4) , the closer the corresponding line gets to what intuitively looks like the "tangent line."

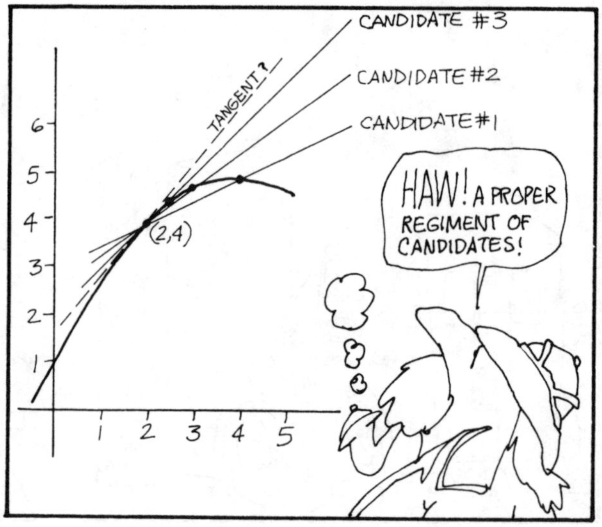

So what we need is some way to look at the actual numbers that are the slopes of these lines and see if they tend toward some limit number. If we can find it, such a limit number would be a good choice for the official slope of the tangent line and once we have the slope, we can easily find the formula for the tangent line. This is where the new "tangent-hunting" function comes in— it reports the <u>slopes</u> of these tangent-line <u>candidates</u> and tells us how to find such a limit number if there is one.

The new function does this reporting by looking at the wedge that props up a typical tangent-line candidate through point $(2, f(2)) = (2, 4)$ and a second point on the graph. A typical second point is traditionally labeled

$$(2+h, f(2+h))$$

so that "h" also gives the length of the "run" of the triangular wedge propping up the line through the two points.

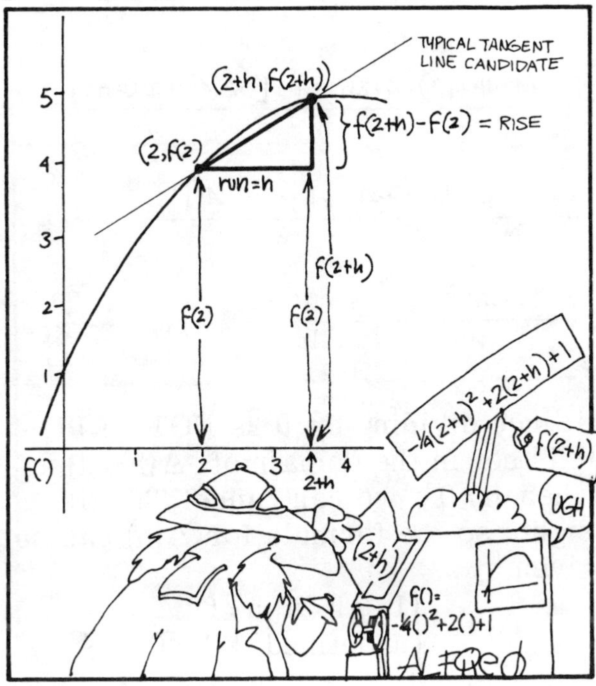

The top vertex of the triangle will be on the graph of f( ) above $x = 2+h$ at point $(2+h, f(2+h))$. Then the SLOPE of this typical tangent-line candidate will be

$$\text{slope} = \frac{\text{rise}}{\text{run}} = \frac{f(2+h) - f(2)}{h}.$$

This is exactly the formula mathematicians used for their new tangent-hunting function! The new function's <u>domain</u> is the set of all <u>h's</u> that represent the "runs" of wedges propping up tangent line candidates and, for each such run "h," it reports the <u>slope</u>

$$\frac{f(2+h) - f(2)}{h}$$

of the corresponding tangent-line <u>candidate</u> through (2, 4) and $(2+h, f(2+h))$. We will use the symbol "$\Delta(h)$" (called "delta of h") for this new tangent-hunting function.

To sum up,

$$\Delta(h) = \left\{ \begin{array}{l} \text{SLOPE OF TANGENT-} \\ \text{LINE \underline{CANDIDATE}} \\ \text{WHEN \underline{RUN} OF PROPPING} \\ \text{WEDGE IS "h".} \end{array} \right\} = \frac{f(2+h) - f(2)}{h}.$$

This new tangent-hunting function just reports the slopes of <u>candidate lines</u> that approximate the tangent line; it doesn't ever get around to reporting the slope of the tangent line itself. But it looks like the candidates get closer and closer to being the tangent line as the "other point $(2+h, f(2+h))$" gets closer and closer to $(2, f(2))$ and the wedges get smaller and smaller.

Though the wedges disappear as h goes to zero, the slopes won't necessarily disappear too, since slopes are <u>ratios</u> of the sides of triangles and don't depend on the size of similar triangles. So, just look at the <u>slopes</u> of the candidates using $\Delta(h) = \frac{f(2+h) - f(2)}{h}$.

As the run "h" gets close to zero, the slope reported by $\Delta(h)$ for any particular run h gets close to what should be the slope of the tangent line.

But, unfortunately, we can't compute $\Delta(0)$, since $h=0$ is NOT in the domain of $\Delta(\ )$ because dividing by 0 is NOT FAIR.

However, if we can figure out what value $\Delta(h)$ wants to take at $h=0$ by figuring out what number is the <u>limit</u> of $\Delta(h)$ as h goes to zero, it seems as if this limit number would be an excellent number to legislate as the <u>slope</u> of the tangent line, even though we won't end up with an actual wedge.

Let's see how this works out for

$$f(\ ) = -\tfrac{1}{4}(\ )^2 + 2\cdot(\ ) + 1:$$

$$\Delta(h) = \frac{f(2+h) - f(2)}{h} =$$

PUT TWO COPIES OF f( ) HERE, AND HERE,

NOTICE HOW 2+h GOES INTO THE FIRST SET OF BLANKS

$$\frac{[-\tfrac{1}{4}(2+h)^2 + 2\cdot(2+h) + 1] - [-\tfrac{1}{4}(2)^2 + 2\cdot(2) + 1]}{h} =$$

$$\frac{[-\tfrac{1}{4}(4+4h+h^2) + 4 + 2h + 1] - [-\tfrac{1}{4}(4) + 2\cdot(2) + 1]}{h} =$$

$$\frac{[-1 - h - \tfrac{1}{4}h^2 + 4 + 2h + 1] - [4]}{h} =$$

$$\frac{h - \tfrac{1}{4}h^2}{h} = \frac{h}{h}\left(1 - \tfrac{1}{4}h\right).$$

Since dividing by 0 is NOT FAIR, 0 is not in the domain of $\Delta(h)$. But when $h \neq 0$, we can cancel "h" all right, so the formula for $\Delta(h)$ can be given by

$$\Delta(h) = \begin{cases} 1 - \tfrac{1}{4}h & \text{if } h \neq 0 \\ \text{undefined} & \text{if } h = 0 \end{cases}$$

Even though 0 is NOT in the domain of $\Delta(h)$, it's pretty clear that as h gets smaller and smaller, $\Delta(h)$ wants to approach $1-\frac{1}{4}(0)=1$ as its limit number.

To make sure that "1" actually is the right limit number, we have to go through the ritual of proving that

$$\lim_{h\to 0} \Delta(h) \text{ exists and equals } 1,$$

where $\Delta(h)=\begin{cases}1-\frac{1}{4}h & \text{if } h\neq 0\\ \text{undefined} & \text{if } h=0\end{cases}$.

Proof: 1. For any $\epsilon>0$, let $\delta_{(\epsilon)}=4\epsilon$.

2. IF $0-4\epsilon < h < 0+4\epsilon$ and $h\neq 0$,

so (times "$-\frac{1}{4}$"): $\epsilon > -\frac{1}{4}h > -\epsilon$

then (add 1) $1+\epsilon > 1-\frac{1}{4}h > 1-\epsilon$,

or $1+\epsilon > \Delta(h) > 1-\epsilon$.

So "1" is officially the limit number all right, and it is the only one possible by Theorem II-1.

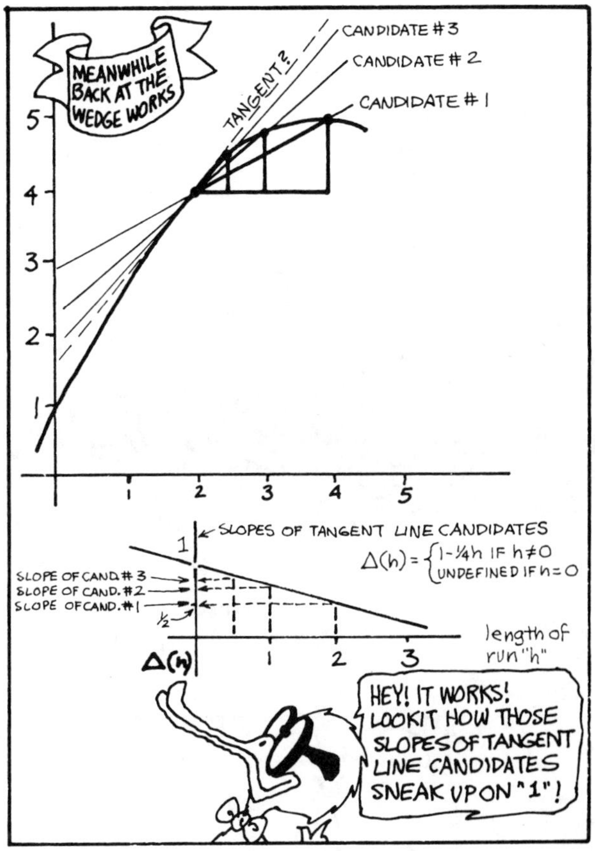

Recall that the tangent-line candidates get closer and closer to what should be a good "tangent line" as h — the length of the "run" of the triangular wedge— gets smaller and smaller. So the limit number "1" of $\Delta(h)$ as h goes to zero is what should be the official slope of the tangent line, since Δ(h) gives the slopes of the tangent-line candidates.

Since "1" is definitely THE limit number for $\Delta(h)$ as h goes to zero, the line through (2, 4) with the limit number "1" as its slope is a good tangent line to the graph of

$$f(\ ) = -\frac{1}{4}(\ )^2 + 2\cdot(\ ) + 1$$

as far as anyone can tell. So mathematicians decided that this sort of argument is enough to show that the graph of f( ) is smooth enough near (2, 4) to officially have a tangent line, with the limit-number "1" of $\Delta(h)$ officially legislated to be its slope.

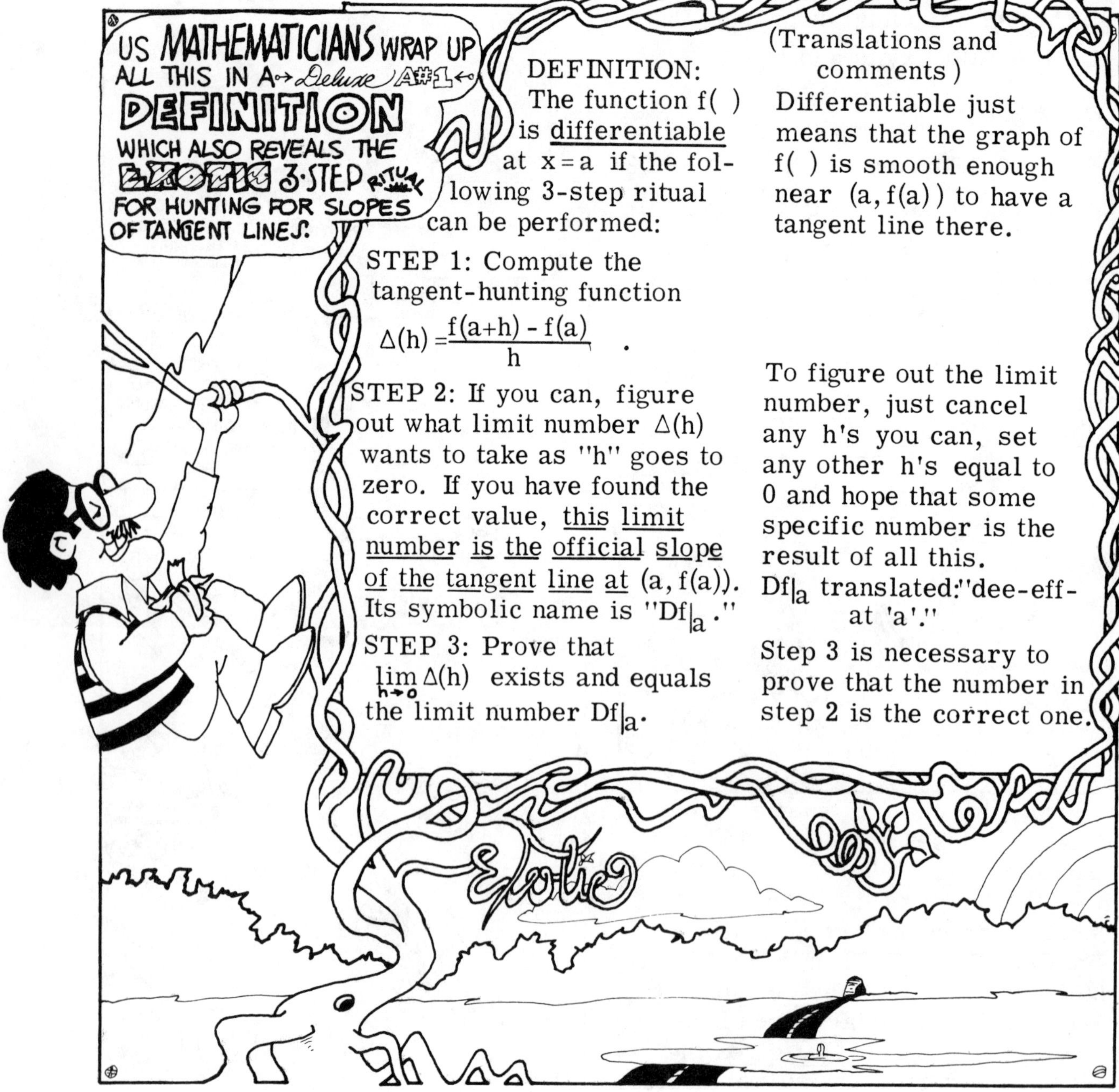

US MATHEMATICIANS WRAP UP ALL THIS IN A → Deluxe A#1 ← DEFINITION WHICH ALSO REVEALS THE EXOTIC 3-STEP RITUAL FOR HUNTING FOR SLOPES OF TANGENT LINES.
Exotic

DEFINITION: The function f( ) is <u>differentiable</u> at x = a if the following 3-step ritual can be performed:

(Translations and comments) Differentiable just means that the graph of f( ) is smooth enough near (a, f(a)) to have a tangent line there.

STEP 1: Compute the tangent-hunting function

$$\Delta(h) = \frac{f(a+h) - f(a)}{h}.$$

STEP 2: If you can, figure out what limit number $\Delta(h)$ wants to take as "h" goes to zero. If you have found the correct value, <u>this limit number is the official slope of the tangent line at</u> (a, f(a)). Its symbolic name is "$Df|_a$."

To figure out the limit number, just cancel any h's you can, set any other h's equal to 0 and hope that some specific number is the result of all this.

$Df|_a$ translated: "dee-eff-at 'a'."

STEP 3: Prove that $\lim_{h \to 0} \Delta(h)$ exists and equals the limit number $Df|_a$.

Step 3 is necessary to prove that the number in step 2 is the correct one.

Finding the number "$Df|_a$" is called "differentiating f( ) at a," and the official slope number "$Df|_a$" is called "the DERIVATIVE of f( ) at a." Other symbols that mean the same thing are

$$Df|_a = f'(a) = D_x f = Df = \frac{df}{dx} = f_x(a).$$

Whenever anybody says that the number $Df|_a$ is the "derivative of f( ) at a" it means that the entire 3-step ritual can be done using the particular f( ) and a.

Here is an example of performing the ritual for a new function and a new "a."

EXAMPLE: Prove that

$$f(x) = 2x^2+5x+3$$

is differentiable at $x=-2$. Find the derivative $Df|_{-2}$.

Proof: Step 1. Compute the tangent-hunting function:

$$\Delta(h) = \frac{f(-2+h)-f(-2)}{h} =$$

$$\frac{[2(-2+h)^2+5(-2+h)+3]-[2(-2)^2+5(-2)+3]}{h} =$$

$$\frac{[2(4-4h+h^2)-10+5h+3]-[8-10+3]}{h} =$$

$$\frac{[8-8h+2h^2+5h-7]-[1]}{h} =$$

$$\frac{-3h+2h^2}{h} = \frac{h}{h}(-3+2h)$$

When $h \neq 0$ we can cancel "h," so the formula for $\Delta(h)$ is

$$\Delta(h) = \begin{cases} -3+2h & \text{if } h \neq 0 \\ \text{undefined} & \text{if } h=0 \end{cases} .$$

Step 2. Unofficially set $h=0$:

$$-3+2(0) = -3 .$$

So it looks like we should have

$$D(2x^2+5x+3)|_{-2} = Df|_{-2} = -3.$$

Step 3. Prove that $\lim_{h \to 0} \Delta(h)$ exists and equals $-3 = Df|_{-2}$, where

$$\Delta(h) = \begin{cases} -3+2h & \text{if } h \neq 0 \\ \text{undefined} & \text{if } h=0 \end{cases} .$$

Proof: 1. For any $\epsilon>0$, let $\delta(\epsilon) = \epsilon/2$.

2. If $0-\frac{\epsilon}{2} < h < 0+\frac{\epsilon}{2}$ and $h\neq 0$,

so $-\epsilon < 2h < \epsilon$

then $-3-\epsilon < -3+2h < -3+\epsilon$

or $-3-\epsilon < \Delta(h) < -3+\epsilon$.

Then $Df|_{-2} - \epsilon < \Delta(h) < Df|_{-2} + \epsilon$.

The ritual is complete! The graph of

$$f(x) = 2x^2+5x+3$$

is officially smooth enough to have a tangent line at point $(-2, f(-2)) = (-2, 1)$, and the official slope for the tangent line through this point is

$$-3 = D(2x^2+5x+3)|_{-2} .$$

EXERCISES

coming up.

III.1.1 Prove that

$$f(x) = 3x^2 - 6x + 2$$

is differentiable at $x = -3$. Find the derivative $Df|_{-3}$.

III.1.2 Prove that

$$f(x) = 4 - 3x - x^2$$

is differentiable at $x=2$. Find the derivative $Df|_2$.

III.1.3 Prove that $f(x) = x$ is differentiable at $x=a$ for any real number a. Verify that $Dx|_a = 1$.

III.1.4 Prove that

$$f(x) = 2x^3 - 3x - 5$$

is differentiable at $x=2$. Find the derivative $Df|_2$.

III.1.5 Prove that $f(x) = x^2$ is differentiable at x=a for any number a. Verify that $Dx^2|_a = 2a$.

III.1.6 Prove that $f(x) = x^3$ is differentiable at x=a for any number a. Verify that $Dx^3|_a = 3a^2$.

III.1.7 In this section we started with the problem of finding the tangent line to the graph of

$$f(x) = -\frac{1}{4}x^2+2x+1$$

at point (2, f(2)). The slope of tangent-line candidate #1 through (2, f(2)) and (4, f(4)) was found to be $\frac{1}{2}$ by constructing an appropriate wedge with "run" = 2. Next the

III.1.7 Continued:
tangent-hunting function turned out to be

$$\Delta(h) = \left\{\begin{array}{c}\text{SLOPE OF TANGENT}\\ \text{LINE CANDIDATE}\\ \text{WHEN RUN OF}\\ \text{PROPPING WEDGE IS } h\end{array}\right\} = \begin{cases} 1-\frac{1}{4}h & \text{if } h\neq 0 \\ \text{undefined} & \text{if } h=0 \end{cases}$$

So $\Delta(2)$ had better be $\frac{1}{2}$ —

$\Delta(2) = 1 - \frac{1}{4}(2) = 1 - \frac{1}{2} = \frac{1}{2}$; it checks. But while deriving the formula

$$\Delta(h) = \frac{f(2+h) - f(2)}{h}$$

for reporting slopes of candidates, we only considered candidates through (2, f(2)) and a second point to the right of (2, f(2)). However, the formula also works for second points to the left of (2, f(2))! Check this out using point (0, f(0)) as the "second point" by first finding the slope of this candidate using a wedge and then check to see that the result obtained by using the formula for $\Delta(h)$ agrees with the first result. (Remember that "runs" are counted negative when wedges are to the left of the point.)

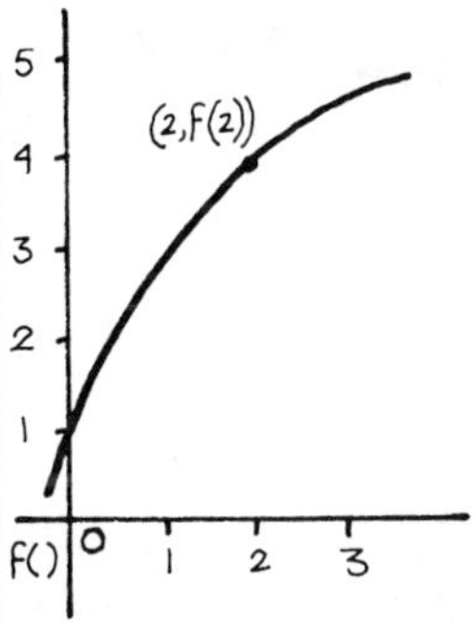

III.2 TWO (2) DERIVATIVE THEOREMS: THE EASY WAY OUT

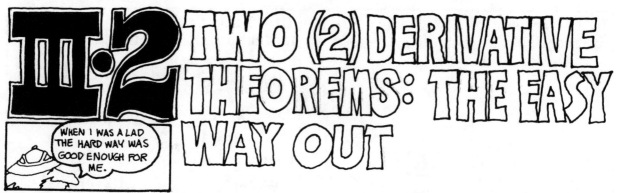

WHEN I WAS A LAD THE HARD WAY WAS GOOD ENOUGH FOR ME.

Doing the entire three-step ritual for any function to find its derivative gets to be quite a chore as functions get more complicated. Computing the tangent-hunting function

$$\Delta(h) = \frac{f(a+h) - f(a)}{h}$$

in step 1 is bad enough and showing that the necessary limits exist in step 3 is even worse. From a practical point of view, all we really use is the official "slope of the tangent line" —the number that is the derivative of f( ) at a. So mathematicians, having in the first place legislated that the three-step ritual is necessary to find a derivative, then hunted around to find a logical way to get out of having to do it.

We will see how they found a way out for polynomial functions. They first discovered a easy way to find the derivative number "$Df|_a$" without going through step 1. Then, using the limit theorems from Chapter II, they argued themselves out of ever having to prove that limits exist in step 3, and so made things easier for everyone.

To get an idea of how the argument will go, start the ritual with

$f(\ ) = 5(\ )^2 - 3(\ )$ and let $a=2$.

In other words, we want to prove that $f(x)=5x^2-3x$ is differentiable at $x=2$ and find $D(5x^2-3x)|_2$.

STEP 1:

$$\Delta(h) = \frac{[5(2+h)^2 - 3(2+h)] - [5(2)^2 - 3(2)]}{h}$$

Now, instead of multiplying everything out, rearrange the terms by taking the $5(\ )^2$ term from each square bracket and putting them together and do the same with the $-3(\ )$ terms:

$$= \frac{[5(2+h)^2 - 5(2)^2] + [-3(2+h) - (-3)(2)]}{h}$$

$$= \frac{5\cdot[(2+h)^2 - (2)^2] + (-3)[(2+h) - 2]}{h}$$

$$= 5\cdot\frac{(2+h)^2 - (2)^2}{h} + (-3)\,\frac{(2+h) - 2}{h}$$

$$= 5\cdot\left[\frac{\Delta(h) \text{ for function } f()=()^2}{\text{at } a=2}\right] + (-3)\cdot\left[\frac{\Delta(h) \text{ for function } f()=()}{\text{at } a=2}\right]$$

STEP 2: Figure out what value $\Delta(h)$ wants to take as its limit number as h goes to zero.

Now, from exercise III.1.5, "$\Delta(h)$ for function $f(\ )=(\ )^2$" wants to take value $Dx^2|_2 = 2\cdot 2 = 4$ as its limit number as h goes to zero and has a limit-machine to show that 4 is the correct value, and from exercise III.1.3, " $\Delta(h)$ for function $f(\ )=(\ )$" wants to take value $Dx|_2 = 1$ as its limit number as h goes to zero and has a limit-machine to show that 1 is the correct value.

We just discovered that, for function $f(x) = 5x^2-3x$,

$$\Delta(h) = 5\cdot\begin{bmatrix}\Delta(h) \text{ for} \\ f(\ )=(\ )^2 \\ \text{at } a=2\end{bmatrix} + (-3)\begin{bmatrix}\Delta(h) \text{ for} \\ f(\ )=(\ ) \\ \text{at } a=2\end{bmatrix}$$

(first bracket) ↑ WANTS TO TAKE VALUE $Dx^2|_2=4$ WHEN $h \to 0$

(second bracket) ↑ WANTS TO TAKE VALUE $Dx|_2=1$ WHEN $h \to 0$

So it looks like $\Delta(h)$ for $f(x)=5x^2-3x$ wants to take value

$$5\cdot(Dx^2|_2) + (-3)\cdot(Dx|_2)$$
$$= 5\cdot(4)+(-3)\cdot(1) = 17$$

for its limit number as h goes to zero, so a good guess is that

$$D(5x^2-3x)|_2 = 5\cdot(Dx^2|_2) + (-3)\cdot(Dx|_2)\ !$$

STEP 3: Prove that $\lim_{h\to 0} \Delta(h)$ exists and equals

$$5\cdot(Dx^2|_2) + (-3)\cdot(Dx|_2)\ .$$

But now we can use the limit theorems to get out of having to do this, since

$$\Delta(h) = 5\cdot\begin{bmatrix}\Delta(h) \text{ FOR} \\ f(\ )=(\ )^2 \\ \text{AT } a=2\end{bmatrix} + (-3)\begin{bmatrix}\Delta(h) \text{ FOR} \\ f(\ )=(\ ) \\ \text{AT } a=2\end{bmatrix}$$

(first bracket) ↑ HAS A LIMIT-MACHINE TO PROVE THAT $Dx^2|_2(=4)$ IS THE CORRECT LIMIT VALUE AS $h \to 0$

(second bracket) ↑ HAS A LIMIT-MACHINE TO PROVE THAT $Dx|_2(=1)$ IS THE CORRECT LIMIT VALUE AS $h \to 0$

SO THE "Kf" LIMIT THEOREM SAYS THAT THERE IS A LIMIT-MACHINE TO PROVE THAT $5\cdot(Dx^2|_2)$ (↑ K) IS THE CORRECT LIMIT VALUE FOR $5\cdot(\Delta(h) \text{ FOR } (\ )^2)$ AS $h \to 0$

SO THE "Kf" LIMIT THEOREM SAYS THAT THERE IS A LIMIT-MACHINE TO PROVE THAT $(-3)\cdot(Dx|_2)$ (↑ K) IS THE CORRECT LIMIT VALUE FOR $(-3)(\Delta(h) \text{ FOR } (\ ))$ AS $h \to 0$

THEN, FINALLY, THE "f+g" LIMIT THEOREM SAYS THAT THERE IS A LIMIT-MACHINE TO PROVE THAT ALL OF THE FUNCTION

$$\Delta(h) = 5[\Delta(h) \text{ FOR } (\ )^2] + (-3)[\Delta(h) \text{ FOR } (\ )]$$

HAS A CORRECT LIMIT VALUE EQUAL TO

$$5\cdot(Dx^2|_2) + (-3)\cdot(Dx|_2)$$

AS $h \to 0$

So the entire three-step ritual for

$$f(x) = 5x^2 - 3x$$

is officially done, and we have proved that $f(\ )$ is differentiable at $x=2$, and

$$D(5x^2-3x)|_2 = 5\cdot(Dx^2|_2) + (-3)\cdot(Dx|_2)\ !$$

It looks like the symbol "D" works like this: If b and c are constant numbers and $f(x)$ and $g(x)$ are functions, then

$$D(b\cdot f(x) + c\cdot g(x))|_a = b\cdot(Df|_a) + c\cdot(Dg|_a)\ ,$$

i.e., the symbol "D" attacks any functions separately that add together to make a more complicated function
SET, 1-2-39...
GRR
g( )
f( )
GRONK
POUNCE
g( )
Grover
f( )
ALFRED
and "D" slides right by constants that multiply functions!
K
f( )
ALFRED
SNARL
WHAP
ZOOM
f( )

Here are two theorems that prove that it is all right to move D around like this. The first one shows that D slides by constants that multiply functions.

THEOREM III-1	Translation:
If g( ) is differentiable at x=a,	If the 3-step ritual can be done for g( ) at x=a, and $Dg\|_a$ is the slope of the tangent line to the graph of g( ) at (a, g(a)),
then K·g( ) is also differentiable at x=a, and $$D(K\cdot g)\|_a = K\cdot(Dg\|_a)$$ where K is any constant number.	then automatically the 3-step ritual can be done for K·g( ) at x=a, and $K\cdot(Dg\|_a)$ is the slope of the tangent line to the graph of K·g( ) at point (a, K·g(a)).

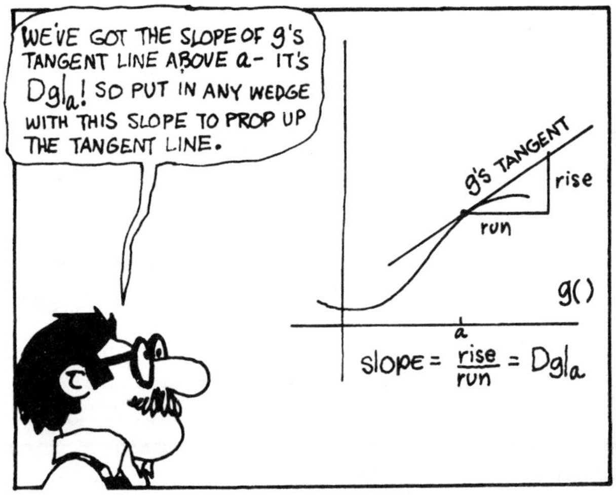

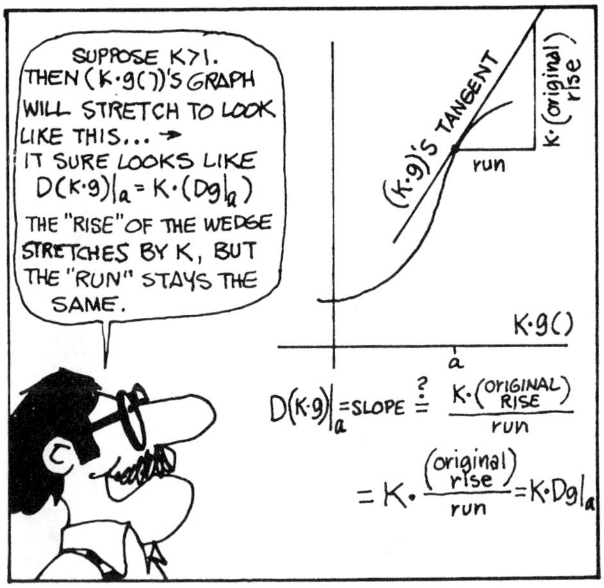

Now for the 3-step ritual proof:

STEP 1: The tangent-hunting function for K·g( ) is

$$\left(\begin{matrix}\Delta(h) \text{ for } K\cdot g() \\ \text{at } x=a\end{matrix}\right) = \frac{[K\cdot g(a+h)] - [K\cdot g(a)]}{h}$$

(FACTOR OUT K)

$$= K\cdot\frac{g(a+h)-g(a)}{h} = K\cdot\left(\begin{matrix}\Delta(h) \text{ for } g() \\ \text{at } x=a\end{matrix}\right)$$

STEP 2: Figure out what limit number "Δ(h) for K·g( )" wants to take as h goes to zero. Easy, since

$$(\Delta(h) \text{ FOR } K\cdot g()) = K\cdot\underbrace{(\Delta(h) \text{ FOR } g())}_{\text{WANTS TO TAKE LIMIT VALUE } Dg|_a \text{ AS } h\to 0}.$$

So it looks like "Δ(h) for K·g( )" wants to take value $K\cdot(Dg|_a)$ as its limit number as h goes to zero.

STEP 3. We have to prove that $\lim_{h\to 0}(\Delta(h) \text{ for } K\cdot g)$ exists and equals $K\cdot(Dg|_a)$.

Step 1. of our ritual proved that $(\Delta(h) \text{ for } K\cdot g) = K\cdot(\Delta(h) \text{ for } g)$. Since the two functions are equal, if we can demonstrate that there is a limit-machine proving that a limit exists for K·(Δ(h) for g), the same machine can automatically be used to prove that (Δ(h) for K·g) has a limit as well.

So, to finish step 3, all we need is to show that there is a limit-machine proving that $\lim_{h\to 0}[K\cdot(\Delta(h) \text{ for } g)]$ exists and equals $K\cdot(Dg|_a)$. Now we assumed at the beginning of the theorem that the three-step ritual can be done for g( ), so we can assume that there is a limit-machine proving that $\lim_{h\to 0}(\Delta(h) \text{ for } g)$ exists and equals $Dg|_a$. Theorem II-2 says that we can then use this machine to construct a new limit-machine to prove that K·(Δ(h) for g) has a limit too, and the correct limit number is $K[(\Delta(h) \text{ for } g)\text{'s limit number}] = K\cdot(Dg|_a)$. Theorem II-2 also makes sure that all the necessary guarantees will hold, and that is enough to wrap up the theorem.

The second theorem shows that the symbol D attacks any functions separately that add together to make a more complicated function.

THEOREM III-2	Translation:
If f( ) and g( ) are differentiable at x = a,	If the 3-step ritual can be done for f( ) and g( ) at x = a,
then "f( ) + g( )" is differentiable at x = a, and	then automatically the 3-step ritual can be done for "f( )+g( )" at x = a.
$D(f+g)\|_a = Df\|_a + Dg\|_a$.	The slope of the tangent line to the graph of "f( )+g( )" at (a, f(a)+g(a)) is $D(f+g)\|_a = Df\|_a + Dg\|_a$.

PROOF:

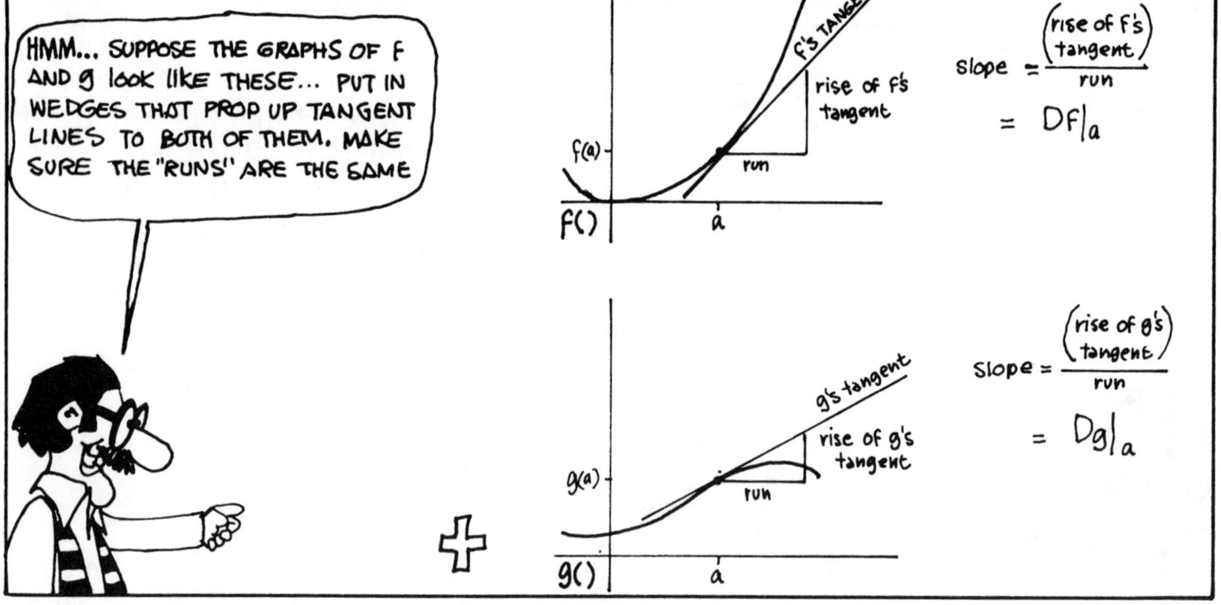

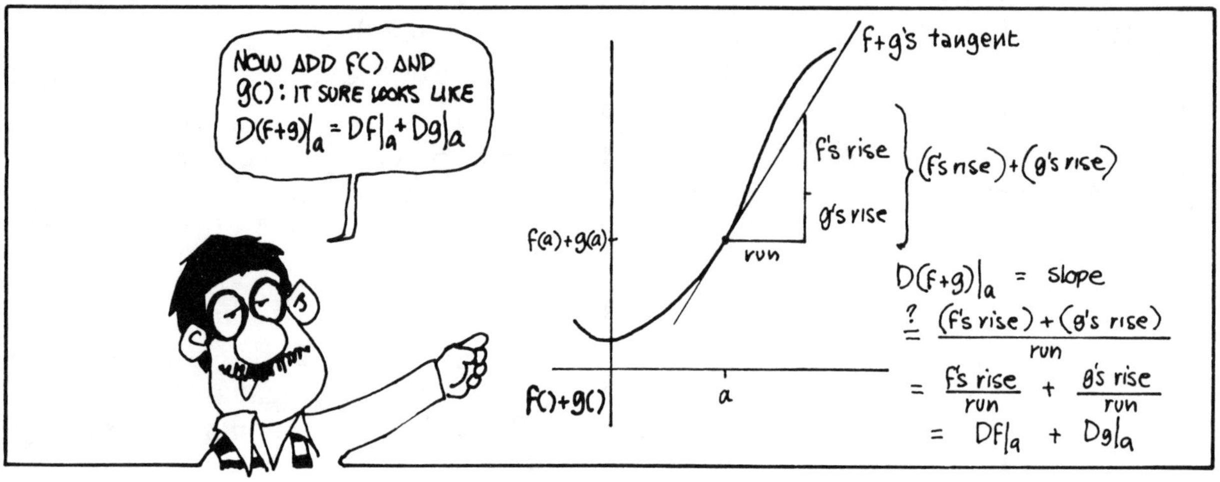

Try the 3-step ritual for $(f(\ )+g(\ ))$:

STEP 1: The tangent-hunting function for "$f(\ )+g(\ )$" is

$$(\Delta(h) \text{ FOR } f+g)$$

$$= \frac{[f(a+h)+g(a+h)] - [f(a)+g(a)]}{h}$$

(NOW, PUT THE f( ) TERMS TOGETHER AND THE SAME WITH THE g( ) TERMS)

$$= \frac{f(a+h)-f(a)}{h} + \frac{g(a+h)-g(a)}{h}$$

$$= [\Delta(h) \text{ FOR } f] + [\Delta(h) \text{ FOR } g].$$

STEP 2: Figure out what limit number "$\Delta(h)$ for $f+g$" wants to take as h goes to zero. Since

$$(\Delta(h) \text{ FOR } f+g)$$

$$= [\Delta(h) \text{ FOR } f] + [\Delta(h) \text{ FOR } g]$$

WANTS TO TAKE VALUE $Df|_a$ AS $h \to 0$ (for $[\Delta(h)$ FOR $f]$); WANTS TO TAKE VALUE $Dg|_a$ AS $h \to 0$ (for $[\Delta(h)$ FOR $g]$)

it's pretty clear that $(\Delta(h)$ for $f+g)$ wants to take value

$$Df|_a + Dg|_a$$

as its limit number as $h \to 0$.

STEP 3: Prove that

$$\lim_{h\to 0}(\Delta(h) \text{ for } f+g)$$

exists and equals $Df|_a + Dg|_a$.

We <u>assumed</u> at the beginning of the theorem that the 3-step ritual can be done for $f(\ )$ and $g(\ )$. The third step of the rituals guarantees that

both $\lim_{h\to 0}(\Delta(h) \text{ for } f(\ ))$

exists and equals $Df|_a$ and

$\lim_{h\to 0}(\Delta(h) \text{ for } g(\ ))$

exists and equals $Dg|_a$. Then the "f+g" <u>limit</u> theorem guarantees that

$$\lim_{h\to 0}\underbrace{((\Delta(h) \text{ for } f(\ )) + (\Delta(h) \text{ for } g(\ )))}_{(\Delta(h) \text{ for } f+g)}$$

exists and equals $Df|_a + Dg|_a$, and this proves that step 3 goes through.

So having derivatives is INHERITED when we multiply a function by a constant or add two functions with derivatives.

Using these two theorems we can now show that any polynomial no more complicated than a cubic is differentiable and we can find the derivative without bothering with the 3-step ritual! The argument goes like this:

In the exercises in III.1 we actually did the ritual for functions

$$x, \ x^2 \text{ and } x^3,$$

proving that they are differentiable for any "a," and the derivatives have formulas

$$Dx|_a = 1 \ ; \ Dx^2|_a = 2 \cdot a \text{ and } Dx^3|_a = 3a^2.$$

Also, the derivative of any constant function should be "0" since any constant function's graph is just a horizontal straight line and thus the graph should be its own tangent line (with slope zero). BUT THE RITUAL MUST BE OBEYED! Here it is:

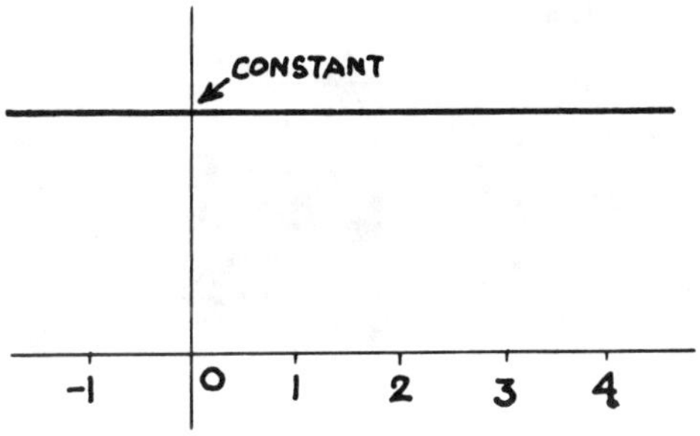

PROOF that the derivative of a constant function is zero:

STEP 1:

$$\left(\Delta(h) \text{ FOR A CONSTANT FUNCTION } F(\;)\right) = \frac{F(a+h)-F(a)}{h}$$

$$= \frac{(\text{CONSTANT})-(\text{SAME CONSTANT})}{h} = \frac{0}{h}$$

$$= 0 \quad (\text{IF } h \neq 0)$$

STEP 2: Since $\Delta(h) = \begin{cases} 0 \text{ if } h \neq 0 \\ \text{undefined if } h=0 \end{cases}$

its limit number should be 0, i.e.,

$$D(\text{constant})|_a = 0.$$

STEP 3: Prove that $\lim_{h \to 0} \Delta(h)$ exists and equals 0, when $\Delta(h) = \begin{cases} 0 \text{ if } h \neq 0 \\ \text{undefined if } h=0 \end{cases}$.

Proof:

For any $\epsilon > 0$, let $\delta(\epsilon)$ = anything positive! (IT DOESN'T MATTER WHAT THIS SAYS

If $0 - \delta(\epsilon) < h < 0 + \delta(\epsilon)$ and $h \neq 0$ SINCE

Then $0 - \epsilon < 0 < 0 + \epsilon$ ← THIS IS ALWAYS TRUE)

or $0 - \epsilon < \Delta(h) < 0 + \epsilon$, and the ritual is complete.

Now suppose we want to argue that a function like

$$f(x) = -5 + 2x - 3x^2 + 15x^3$$

is differentiable at "a" and find its derivative. Having actually done the 3 step ritual for functions

(constant), x, x^2, and x^3,

theorem III-1 guarantees that we could do the ritual for functions

(-5), $2x$, $-3x^2$ and $15x^3$

if we had to, so we don't have to bother with it. Furthermore, theorem III-1 says that we can compute the derivatives as follows:

$D(-5)|_a = 0$,

$D(2x)|_a = 2\cdot(Dx|_a) = 2\cdot 1$,

$D(-3x^2)|_a = (-3)\cdot(Dx^2|_a) = (-3)\cdot(2a)$ and

$D(15x^3)|_a = (15)\cdot(Dx^3|_a) = (15)\cdot(3a^2)$.

Now, knowing that the 3-step ritual could without fail be done for

(-5), $2x$, $-3x^2$, and $15x^3$,

theorem III-2 (used several times) guarantees that the ritual could be done for

$$[(-5 + 2x) + (-3x^2)] + [15x^3],$$

so we don't have to bother with it, and furthermore,

$$D(-5+2x+(-3)x^2+15x^3)|_a$$

$$= D(-5)|_a + D2x|_a + D(-3)x^2|_a + D15x^3|_a$$

$$= 0 + 2\cdot Dx|_a + (-3)Dx^2|_a + 15\cdot Dx^3|_a$$

$$= 0 + 2\cdot 1 + (-3)\cdot 2a + 15\cdot 3a^2$$

$$= 2 - 6a + 45a^2.$$

EXERCISES

coming up.

III.2.1 Compute $D(3x^2-2x-17)|_a$.

Find a formula for the line tangent to the graph of $f(x)=3x^2-2x-17$ at point $(3,f(3))$.

III.2.2 Compute the derivatives of the following functions at any point "a."

(i) $f(x)=x^3+3(x-1)^2+2x-1$

(ii) $f(x) = \frac{x^2+2x}{5}$

III.2.3 If $f(x) = 2x^3+3x^2-36x+5$, find the points where the tangent line to the graph of f( ) is flat (has slope zero).

III.2.4 Find a formula for the line tangent to the graph of

$$f(x)=-3x^2+12x-5$$

at point $(2,f(2))$.

III.2.5 If $f(x) = x^3+3x^2-6x-9$, find the points where the line tangent to the graph of f( ) will have slope equal to 3.

II·3 DIFFERENTIATING POLYNOMIALS AND FINDING MAXIMA AND MINIMA...

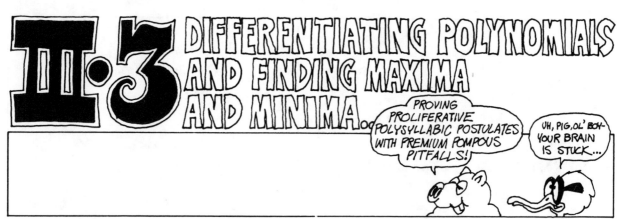

The theorems in the last section prove that the differentiating symbol "D" slides by constants that multiply functions and attacks any functions separately that add together to make more complicated functions. IF we knew that we could do the three-step ritual for

x^4, x^5, x^6, and so forth,

then the theorems would guarantee that we could do the three-step ritual for any specific polynomial, and, furthermore, we could compute like this: $D(5x^9 + 3x^5 - 2x^2 - 2)|_a$

$= 5Dx^9|_a + 3Dx^5|_a - 2Dx^2|_a + D(-2)|_a$.

So to wrap this up we need to show that $Dx^n|_a$ exists and find its value, where n is any positive integer. Try the three-step ritual for x^n — we can't get all the way through, but we can get as far as step 2 and figure out what $Dx^n|_a$ ought to be. The argument goes like this:

STEP 1: $\Delta(h)$ for x^n $= \dfrac{(a+h)^n - (a)^n}{h}$.

Instead of now trying to multiply all this out we will use the following handy formula from algebra:

For any numbers a and b ,

$$b^n - a^n$$

$$= (b-a)(b^{n-1} + b^{n-2}a + b^{n-3}a^2 + \dots + ba^{n-2} + a^{n-1})$$

HERE IS ONE TERM AND HERE ARE (n-1) TERMS, GIVING A TOTAL OF 1+(n-1)=n TERMS.

To make the formula plausible, check it out for n=5 :

$$(b-a)(b^4 + b^3a + b^2a^2 + ba^3 + a^4)$$
$$= b(b^4 + b^3a + b^2a^2 + ba^3 + a^4)$$
$$- a(b^4 + b^3a + b^2a^2 + ba^3 + a^4)$$
$$= b^5 + b^4a + b^3a^2 + b^2a^3 + ba^4$$
$$- b^4a - b^3a^2 - b^2a^3 - ba^4 - a^5$$
$$= b^5 - a^5 .$$

We need to figure out the tangent-hunting function

$$\frac{(a+h)^n - (a)^n}{h} ,$$

so plug in (a+h) for "b" in the handy formula and we get

$$\Delta(h) \text{ FOR } x^n = \frac{(a+h)^n - (a)^n}{h}$$

$$= \frac{\overbrace{((a+h)-a)}^{h}\left((a+h)^{n-1}+(a+h)^{n-2}a+\ldots+(a+h)a^{n-2}+a^{n-1}\right)}{h}$$

(CANCEL h)

$$= \begin{cases} \overbrace{\left((a+h)^{n-1}+(a+h)^{n-2}a+\ldots+(a+h)a^{n-2}+a^{n-1}\right)}^{n \text{ TERMS}} & \text{IF } h \neq 0 \\ \text{UNDEFINED IF } h=0. \end{cases}$$

STEP 2: Figure out what limit number $\Delta(h)$ wants to take as h goes to zero. (Unofficially make any h's equal to 0.) It looks like " $\Delta(h)$ for x^n " wants to take value

$$Dx^n|_a \stackrel{?}{=} \overbrace{\left((a+0)^{n-1}+(a+0)^{n-2}a+(a+0)^{n-3}a^2+\ldots+(a+0)^{n-2}a+a^{n-1}\right)}^{n \text{ TERMS}}$$

$$= \overbrace{\left(a^{n-1}+a^{n-2}a+a^{n-3}a^2+\ldots+a\cdot a^{n-2}+a^{n-1}\right)}^{n \text{ TERMS}}$$

THESE ALL MULTIPLY OUT TO BE a^{n-1}

$$= \overbrace{\left(a^{n-1}+a^{n-1}+\ldots+a^{n-1}+a^{n-1}\right)}^{n \text{ TERMS}}$$

$$= n\cdot a^{n-1} \quad .$$

as its limit number as h goes to 0.

STEP 3: Prove that

$$\lim_{h\to 0}\left((a+h)^{n-1}+(a+h)^{n-2}a+\ldots+(a+h)a^{n-2}+a^{n-1}\right)$$

exists and equals $n{\cdot}a^{n-1}$. Even using the limit theorems we already have, we can't officially prove this, since we haven't ever proved that

$\lim_{h\to 0}(a+h)^k$ exists and equals a^k, but mathematicians have checked it out, and it can be done using more limit theorems. So, leaving the completion of the ritual to the mathematicians, we will take it that x^n is officially differentiable and

$$Dx^n|_a = n{\cdot}a^{n-1} \quad .$$

This formula is easy to remember:

We use the formula a bit:

$$Dx^9|_a = 9a^{9-1} = 9a^8 \quad ;$$

$$Dx^7|_4 = 7{\cdot}(4)^{7-1} = 7{\cdot}(4)^6 \;;$$

$$Dx^4|_{(-2)} = 4(-2)^{4-1} = 4(-2)^3 = -32.$$

Now we can finish up the computing that we did earlier:

$$D(5x^9 + 3x^5 - 2x^2 - 2)|_a$$

$$= 5Dx^9|_a + 3Dx^5|_a - 2Dx^2|_a + D(-2)|_a$$

$$= 5{\cdot}9a^8 + 3{\cdot}5a^4 - 2{\cdot}2a + 0$$

$$= 45a^8 + 15a^4 - 4a \;.$$

Back in Chapter I (I.6), when talking about drawing parabolas, we used a convenient formula for finding the point that is the "nose" of the graph of a typical parabola with formula $f(x) = Ax^2 + Bx + C$.

The formula said that the "nose" is always directly above or below the number

$$x = \frac{-B}{2A}$$

on the x-axis.

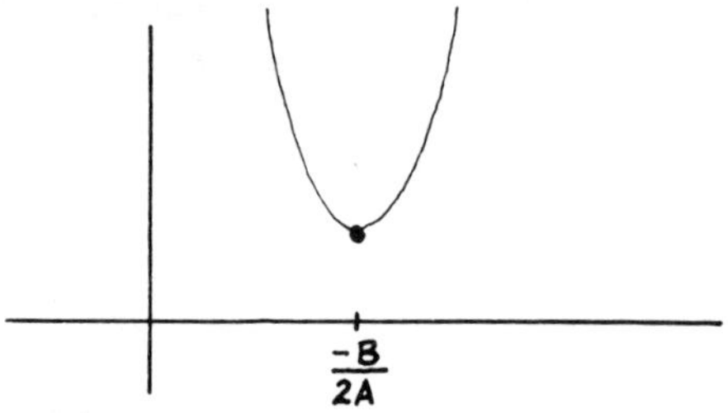

Now, using derivatives, we can at last see how to get this formula.

The line tangent to the graph of parabola $f(x) = Ax^2 + Bx + C$ at the "nose" point is clearly flat,

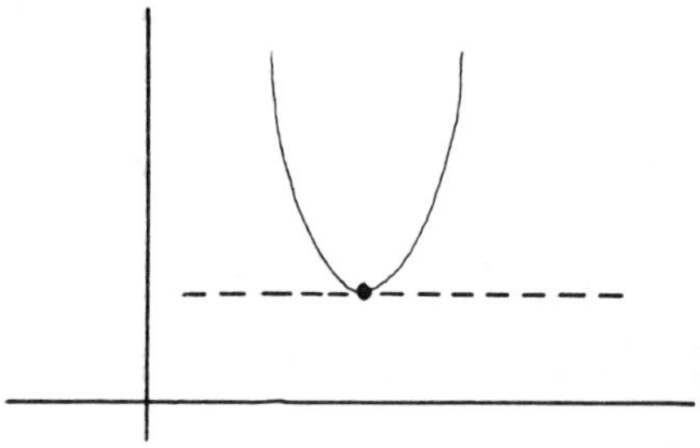

parallel to the x-axis, so its slope must be zero. And furthermore, the nose is the only point on the graph where the tangent line is flat, with slope zero. So, since the slope of the tangent line is given by the derivative, if we find out where the derivative is zero, we will find out where the nose is !

Now $Df(x)|_a = D(Ax^2+Bx+C)|_a$

$= A{\cdot}Dx^2|_a + B{\cdot}Dx|_a + D(C)|_a = A{\cdot}2a+B{\cdot}1+0.$

We want the value of "a" that makes the derivative of $f(x)$ zero, i.e., we want to find the value "a" such that

$$2A{\cdot}a + B = 0$$

or

$$2A{\cdot}a = -B$$

or

$$a = \frac{-B}{2A},$$

and this is exactly the formula we used in Chapter I to find the x-value of the nose point.

The same sort of reasoning can be used with cubics to find the top or bottom of humps (if any) on their graphs. If a hump looks like this:

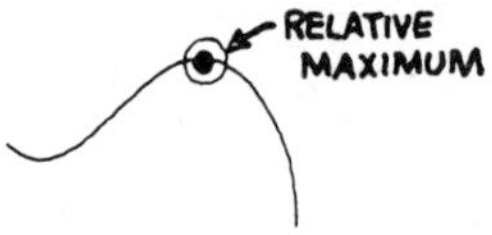

the top is called a "relative maximum". If the hump looks like this:

the bottom is called a "relative minimum." Again, at both of these points, the tangent line is flat, so the derivative will be zero there.

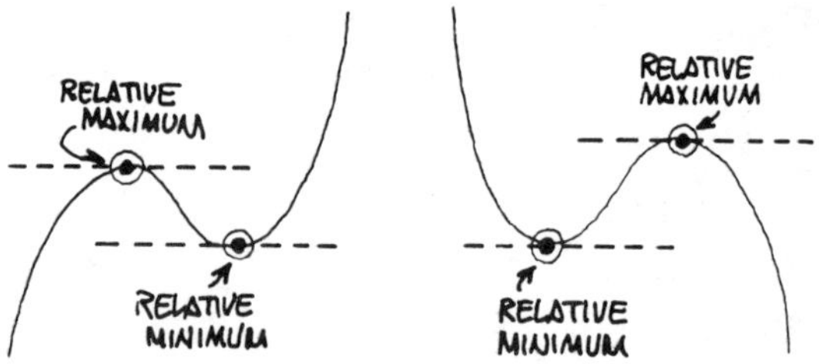

Finding such "relative maxima and minima" turns out to be handy in practical problems where we want to find the high (or low) points of whatever the function represents.

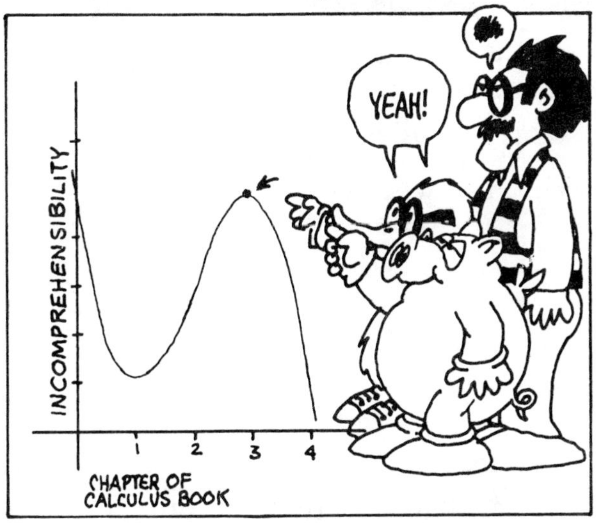

For example, suppose we want to find the relative maximum and minimum of function

$$f(x) = -x^3 + 6x^2 - 9x + 5.$$

The function is a cubic. Since the coefficient of x^3 is -1, we know from Chapter I that the graph will look like  or .

The "point of inflection" (from page 65) will be above (or below)

$$x = \frac{-6}{3\cdot(-1)} = 2 .$$

To find the humps (if any), first compute the derivative of f(x):

$$D(-x^3+6x^2-9x+5)|_a$$

$$= -Dx^3|_a + 6Dx^2|_a - 9Dx|_a + D5|_a$$

$$= -3a^2 + 12a - 9 + 0.$$

Now, since the tangent line will be flat (and so have slope = 0) at the top (or bottom) of any humps, the derivative of f( ) will have to be zero at these points. So we hunt for values of "a" where

$$D(-x^3+6x^2-9x+5)|_a = -3a^2+12a-9 = 0.$$

We can factor $-3a^2+12a-9$:

$$-3a^2+12a-9 = (-3)\cdot(a^2-4a+3)$$
$$= (-3)\cdot(a-1)\cdot(a-3).$$

(If $Df|_a$ doesn't factor, there will be no points where the tangent line is flat, so the cubic would have to look like  .)

The factored $Df|_a = (-3)\cdot(a-1)\cdot(a-3)$. So $Df|_a$ will be zero if a=1 or a=3 and this means that the tangent line will be flat if a=1 or a=3. Since the cubic looks like  , we must have a

relative minimum above a=1 at point $(1, f(1)) = (1, 1)$ and a

relative maximum above a=3 at point $(3, f(3)) = (3, 5)$.

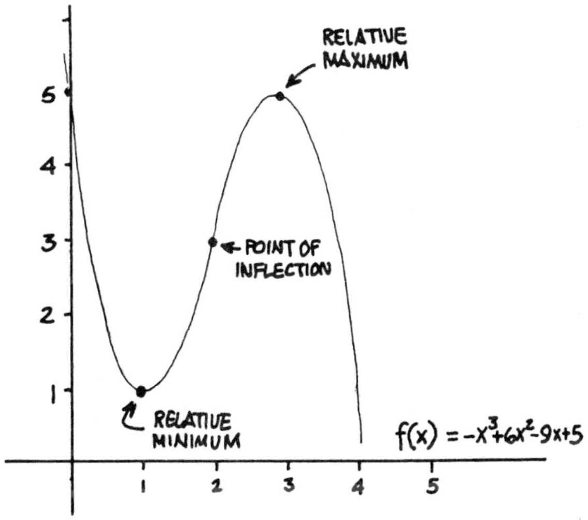

EXERCISES

III.3.1 Compute the following

$D(x^5-4x^3+2x+97)|_2 =$

$D(x^4-3x^9+2x)|_{-1} =$

$D(\frac{2}{3}x^3-x^{14}+3(x+1)^2)|_1 =$

$D(22x^{47} - 3x^{17})|_a =$

III.3.2 Find any relative maxima or minima of

$f(x) = -x^3-6x^2-9x-3$.

Which is which?

III.3.3 Find any relative maxima or minima of

$f(x) = x^3-3x+5$.

Which is which?

III.3.4 The fact that the tangent line to the graph of a function has slope zero at a point doesn't <u>necessarily</u> mean that the point is a relative maximum or minimum. To see what else can happen, let $f(x) = x^3-6x^2+12x-6$. Find where the derivative is zero and compare your answer with the location of the point of inflection. Carefully sketch the graph.

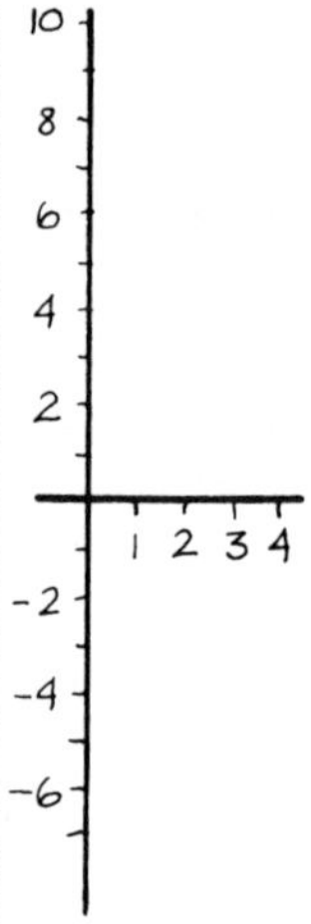

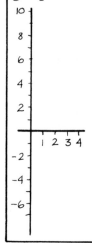

III.3.5 The O. K. Corral Construction Co. specializes in rectangular corrals. Fence running north-south costs $2.00 per running foot while fence running east-west costs $3.00 per foot. Find the dimensions of the largest corral you can get for $960.

ANSWER: IF ℓ = (NORTH-SOUTH LENGTH) AND W = (EAST-WEST WIDTH), THEN, REQUIRING THAT THE COST BE $960 TRANSLATES INTO MATHEMATICS AS

$$\$960 = \ell\cdot\$2 + W\cdot\$3 + \ell\cdot\$2 + W\cdot\$3$$
$$= 2\ell\cdot\$2 + 2W\cdot\$3, \text{ OR}$$
$$960 = 4\ell + 6W.$$

WE NEED TO INVENT A FUNCTION THAT REPORTS THE AREA FOR VARIOUS CHOICES OF ℓ AND w AND THEN FIND THE FUNCTION'S MAXIMUM. A FORMULA FOR THE AREA IS EASY ENOUGH: AREA = (LENGTH)·(WIDTH) = $\ell\cdot w$. BUT WE CAN'T USE CALCULUS TO FIND THE MAXIMUM AREA UNLESS WE HAVE ONLY ONE "UNKNOWN," LIKE ℓ OR w.

THE TRICK IS TO USE THE REQUIREMENT THAT THE COST BE $960, I.E., "$960 = 4\ell + 6W$" TO EXPRESS ℓ IN TERMS OF w (OR VICE VERSA). TRY A LITTLE ALGEBRA: $960 - 6w = 4\ell$ OR $\ell = \frac{1}{4}(960 - 6w)$. NOW WE CAN SUBSTITUTE FOR ℓ TO GET AREA $= \ell\cdot w = \frac{1}{4}(960 - 6w)\cdot w = 240w - \frac{3}{2}w^2$

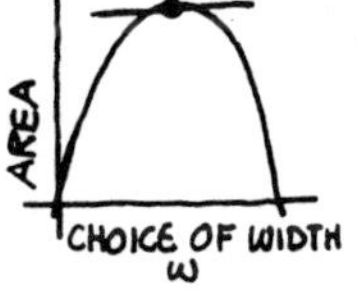

THIS IS A PARABOLA LOOKING LIKE THIS, SINCE THE COEFFICIENT OF w^2 IS NEGATIVE ($-\frac{3}{2}$ HERE). SO FINDING WHERE THE AREA IS A MAXIMUM IS THE SAME AS FINDING WHERE THE TANGENT LINE IS FLAT: FIND THE DERIVATIVE:

$D(\text{AREA})|_w = 240 - \frac{3}{2}\cdot(2w)$. SET IT EQUAL TO ZERO: $0 = 240 - 3w$: SOLVE FOR w:

$w = 80$;
$\ell = \frac{1}{4}(960 - 6\cdot 80) = 120$
ANSWER

III.3.6 The Classy Carton Corp. has just bought a load of square cardboard 12 inches on a side. Find the dimensions of the open-topped box of greatest volume that they can make out of the cardboard by cutting square pieces from the corners. (Hint: Let "x" be the length of the side of one of the pieces to be cut out. Then invent a function that reports the volume for any choice of "x" and find its maximum.)

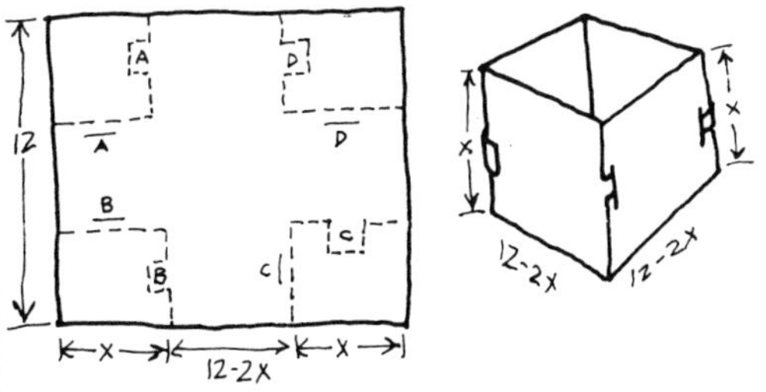

III·4 VELOCITIES & DERIVATIVES

Since differential calculus was originally invented to handle velocities, it's about time we figured out what velocities have to do with derivatives.

To start, suppose a point is moving along a wire.

The point has a speedometer to tell how fast he is going. (The speedometer registers inches-per-minute instead of the usual miles-per-hour.)

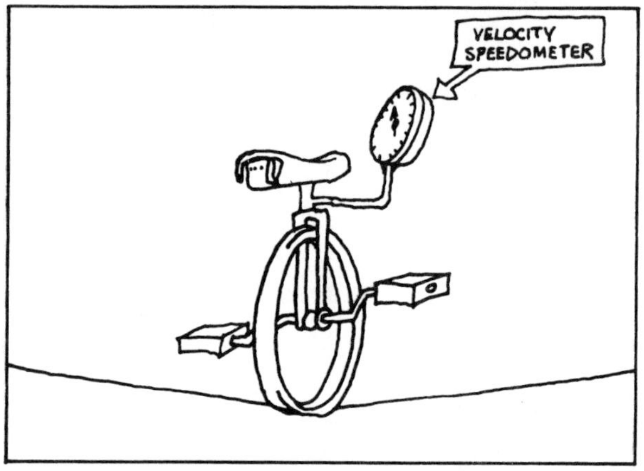

This speedometer is a special custom job that reports velocities, which just means that when the point moves forward, the speedometer calls the speed positive and when the point moves backward, the speedometer reports the speed backward as negative. (The only mathematical difference between speed and velocity (when the motion is in a straight line) is that "speed" is usually considered positive no matter what direction the motion is, whereas "velocity" counts forward speed as positive and backward speed as "negative.")

We have to figure out what the speedometer will read at any particular instant as the point moves along. First we need some sort of way to record what the point does.

One way to keep track of where the point has been in a rough sort of way is to label its location on the wire at the end of every minute.

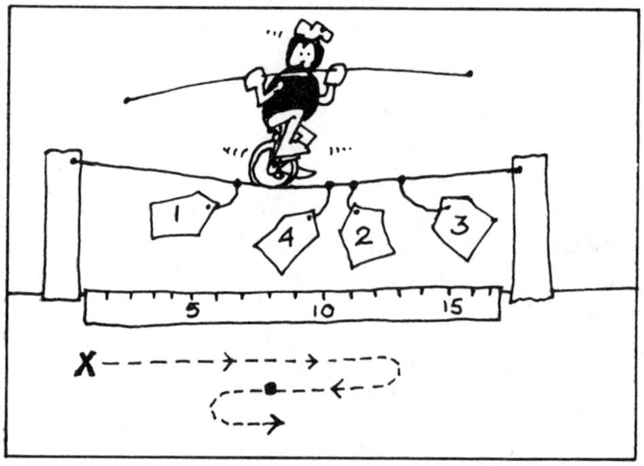

But this doesn't indicate what happens in enough detail to figure out "instantaneous velocity," which is what speedometers are supposed to report.

So scientists devised a clever new scheme that records exactly where the point is at every instant of time, not just at the end of every minute. It goes like this:

Put a sheet of graph paper under the wire and move the paper at a <u>steady</u> speed to the left as the point moves up and down the wire. (The wire is held fixed above the paper.) Arrange things so that at the beginning of each minute one of the ruled lines on the graph paper is directly under the wire. Finally, the point now has a pencil to trace on the paper as both the point and the paper move.

Our scientists will now demonstrate what will happen:

Now, when the point stops, a complete record of the motion of the point will be given by the curve drawn on the graph paper! Of course, the curve is NOT the actual path of the point — (remember that the point just moves up and down the wire). But if we pull the paper back so that the line through "start" is under the wire, we can use the traced curve to find out exactly where the point was at any particular time during the experiment.

For example, to find out where the point was at $2\frac{1}{2}$ minutes, go straight up to the traced curve above $2\frac{1}{2}$ and make a right angled turn over to the wire. We will end up exactly where the point was at $2\frac{1}{2}$ minutes!

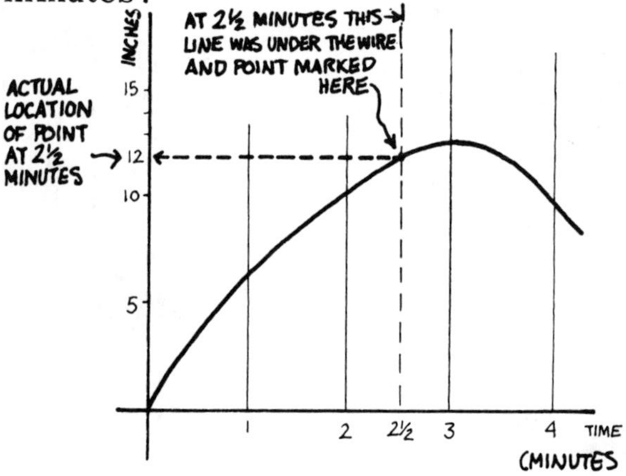

So we can use the traced curve as the graph of a function whose arrows start at any particular time and end up pointing to the position of the point at that particular time. Call this function p(t): it reports the position of the point as a function of time.

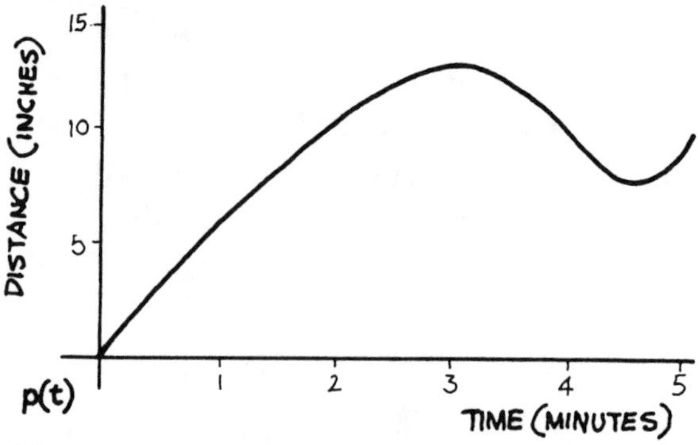

Now that we have the function p(t) to provide a record of where the point has been, we ought to be able to figure out how fast it was moving at any particular time along the way. Try the end of the second minute: what was the speedometer reading as the point moved past $p(2) = 10$? To answer this, we first need a closer look at what "velocity" means.

Velocity is described here in terms like "3 inches per minute" instead of the usual "miles per hour." The usual formula linking distance, time and velocity is

$$(\text{velocity})\cdot(\text{time}) = (\text{distance})$$

or, equivalently,

$$(\text{velocity}) = \frac{(\text{distance})}{(\text{time})} ,$$

but these formulas only work if the velocity is constant.

For example, if the point moves along and keeps his speedometer constantly at "3 inches per minute" for 2 minutes, the point will move a total of 6 inches — the formula works just fine:

$$(\text{velocity})\cdot(\text{time}) = (\text{distance})$$

$$(3\text{ in./min.})\cdot(2\text{ min.}) = (6\text{ in.})\ .$$

But if we suppose that the point speeds up and slows down and even reverses direction as he moves along, the formula doesn't work any more: writing

$$(\text{VELOCITY}) \cdot (\text{TIME}) = (\text{DISTANCE})$$

$$\left(\begin{array}{l}\text{SOMETIMES } 3''/\text{MIN.}\\ \text{BUT OTHER TIMES } 2''/\text{MIN.}\\ \text{AND OCCASIONALLY EVEN}\\ -1''/\text{MIN. AND OTHER}\\ \text{VELOCITIES INBETWEEN.}\end{array}\right)\cdot(\text{TIME}) = ?$$

doesn't make much sense.

On the other hand, if we know the distance that the point moves and the time it takes to move, it does make sense to define "average velocity" as

$$(\text{AVERAGE VELOCITY}) = \frac{(\text{DISTANCE})}{\left(\begin{array}{l}\text{TIME IT TAKES TO}\\ \text{TRAVEL THE DISTANCE}\end{array}\right)}$$

For example, if we know that the point traveled 5 inches in, say, 2 minutes, the average velocity of the point during those 2 minutes is

$$\left(\begin{array}{l}\text{AVERAGE}\\ \text{VELOCITY}\end{array}\right) = \frac{(\text{DISTANCE})}{\left(\begin{array}{c}\text{TIME IT TAKES}\\ \text{TO TRAVEL THE}\\ \text{DISTANCE}\end{array}\right)} = \frac{5\text{ IN.}}{2\text{ MIN.}} = \tfrac{5}{2}\text{ IN./MIN.}$$

even though the point may not keep his velocity constant as he moves.

This idea of "average velocity" turns out to be the key to figuring out what the speedometer will read at any particular time.

So, back to the problem of trying to figure out what the velocity of the point should be at the end of the second minute as it moved past p(2)=10 in. From minute 2 to minute 3 , the point covers distance

$$p(3) - p(2) = 13 - 10 = 3\text{ in.}$$

So the average velocity during that minute is

$$\left(\begin{array}{l}\text{AVERAGE VELOCITY}\\ \text{DURING TIME FROM}\\ \text{2 MIN. TO 3 MIN.}\end{array}\right) = \frac{(\text{DISTANCE})}{\left(\begin{array}{c}\text{TIME TO TRAVEL}\\ \text{THE DISTANCE}\end{array}\right)} = \frac{P(3)-P(2)}{1\text{ MIN.}} = 3\text{ IN./MIN.}$$

But it looks like the point moved faster during the $\frac{1}{2}$-minute from minute 2 to minute $2\frac{1}{2}$ (covering 2 inches) than it did during the $\frac{1}{2}$-minute from minute $2\frac{1}{2}$ to minute 3 (covering only 1 inch).

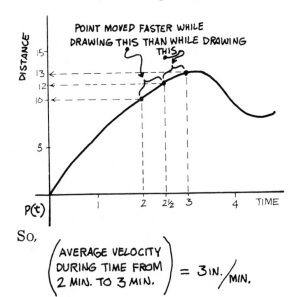

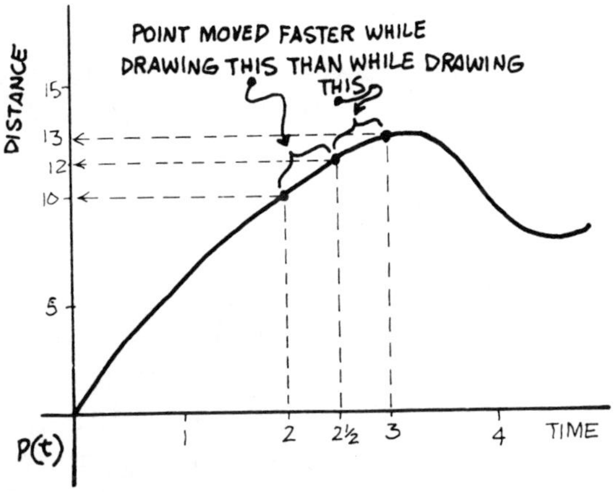

So,

$$\left(\begin{array}{l}\text{AVERAGE VELOCITY}\\ \text{DURING TIME FROM}\\ \text{2 MIN. TO 3 MIN.}\end{array}\right) = 3\text{ IN./MIN.}$$

won't actually be what the point's speedometer reads as passes p(2) = 10 in. on the wire.

It looks like a better guess for the speedometer reading at time "2 minutes" could be found by figuring out the point's average velocity during the time from 2 minutes to $2\frac{1}{2}$ minutes. The point moves from

$$p(2)=10 \text{ in. to } p(2+\tfrac{1}{2})=12 \text{ in.},$$

so the distance covered is

$$p(2+\tfrac{1}{2}) - p(2) = 12 - 10 = 2 \text{ in.}$$

and the time used to cover the distance is $\frac{1}{2}$ minute. The formula for average velocity then gives

$$\left(\begin{array}{l}\text{AVERAGE VELOCITY DURING}\\ \text{TIME FROM 2 MIN. TO } 2\tfrac{1}{2} \text{ MIN.}\end{array}\right) = \frac{\text{DISTANCE}}{\left(\begin{array}{c}\text{TIME TO TRAVEL}\\ \text{THE DISTANCE}\end{array}\right)}$$

$$= \frac{P(2+\frac{1}{2})-P(2)}{\frac{1}{2}} = \frac{2}{\frac{1}{2}} = 4 \text{ IN./MIN.}$$

and this should be closer to what the speedometer should report at 2 minutes.

In fact, for any small amount of time — call it "h" — the formula gives

$$\left(\begin{array}{l}\text{AVERAGE VELOCITY DURING}\\ \text{TIME FROM 2 MIN. TO } (2+h) \text{ MIN.}\end{array}\right) = \frac{\text{DISTANCE}}{\left(\begin{array}{c}\text{TIME TO TRAVEL}\\ \text{THE DISTANCE}\end{array}\right)}$$

$$= \frac{P(2+h)-P(2)}{h} \text{ in./min.}$$

It's pretty clear that if the "average velocity" gets closer and closer to some limit number as the small amount of time "h" gets smaller and smaller, then this limit number should be what scientists call the "instantaneous velocity" of the point at minute 2 as the point moves past $p(2)=10$ inches on the wire. So we define "instantaneous velocity" by

$$\left(\begin{array}{l}\text{INSTANTANEOUS VELOCITY}\\ \text{AT TIME 2 MINUTES}\end{array}\right) = \lim_{h\to 0} \frac{P(2+h)-P(2)}{h}.$$

But

$$\lim_{h\to 0} \frac{p(2+h)-p(2)}{h}$$

is also the derivative $Dp|_2$ of function p( ) at time 2 minutes! So the slope $Dp|_2$ of the tangent line to the graph of p( ) at point $(2, p(2))=(2,10)$ is also the instantaneous velocity at time 2 min.; the derivative $Dp|_2$ is precisely the number that the point's speedometer is supposed to report as it moves up past $p(2)=10$ in. on the wire at the end of the second minute.

All this theory sums up as follows:

If the function $p(t)$ gives the position of a point at any time "t," then the velocity of the point at any particular time "t" is

$$Dp|_t \; .$$

A word about notation: When a function reports something that depends on time "t" instead of distance "x" as $p(t)$ does here, scientists often use notation "$\dot{p}(t)$" (translated "p-dot") instead of $Dp|_t$.

The theory above doesn't result in being able to compute anything unless we actually have a formula for $p(t)$. Fortunately, however, the laws of physics do often result in motion being described by a familiar function and once the motion of a point is described by an actual formula, computing velocities using derivatives turns out to be pretty useful.

For example, suppose the motion of a point is described by the formula $p(t) = \frac{1}{3}t^3 - 4t^2 + 15t$; $0 \le t \le 6$
= location of point in inches at time t min. between 0 min. and 6 min.

We can actually compute the answers to questions like these:

1. How fast was the point moving at the end of the fourth minute?

2. When did the point reverse direction?

3. What was the maximum distance from "start" that the point reached during the 6 minutes?

4. When was the point traveling most rapidly backward ?

Here are some answers to these questions. First, draw the graph of p(t) and compute the velocity:

$$Dp|_t = \dot{p}(t) = t^2-8t+15 \text{ in./min.}$$

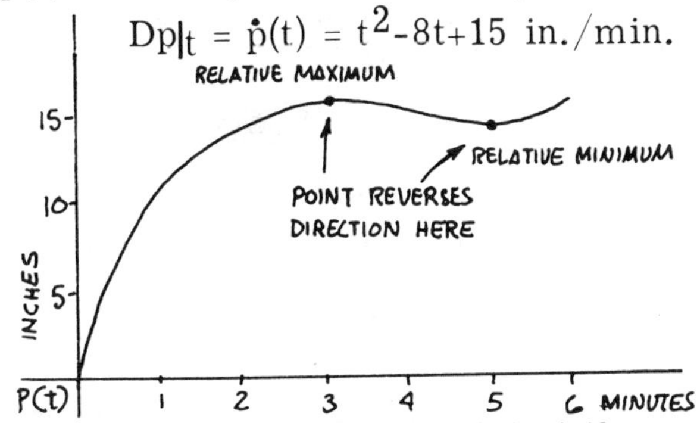

1. The velocity of the point at the end of the 4th minute will be

$Dp|_4 = \dot{p}(4) = (4)^2 - 8(4) + 15 = -1$ in./min.

Since the velocity is <u>negative</u>, the point will be moving backward.

2. A look at the graph of p(t) shows that when the point reverses direction the graph will have a relative maximum or minimum at these times, so the derivative $\dot{p}(t)$ must be zero at these times. This seems reasonable, since the point's velocity is also $\dot{p}(t)$, and the velocity must be zero momentarily to be able to reverse direction smoothly.

So, to find the relative maximum or minimum, we find out the times that $\dot{p}(t)=0$:

$$\dot{p}(t) = t^2-8t+15 = (t-5)(t-3) \quad .$$

So $\dot{p}(t)=0$ only when t=3 min. or t=5 min., and these are exactly the times that the point will reverse direction.

3. It looks like the point reached its maximum distance from start when it first reversed direction at t=3 min., reaching point

$$p(3) = \tfrac{1}{3}(3)^3 - 4(3)^2 + 15(3) = 18 \text{ in.}$$

But to be sure that this is the right answer we should compare this distance with the distance reached by the point at the end of the 6 min., after it had reversed direction again and started forward once more:

$$p(6) = \tfrac{1}{3}(6)^3 - 4(6)^2 + 15(6) = 18 \text{ in.},$$

the same distance! Either way the answer is 18 in.

4. To find out when the point was travelling most rapidly backward, have a look at the graph of the <u>velocity</u> $Dp|_t = \dot{p}(t)$:

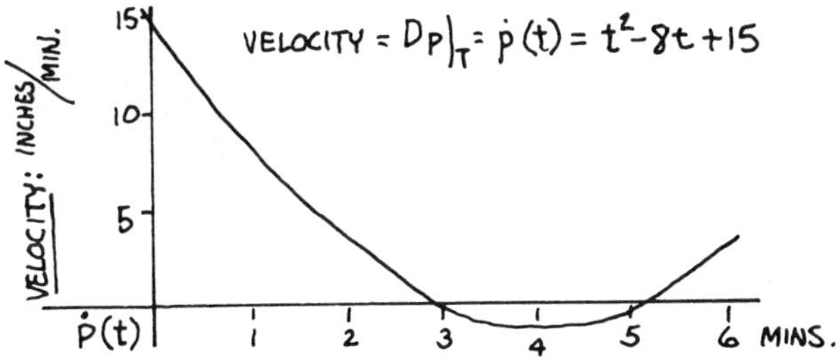

Now backward speed corresponds to negative velocity and, from the looks of the graph of $\dot{p}(t)$, it's pretty clear that the velocity was most <u>negative</u> at the bottom of the graph of $\dot{p}(t)$, where a tangent line will have slope zero. To find the time when this happens, compute the <u>derivative of $\dot{p}(t)$</u> this time, set it equal to zero and solve for t:

$$D\dot{p}(\ )|_t = 2t-8 = 0 \text{ when } t=4 \text{ min.}$$

So the greatest backward speed happens at 4 min., and, from the answer to the first question, the velocity at t=4 is

$$\dot{p}(4) = -1 \text{ in./min.}$$

EXERCISES

III.4.1 If the motion of a point is described by the formula

$p(t) = t^3 - 6t^2 + 9t$ inches; $0 \leq t \leq 5$ (min.)

1. How fast was the point moving at the end of the fourth minute?

2. When did the point reverse direction?

3. What was the maximum distance from "start" that the point reached during the five minutes?

4. When was the point traveling most rapidly backward?

Any object tossed straight up (or down) obeys the formula for position

$p(t) = s + vt - 16t^2$ feet ,

where t = seconds;
s = starting height
and v = starting velocity (positive if directed upward).

III.4.2 A coin is dropped in "Ye Olde Wishinge Welle." Three seconds later a "plink" is heard as it hits the bottom. How deep is the well? How fast was the coin going when it hit bottom?

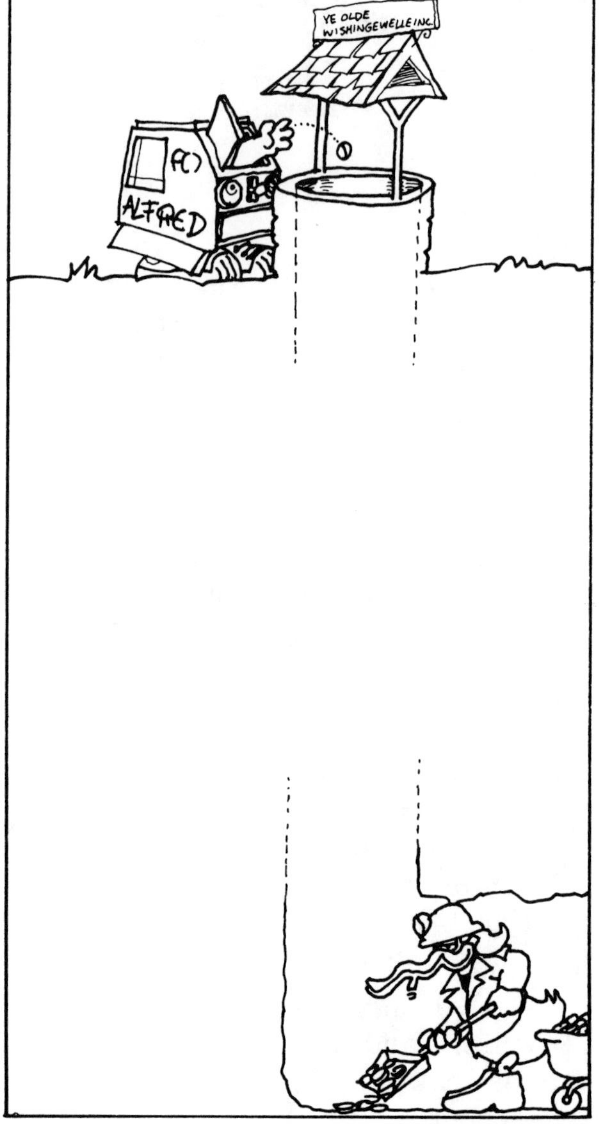

III.4.3 An avid birdwatcher is on the top of a cliff 48 feet above the water. She tosses a chunk of bread up to a seagull. The gull misses and the bread falls back down to the water. If the initial velocity of the bread is 32 ft./sec. ,

1. How many seconds before the bread hits the water?

2. How high does the bread get before it falls in the water?

3. When does the bread pass its starting point on the way back down? Compare the velocity at this time with the starting velocity.

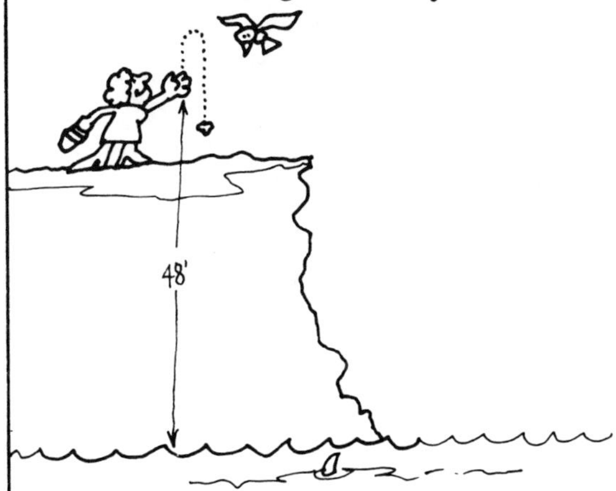

III.4.4 El Zippo, the Human Cannonball, is to be projected straight up to go through a flaming hoop at the top of his flight. He then falls back down to end his flight by diving into a tub of water. If the center of the hoop is 25 feet above the cannon's mouth, what must the cannon's muzzle velocity be so that Zippo will go through the hoop? If the tub is 49 feet below the hoop, what will Zippo's velocity be when he hits the water?

EXTRA EXERCISES & EXCURSIONS

Extra exercises are given for each section in chapter 3, and they are numbered, for example, like EIII.1.1 to indicate that they belong to section III.1. The excursions for section III.2 need the results discovered in the excursions extending section II.10. Selected answers and solutions can be found at the end of the book.

For Section III.1:

Exercise EIII.1.1.
Prove that $f(x) = 7-3x-\frac{1}{2}x^2$ is differentiable at $x = -2$. Find the derivative $Df|_{-2}$.

Exercise EIII.1.2.
Prove that $f(x) = 4-6x^2+3x^3$ is differentiable at $x = -1$. Find the derivative $Df|_{-1}$.

Exercise EIII.1.3.
Prove that $f(x) = x^2-4x+5$ is differentiable at $x = 2$. Find the derivative $Df|_2$.

Exercise EIII.1.4.
Prove that $f(x) = -2x^3-12x^2-24x+5$ is differentiable at $x = -2$. Find the derivative $Df|_{-2}$.

Exercise EIII.1.5.
For any constants m and b, prove that $f(x) = mx+b$ is differentiable at $x=a$ for any a and show that $Df|_a=m$. This is to be expected, for $y = mx+b$ is the equation for a straight line with slope m; the line is its own tangent line.

For Section III.2:

Exercise EIII.2.1.
Compute $D(5-3x^2+2x)|_a$. Find a formula for the line tangent to the graph of $f(x) = 5-3x^2+2x$ at point $(-2,f(-2))$.

Exercise EIII.2.2.
If $f(x) = 2x-3+\frac{1}{2}x^2$, find the point where the line tangent to the graph of $f(x)$ at $(2,f(2))$ crosses the x-axis.

Exercise EIII.2.3.

If $f(x) = x^3-3x+1$, sketch its graph. For any point $(x_1,f(x_1))$ on the graph, is there a different point $(x_2,f(x_2))$ on the graph where the slope of the tangent line is the same as the slope of the tangent line at $(x_1,f(x_1))$?

If this is the case, given $(x_1,f(x_1))$, find $(x_2,f(x_2))$.

EXCURSION!

Here we discover rules for finding the derivative of the product and quotient of functions. To prove that the rules are correct, we will need another theorem.

THEOREM III-3. If f(x) is differentiable at x=a, then f(x) is continuous at x=a.

Proof: We must show that

$$\lim_{x\to a} f(x) = f(a).$$

We are assuming that the three-step ritual for finding a derivative can be done for f(x) at point a. So there is some number $Df|_a$ such that $\lim_{h\to 0} \frac{f(a+h)-f(a)}{h}$ exists and equals $Df|_a$.

We also have $\lim_{h\to 0} h = 0$. By the product limit theorem II-4, function

$$\frac{f(a+h)-f(a)}{h} \cdot h = f(a+h)-f(a)$$

will inherit limit $Df|_a \cdot 0 = 0$ from these two limits. So we are sure that $\lim_{h\to 0} f(a+h)-f(a)$ exists and equals 0.

This means that for any $\epsilon>0$, there will be some $\delta(\epsilon)>0$ such that

if $0<|h|<\delta(\epsilon)$ then $|f(a+h)-f(a)| = |f(a+h)-f(a)-0| < \epsilon$.

On the other hand, to show that $\lim_{x\to a} f(x) = f(a)$, for any $\epsilon>0$, we need to find some $\delta^f(\epsilon)>0$ with the guarantee that

if $0<|x-a|<\delta^f(\epsilon)$,
then $|f(x)-f(a)| < \epsilon$.

This is very close to what we already have: we can use the same $\delta(\epsilon)$. Just replace h by (x-a). Then the guarantee on $\delta(\epsilon)$ says that if $0<|x-a|<\delta(\epsilon)$, then

$$|f(a+(x-a)) - f(a)| < \epsilon.$$

But $a+(x-a) = x$, so $|f(a+(x-a))-f(a)| = |f(x)-f(a)|$; we are done.

Next, we show that if two functions have derivatives at some point, then the product of the two functions will inherit the property of being differentiable. The rule for finding the number that is the derivative of the product of the functions is more complicated than the other two rules.

THEOREM III-4.
(The product derivative theorem.)

If f(x) and g(x) are differentiable at x=a, then f(x)·g(x) is differentiable at x=a and

$$D(f\cdot g)|_a = f(a)Dg|_a + g(a)Df|_a.$$

Proof: We are assuming that the three step ritual can be done for both f(x) and g(x) at x=a. So

$$\lim_{h\to 0} \frac{f(a+h)-f(a)}{h} \text{ exists}$$

and equals $Df|_a$

$$\text{and } \lim_{h\to 0} \frac{g(a+h)-g(a)}{h} \text{ exists}$$

and equals $Dg|_a$.

Construct the tangent hunting function for f(x)·g(x): Δ(h) for f(x)·g(x) is

$$\Delta(h) = \frac{f(a+h)g(a+h)-f(a)g(a)}{h} .$$

If we subtract and add f(a+h)·g(a) and do a little algebra, we can write this as

$$\frac{f(a+h)g(a+h)-f(a)g(a)}{h} =$$

$$\frac{f(a+h)g(a+h)-f(a+h)g(a)+f(a+h)g(a)-f(a)g(a)}{h}$$

$$= \frac{f(a+h)[g(a+h)-g(a)] + [f(a+h)-f(a)]g(a)}{h}$$

$$= f(a+h)\frac{g(a+h)-g(a)}{h} + g(a)\frac{f(a+h)-f(a)}{h} .$$

Now we use the limit theorems:

$$\lim_{h\to 0} \frac{f(a+h)-f(a)}{h} \text{ exists}$$

and equals $Df|_a$ and g(a) is a constant, so the 'Kf' limit theorem II-2 can be used to show that

$$\lim_{h\to o} g(a)\frac{f(a+h)-f(a)}{h} \text{ exists}$$

and equals $g(a)Df|_a$.

$\lim_{h\to 0} \frac{g(a+h)-g(a)}{h}$ exists and equals $Dg|_a$ and $\lim_{h\to 0} f(a+h)$ exists and equals f(a) by theorem III-3, so the product limit theorem II-4 tells us that

$$\lim_{h\to 0} f(a+h)\frac{g(a+h)-g(a)}{h} \text{ exists}$$

and equals $f(a)Dg|_a$.

Finally, the 'f+g' limit theorem II-3 can be used to prove that

$$\lim_{h\to o} f(a+h)\frac{g(a+h)-g(a)}{h} + g(a)\frac{f(a+h)-f(a)}{h}$$

exists and equals

$$f(a)Dg|_a + g(a)Df|_a.$$

We are done, for we have shown that the tangent-hunting function Δ(h) for f(x)·g(x) at x=a has inherited the limit

$$f(a)Dg|_a + g(a)Df|_a$$

as h→0; this will be the derivative of f(x)·g(x) at x=a;

$$D(f\cdot g)|_a = f(a)Dg|_a + g(a)Df|_a.$$

Exercise EIII.2.4 Sample.

If f(x) = 3x+2 and g(x) = 4-5x, implement the methods of theorem III-4 to find the derivative of (3x+2)(4-5x) at x=2 and prove that your answer is correct. Check that your answer is the same as that given by the rule for finding the number that is the derivative of the product of two functions.

Solution. First we construct the tangent hunting function Δ(h) for the product f·g at x=2:

$$\Delta(h) = \frac{(3(2+h)+2)(4-5(2+h)) - (6+2)(4-10)}{h}$$

$$= \frac{f(2+h)g(2+h)-f(2)g(2)}{h}$$

$$= f(2+h)\frac{g(2+h)-g(2)}{h} + g(2)\frac{f(2+h)-f(2)}{h}$$

$$= (3(2+h)+2)\frac{(4-5(2+h))-(-6)}{h}$$

$$+ (-6)\frac{(3(2+h)+2)-8}{h}$$

$$= (8+3h)\frac{-5h}{h} + (-6)\frac{3h}{h}$$

$= (8+3h)(-5) + (-6)(3) = -58 - 15h$ if $h \neq 0$.

Next we must prove that $\lim_{h\to 0} \Delta(h) = \lim_{h\to 0} (-58-15h)$ exists and equals (guess:) -58.

From II-4, a good δ for $\Delta(h)$ should be $\delta(\epsilon) = \epsilon/15$. We check that it works: If $0<|h|<\epsilon/15$, then

$$|\Delta(h)-(-58)|=|-58-15h - (-58)|$$

$= |-15h| = 15|h| < 15(\epsilon/15) = \epsilon$; the proof is complete.

We have proved that the derivative of $(3x+2)(4-5x)$ at $x=2$ is (-58). The number that is the derivative should be the same as that provided by the product formula for finding derivatives:

$$D(f\cdot g)|_a = f(a)Dg|_a + g(a)Df|_a.$$

Here, $a=2$; $f(2)=8$; $g(2)=-6$;

$Dg|_2= D(4-5x)|_2=-5$;

$Df|_2=D(3x+2)|_2=3$.

We should have

$-58 = 8\cdot(-5) + (-6)\cdot 3 = -40-18$; we have verified that the formula produces the correct answer in this case.

Exercise EIII.2.5

If $f(x) = 5x-6$ and $g(x) = (x/2)-3$, implement the methods of theorem III-4 to find the derivative of $(5x-6)((x/2)-3)$ at $x=2$ and prove that your answer is correct. Check that your answer is the same as that given by the rule for finding the number that is the derivative of the product of two functions.

Exercise EIII.2.6

If $f(x) = 3x+2$ and $g(x) = 4-5x$, implement the methods of theorem III-4 to find the derivative of $(3x+2)(4-5x)$ at $x=-1$ and prove that your answer is correct. Check that your answer is the same as that given by the rule for finding the number that is the derivative of the product of two functions.

EXCURSION!

To find the derivative of the quotient of two functions, we first discover how to find the derivative of function $\frac{1}{g(x)}$ if we know the derivative of $g(x)$.

THEOREM III-5. If g(x) is differentiable at x=a and g(a)≠0, then $\frac{1}{g(x)}$ is differentiable at x=a and

$$D\left(\frac{1}{g(x)}\right)\Big|_a = \frac{-Dg|_a}{g(a)^2}$$

Proof: We are assuming that the three step ritual can be done for g(x) at x=a. So

$$\lim_{h\to 0} \frac{g(a+h)-g(a)}{h} \text{ exists}$$

and equals $Dg|_a$

Construct the tangent hunting function for $\frac{1}{g(x)}$.

$\Delta(h)$ for $\frac{1}{g(x)}$ is $\dfrac{\frac{1}{g(a+h)} - \frac{1}{g(a)}}{h}$.

We do a little algebra on

$$\frac{1}{g(a+h)} - \frac{1}{g(a)}:$$

$$\frac{1}{g(a+h)} - \frac{1}{g(a)} = \frac{g(a)}{g(a+h)g(a)} - \frac{g(a+h)}{g(a)g(a+h)}$$

$$= \frac{g(a)-g(a+h)}{g(a+h)g(a)} = \frac{-[g(a+h)-g(a)]}{g(a+h)g(a)}.$$

We have to divide this by h to get the tangent hunting function:

$$\Delta(h) = \frac{1}{h} \cdot \frac{-[g(a+h)-g(a)]}{g(a+h)g(a)} =$$

$$-\left[\frac{g(a+h)-g(a)}{h}\right] \cdot \frac{1}{g(a+h)g(a)}.$$

Now we can use the limit theorems:

g(a) is a constant. Since $Dg|_a$ exists, theorem III-3 tells us that $\lim_{h\to o} g(a+h)$ exists and equals g(a). So $\lim_{h\to o} g(a+h)\cdot g(a)$ exists and equals $g(a)\cdot g(a)$ by the 'Kf' limit theorem II-2.

Since g(a)≠0 by assumption, the quotient limit theorem II-5 tells us that $\frac{1}{g(a+h)g(a)}$ inherits the limit $\frac{1}{g(a)g(a)}$ as $h\to 0$.

We are assuming that

$$\lim_{h\to 0} \frac{g(a+h)-g(a)}{h} \text{ exists}$$

and equals $Dg|_a$. So the product limit theorem II-4 tells us that

$$-\left[\frac{g(a+h)-g(a)}{h}\right] \cdot \frac{1}{g(a+h)g(a)}$$

inherits the limit

$$-Dg|_a \cdot \frac{1}{g(a)g(a)} = \frac{-Dg|_a}{g(a)^2}$$

as $h\to 0$. We have proved that the limit of the tangent-hunting function $\Delta(h)$ for function 1/g(x) exists and equals $\frac{-Dg|_a}{g(a)^2}$; the proof is complete.

> Exercise EIII.2.7 Sample.
> If f(x) = 3x+2, implement the methods of theorem III-5 to find the derivative of $\frac{1}{3x+2}$ at x=2 and prove that your answer is correct. Use any theorems you need. Check that your answer is the same as that given by the rule for finding the number that is the derivative of $\frac{1}{3x+2}$ in theorem III-5.

Solution: First we construct the tangent hunting function $\Delta(h)$ for the $1/f(x)$ at $x=2$:

$$\frac{\frac{1}{f(2+h)} - \frac{1}{f(2)}}{h} = \frac{\frac{1}{3(2+h)+2} - \frac{1}{8}}{h}$$

$$= \frac{1}{h}\left(\frac{8}{(8+3h)8} - \frac{8+3h}{8(8+3h)}\right)$$

$$= \frac{1}{h}\left(\frac{8 - (8+3h)}{(8+3h)8}\right) = \frac{1}{h}\left(\frac{-3h}{(8+3h)8}\right)$$

$$= \frac{-3}{(8+3h)8} \text{ if } h \neq 0.$$

Since $\lim_{h\to 0} h = 0$ and $\lim_{h\to 0} 8 = 8$, the 'Kf' limit theorem II-2 (used twice) and the 'f+g' limit theorem II-3 tell us that $\lim_{h\to 0} (8+3h)8$ exists and equals 64.

Then theorem II-5 assures us that $\lim_{h\to 0} \Delta(h) = \lim_{h\to 0} \frac{-3}{(8+3h)8}$ exists and equals -3/64. So we should have $D\left(\frac{1}{3x+2}\right)\Big|_2 = -3/64$.

Let's check to see if this agrees with the answer given by theorem III-5. $f(2) = 8$; $Df|_2 = D(3x+2)|_2 = 3$. Theorem III-5 tells us that

$$D\left(\frac{1}{f(x)}\right)\Big|_2 = \frac{Df|_2}{f(2)^2}.$$

So the answer should be $\frac{-3}{8^2} = -3/64$. We have verified the theorem in this case.

Exercise EIII.2.8

If $f(x) = 4-5x$, implement the methods of theorem III-5 to find the derivative of $\frac{1}{4-5x}$ at $x=2$ and prove that your answer is correct. Use any theorems you need. Check that your answer is the same as that given by the rule for finding the number that is the derivative of $\frac{1}{4-5x}$ in theorem III-5.

Exercise EIII.2.9

If $f(x) = 4-5x$, implement the methods of theorem III-5 to find the derivative of $\frac{1}{4-5x}$ at $x=-1$ and prove that your answer is correct. Use any theorems you need. Check that your answer is the same as that given by the rule for finding the number that is the derivative of $\frac{1}{4-5x}$ in theorem III-5.

We can use theorem III-5 and theorem III-4 about computing the derivative of the product of two functions to find the derivative of the quotient of two functions. Here is the theorem:

THEOREM III-6.
(The quotient derivative theorem.)

If f(x) and g(x) are differentiable at x=a, and $g(a) \neq 0$, then $\frac{f(x)}{g(x)}$ is differentiable at x=a, and

$$D\left(\frac{f}{g}\right)\bigg|_a = \frac{g(a)\cdot Df|_a - f(a)\cdot Dg|_a}{g(a)^2}.$$

Proof: We are assuming that $Df|_a$ exists and, from the previous theorem III-5, we know that $D(\frac{1}{g(x)})|_a$ exists and equals $\frac{-Dg|_a}{g(a)^2}$.

If we write $\frac{f(x)}{g(x)} = f(x)\cdot\frac{1}{g(x)}$, we can use the product derivative theorem III-4 to prove that $D\frac{f(x)}{g(x)}\big|_a = D\left(f(x)\cdot\frac{1}{g(x)}\right)\big|_a$ exists and equals $f(a)D(\frac{1}{g(x)})|_a + \frac{1}{g(a)}Df|_a$

$$= f(a)\frac{-Dg|_a}{g(a)^2} + \frac{1}{g(a)}Df|_a$$

$$= \frac{-f(a)Dg|_a}{g(a)^2} + \frac{g(a)Df|_a}{g(a)^2}$$

$$= \frac{g(a)\cdot Df|_a - f(a)\cdot Dg|_a}{g(a)^2}.$$

*FOR USE IN BETTER MATH CLASSES EVERYWHERE

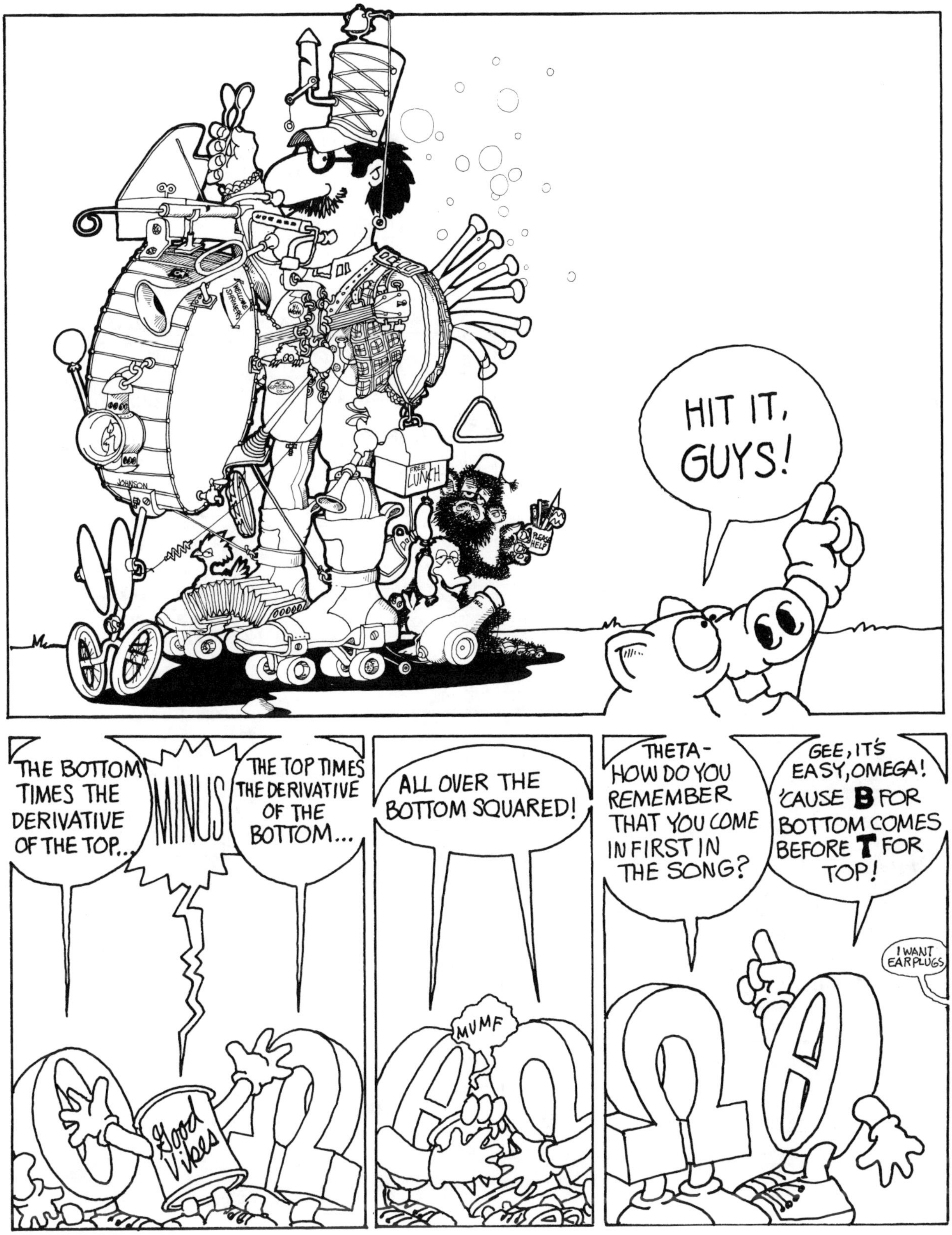
HIT IT, GUYS!
FREE LUNCH
JOHNSON
THE BOTTOM TIMES THE DERIVATIVE OF THE TOP...
MINUS
THE TOP TIMES THE DERIVATIVE OF THE BOTTOM...
Good Vibes
ALL OVER THE BOTTOM SQUARED!
MUMF
THETA- HOW DO YOU REMEMBER THAT YOU COME IN FIRST IN THE SONG?
GEE, IT'S EASY, OMEGA! 'CAUSE B FOR BOTTOM COMES BEFORE T FOR TOP!
I WANT EARPLUGS

When 'a' is any x, we shorten the notation $Df|_x$ to be just Df. Other symbols for Df are f′(x) or $\frac{d}{dx}f$ or $\frac{df}{dx}$.

To do the exercises coming up recall that we have shown that DK=0, where K is any constant; Dx=1; $Dx^2=2x$ and $Dx^3=3x^2$.

Exercise EIII.2.10. Sample.
Compute:

(a) Dx^4.

(b) Dx^{-4}.

(c) If $f(x) = \frac{x^2+5x}{(x^2+1)(x^3-3x+4)}$, compute f′(2).

(d) If $g(x) = \frac{\frac{1}{5x-2}+7}{2x^2+3}$, find $\frac{dg}{dx}$.

Solution:

(a) $x^4=x\cdot x^3$, so we can use the product rule: $Dx^4 = D(x\cdot x^3) = x\cdot Dx^3+x^3Dx = x(3x^2)+x^3\cdot 1 = 3x^3+x^3 = 4x^3$.

(b) $Dx^{-4} = D\frac{1}{x^4} = \frac{-Dx^4}{(x^4)^2} = \frac{-4x^3}{(x^4)^2} = -4x^3\cdot x^{-8} = -4x^{-5}$.

(c) Use both the product and the quotient rule on this one.

$$f'(x) = D\frac{x^2+5x}{(x^2+1)(x^3-3x+4)} =$$

$$\frac{(x^2+1)(x^3-3x+4)D(x^2+5x)-(x^2+5x)D\big((x^2+1)(x^3-3x+4)\big)}{((x^2+1)(x^3-3x+4))^2}$$

$$= \frac{(x^2+1)(x^3-3x+4)(2x+5)}{((x^2+1)(x^3-3x+4))^2} - \frac{(x^2+5x)\big((x^2+1)D(x^3-3x+4)+(x^3-3x+4)D(x^2+1)\big)}{((x^2+1)(x^3-3x+4))^2}$$

$$= \frac{(x^2+1)(x^3-3x+4)(2x+5)}{((x^2+1)(x^3-3x+4))^2} - \frac{(x^2+5x)\big((x^2+1)(3x^2-3)+(x^3-3x+4)(2x)\big)}{((x^2+1)(x^3-3x+4))^2}.$$

To find $f'(2) = Df|_2$, just substitute 2 for x in the mess above to get f′(2) =

$$\frac{(2^2+1)(2^3-3(2)+4)(2(2)+5)}{((2^2+1)(2^3-3(2)+4))^2} - \frac{(2^2+5(2))\big((2^2+1)(3(2)^2-3)+(2^3-3(2)+4)(2(2))\big)}{((2^2+1)(2^3-3(2)+4))^2}$$

$$= \frac{270}{900} - \frac{(14)(69)}{900}.$$

(d) $$\frac{dg}{dx} = D\left(\frac{\frac{1}{5x-2}+7}{2x^2+3}\right) =$$

$$\frac{(2x^2+3)\,D\left(\frac{1}{5x-2}+7\right)-\left(\frac{1}{5x-2}+7\right)D(2x^2+3)}{(2x^2+3)^2}$$

$$= \frac{(2x^2+3)\left(\frac{-5}{(5x-2)^2}\right) - \left(\frac{1}{5x-2}+7\right)(4x)}{(2x^2+3)^2}.$$

Exercise EIII.2.11.

(a) Compute $D(x^2 \cdot x^3)$.

(b) Compute $D(x \cdot x^4)$. The answers to (a) and (b) should be the same; both will be Dx^5.

(c) Compute Dx^9.

(d) Compute $Dx^{-9} = D\left(\frac{1}{x^9}\right)$.

(e) If $f(x) = (x^2+7x-6)(x^9+3x^5-2x)$, compute $f'(-1)$.

(f) Compute $D\,\frac{x^9+7x^5-2x}{x^2+3x-6}$.

(g) Compute $\frac{d}{dx}\left(\frac{(7x-x^2+2)(3-8x^2+6x^5)}{x^5-3x^3+5}\right)$

(h) Compute $\frac{d}{dx}\left(\frac{3x-1}{(x^2-5x+4)^2}\right)$.

For Section III.3.

Exercise EIII.3.1. (For those who know theorem III-5.) For any positive integer k, compute Dx^{-k}. How does your answer compare with the rule for computing Dx^n in section III.3?

Exercise EIII.3.2. Find any relative maximum or minimum for $f(x)=2x^3-3x^2-12x+5$. Which is which?

EXCURSION!

MORE on MAXIMA and MINIMA.

Here we have a closer look at how to use calculus to solve problems that require that we find how to do things so that we get, for example, the maximum value or the minimum cost. Two examples of such problems are the corral fencing problem in III.3.5 and the box problem in III.3.6.

The following steps usually lead to a solution of such problems.

First, we draw a picture if we can.

Next, we construct a function (which we will call the objective function) that reports what we want to maximize or minimize. In the problems cited above, we wanted to maximize the area of a rectangular corral; we wanted to maximize the volume of the box.

Although we may have to invent it, there will be some formula for the objective function. Such a formula will be expressed in terms of one or more variables.

For example, the area of a rectangle = (length)·(width); the variables here are length and width.

The volume of a box is (area of the base)·height = [(length)·(width) of the base]·height.

Choose symbols for the variables and label the picture.

For the box problem, we have a square of cardboard 12 in. on a side:

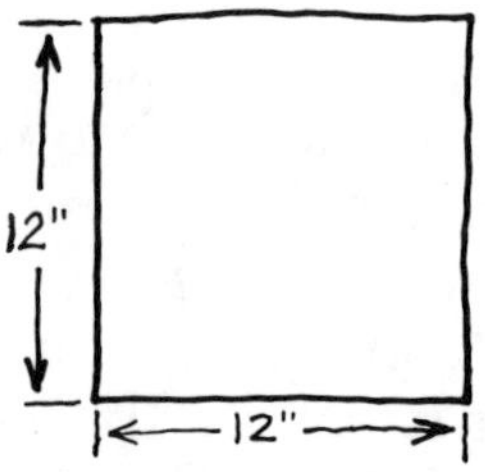

We want to cut square pieces from the corners and make a box with the largest possible volume.

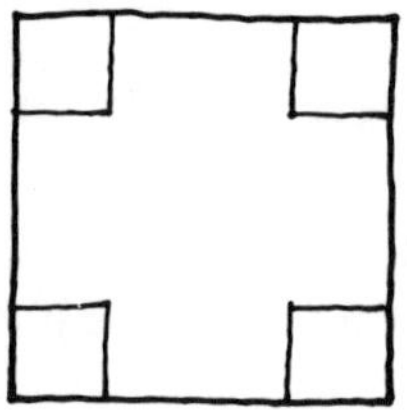

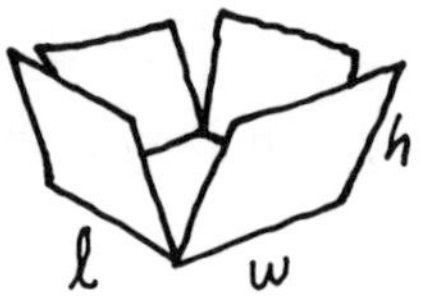

We let ℓ and w stand for the length and width of the base of the box and denote the height as h.

The formula for the objective function, the volume of the box, is $V = \ell \cdot w \cdot h$.

Now we figure out how we would have to cut up the cardboard to make the box:

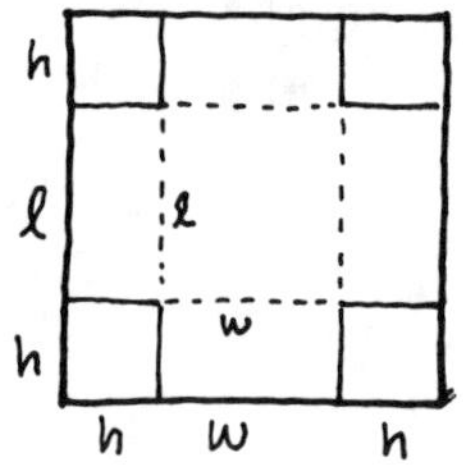

There is usually more than one variable in the formula for the objective function. When this happens, the problem will have constraints that will connect the variables and can be expressed in the form of equations involving the variables.

For the box problem, since the cardboard is a square, we will have $h+\ell+h = h+w+h$, so $\ell=w$.

Furthermore, since a side of the square is 12 in., we will have $12 = h+\ell+h = \ell+2h$.

Finally, we must have

$$\ell>0,\ w>0 \text{ and } h>0$$

for the box to have positive volume, and a look at how the cuts can be made shows us that $\ell=w < 12$ and $h < 6$.

We must use the equations that express the constraints to reduce the number of variables in the objective function to one variable, which we will call the choice variable. (There is usually more than one way to do this; choose the simplest way.)

For the box problem, since $\ell=w$, the objective function is

$$V = \ell \cdot \ell \cdot w = \ell^2 w,$$

which still has two variables. The constraint $12 = \ell+2h$ lets us write $\ell = 12-2h$, so

$$V = (12-2h)^2 h.$$

We have to express the objective function in terms of only one variable so we can use our calculus to find the maximum or minimum of the objective function.

Continuing with the box problem, we now have

$$V = V(h) = (12-2h)^2 h$$

with $0 < h < 6$. We need to find the maximum for $V(h)$.

$$V(h) = (12-2h)^2 h = (144-48h+4h^2)h$$
$$= 144h - 48h^2 + 4h^3.$$

$V(h)$ is a cubic function, and it will have a graph like

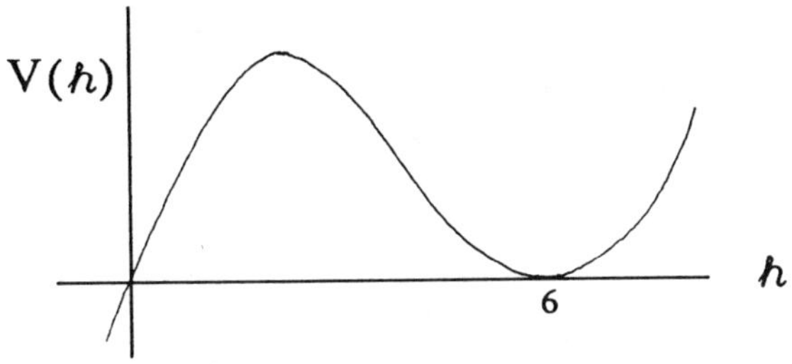

We are only interested in the part of the graph above interval [0,6], since we must have our choice variable h between 0 and 6.

To find any relative maximum for $V(h)$, we compute its derivative and set it equal to zero and see if there are any values for h between 0 and 6 where $DV|_h=0$:

$$DV|_h = 144 - 48 \cdot 2h + 4 \cdot 3h^2 =$$
$$12(12-8h+h^2) = 12(h-6)(h-2) = 0$$

only if $h = 6$ or $h = 2$.

We can't have a maximum when $h=6$, for the volume of a box with $h=6$ would be $V(6) = 0$. The shape of the graph of $V(h)$ tells us that the maximum will occur at the other value of h where

$$DV|_h=0; \text{ at } h=2.$$

So, to make a box out of the cardboard that has the largest possible volume, we must have $h=2$; we must cut squares of side 2 in. out of each corner. Then $w = \ell = 12-2h = 12-2 \cdot 2 = 8$ and the box will have volume

$$V = \ell \cdot w \cdot h = 8 \cdot 8 \cdot 2$$
$$= 128 \text{ cubic inches.}$$

Exercise EIII.3.3. Sample.

A rancher wants to build a rectangular corral that uses all of the side of his barn as part of one side of the corral. The barn is 50′ long. He has 130′ of fence available. What are the dimensions of the largest corral he can build with 130′ of fence?

Solution:

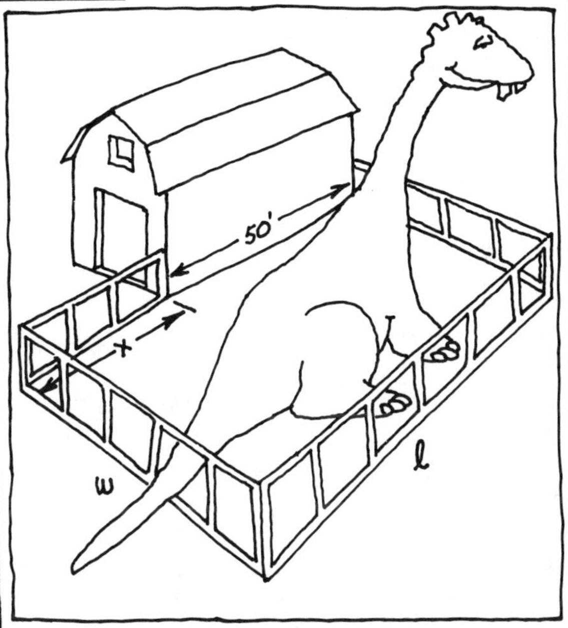

If we denote the length and width of the proposed corral by ℓ and w, our objective function is Area = A = $\ell \cdot w$. We want to maximize the area.

There are two constraints. The fence may be longer than the side of the barn. If x is the length of the fence that may be needed to complete the side of the corral containing the side of the barn, we will have $\ell = 50+x$. We must use all the side of the barn, so we must have $x \geq 0$.

The total amount of fence available is

$$130' = x+2w+\ell = x+2w+(50+x) = 2x+2w+50.$$

If we make x our choice variable, we can use this last constraint equation to solve for w in terms of x: $2w = 130\text{-}50\text{-}2x$ or $w = 40\text{-}x$. We must have $w=40\text{-}x \geq 0$ also, or $40 \geq x$.

Then the area of the corral can be written as

$$A = A(x) = \ell \cdot w = (50+x)\cdot(40\text{-}x) = 2000 - 10x - x^2.$$

$A(x)$ is a parabola with a graph like

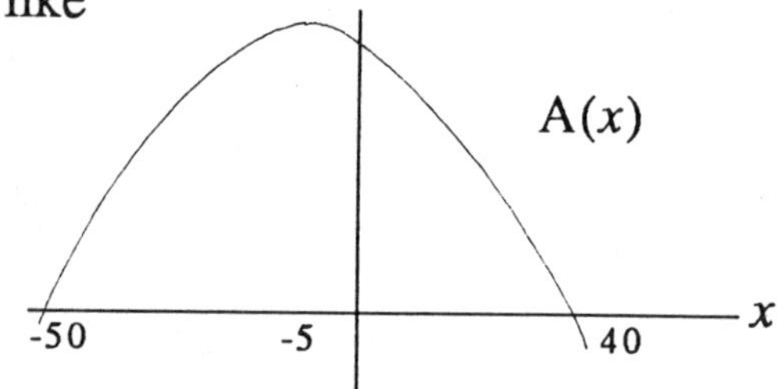

There is a relative maximum for $A(x)$ where $DA|_x = -10-2x = 0$, or $x=-5$. BUT we must have $x \geq 0$ to use all of the side of the barn for the corral. So a look at the graph of $A(x)$ shows us that the maximum for $A(x)$ (with $x \geq 0$) is achieved when $x=0$. Then $\ell = 50\text{-}0 = 50$ and $w = 40\text{-}0 = 40$; these are the dimensions for the largest rectangular corral that uses all of the side of the barn.

In this last example, the maximum for the objective function was achieved at one of the endpoints of the interval [0,40] that is the range for acceptable values for the choice variable.

To be absolutely sure we have the maximum or minimum for the objective function, we must do the following:

(a) Determine the interval containing acceptable values for the choice variable and compute the values for the objective function at the endpoints of this interval.

(b) Find the values of the choice variable where the derivative of the objective function is zero. IF these values are in the interval containing acceptable values for the choice variable, compute the value of the objective function at these points.

(c) If we are seeking the maximum for the objective function, it will be the largest of the values computed in (a) and (b).If we are seeking the minimum for the objective function, it will be the smallest of the values computed in (a) and (b).

Exercise EIII.3.4.

Assuming that the rancher in the previous problem has 210′ of fence available, find the dimensions of the largest corral he can build.

Exercise EIII.3.5.

Find positive numbers x and y satisfying $2x+y = 30$ so that x^2y is a maximim.

Exercise EIII.3.6.

Find numbers x and y satisfying $2x-y = 120$ so that xy is a minimum.

Exercise EIII.3.7.

Prove that the largest rectangular area that can be enclosed by a given amount of fence will be a square.

Exercise EIII.3.8. Sample. (For those who know the quotient derivative theorem III-6, proved in an earlier excursion.)

Suppose we need a cardboard box with an open top and a square base that has a volume of 32 cubic inches. If we wish to use the least amount of cardboard, what should be the dimensions of the box?

Solution: If a side of the base of the box has length ℓ and the height of the box is h,

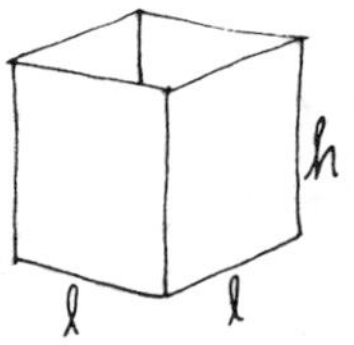

then the objective function, the area of the bottom and sides of the box, will be

$$A = \ell^2 + 4\ell h.$$

The constraint is that the volume be 32 cubic inches; that

$$\ell^2 h = 32.$$

Taking ℓ to be the choice variable, we can solve the constraint equation to get $h = 32/\ell^2$, and express the objective function A as

$$A = A(\ell) = \ell^2 + 4\ell\left(\frac{32}{\ell^2}\right) = \ell^2 + \frac{128}{\ell}$$

with $\ell > 0$. We want to find a minimum for A.

$$DA|_1 = 2\ell + 128\cdot(-1)/\ell^2 = \frac{2\ell^3 - 128}{\ell^2}.$$

$DA|_1 = 0$ only if $2\ell^3 - 128 = 0$ or $\ell^3 = 64$ or $\ell = 4$. Then $h = 32/\ell^2 = 32/16 = 2$. So we have found the dimensions of the open box with volume equal to 32 cubic inches and least surface area: The base will be 4″×4″ and the height will be 2″.

Exercise EIII.3.9. (For those who know the quotient derivative theorem III-6.)

Suppose we need a cardboard box with a top and a square base with a volume of 27 cubic inches. If we wish to use the least amount of cardboard, what should be the dimensions of the box?

Exercise EIII.3.10. (For those who know the quotient derivative theorem III-6.)

A farmer needs to fence in 12,800 square feet of rectangular pasture along a river; she doesn't need to fence along the river. What are the dimensions of the pasture that will require the least amount of fence?

Exercise EIII.3.11.
A farmer has 3200′ of fence available and wishes to fence a rectangular pasture along a river and doesn't need to fence along the river. What are the dimensions of the largest pasture he can obtain?

Exercise EIII.3.12.
A farmer has budgeted $2,880 to fence in a rectangular field. The fencing along one side costs $6 a foot, while the fencing on the other three sides is $3 a foot. What are the dimensions of the largest field she can fence for the $2,880?

Exercise EIII.3.13.
(For those that know the quotient derivative theorem.)
A farmer needs to fence in part of his land to make a rectangular pasture with 240,000 square feet. The fencing along one side costs $6 a foot, while the fencing on the other three sides is $3 a foot. What dimensions should he use so that the fence for the 240,000 square feet costs the least amount of money?

Exercise EIII.3.14.
We need to build a tank with a square base that contains 343 cubic feet of water. Material for the bottom costs $10/sq.ft.; material for the sides costs $15/sq.ft. and the top costs $20/sq.ft. Find out how to design the tank so that the cost is the least.

Exercise EIII.3.15.
A rock band plays in an arena that can hold 12,000 people. When a ticket costs $14, the usual attendence is 8,000. The promoters have found that for each dollar that is taken off the ticket price, the average attendence will increase by 1000. What should the price of a ticket be so that the promoters maximize the revenue obtained from ticket sales?

Exercise EIII.3.16.
What is the solution of the problem in EIII.3.15. if the arena only holds 10,000 people?

Exercise EIII.3.17.
A motel with 50 units is filled every night when the price for renting a unit is $20. Management has found that for each dollar increase in the rent, two fewer units are rented. If it costs $4 to service each rented unit, what price should be charged for each unit to maximize gross profit?

Exercise EIII.3.18.
Does the result in the previous problem still hold if the motel has only 30 units?

For Section III.4.

Exercise EIII.4.1.

Suppose the motion of a point is described by the formula

$$p(t) = -2t^3+21t^2-60t+52 \text{ feet}$$

for $0 \leq t \leq 6$ seconds.

(1) How fast was the point moving at the end of the third second? Was it moving forward or backward?

(2) When did the point reverse direction?

(3) What was the maximum distance the point travelled away from its starting point?

(4) When was the point moving most rapidly forward?

Exercise EIII.4.2.

A skydiver jumps out of an airplane at 5,000 feet. He wants to open his parachute at 1,400 feet. How many seconds should he wait before opening his parachute? What will be his velocity when his parachute opens?

CON'T. NEXT COLUMN...

(We assume that the parachutist's motion follows the formula

$$p(t) = s+vt-16t^2$$

for objects moving according to gravitational force:

t = time in seconds.
s = initial position in feet.
v = initial velocity in feet/sec.
p(t) = position in feet at time t seconds.)

Exercise EIII.4.3.

A cat moves with a velocity of 12′/sec. and a mouse can run at a velocity of 10′/sec.

If a cat sees a mouse 8′ away, how long will it take the cat to catch the mouse if both move in a straight line? How far did the cat go before catching the mouse?

Exercise EIII.4.4.

A train engineer keeps track of his speed during a two hour trip and finds that it fits the formula $v(t) = 120t-60t^2$ mi./hr. What was the maximum speed of the train? How far did the train go?

(Hint: If p(t) is the position of the train and $Dp|_t=v(t)=120t-60t^2$, what must be a formula for p(t)?)

Exercise EIII.4.5.

Our scientists have figured out that the velocity of R.D.'s yo-yo is given by the formula

$$v(t) = -3t^3+9t^2-6 \text{ feet/sec.}$$

during the two seconds it takes to go down and back.

Draw a graph of the velocity function. When does the yo-yo reverse direction?

Our scientists neglected to measure the length of the yo-yo's string. From the information above, can you figure out the length of the string?

AND NOW, THE
INTERGALACTIC EPILOGUE
GUESSING! HE ADMITS IT! ALL THAT SWEET TALK AND HE ADMITS HE'S JUST GUESSING LIMITS! HE CLAIMS TO KNOW HE'S RIGHT WHEN HE COMES UP WITH ONE OF THOSE @[#★] LIMIT MACHINES!
OOO
GRONK?
WE'VE BEEN HAD
OH, MY
SOFT SOAP
HE MERELY GUESSES AND THEN LEGISLATES TO PROVE IT—HOGWASH! LET'S GET HIM!
OH, BOY, HOGWASH! BATH TIME!
GRR

MEANWHILE BACK AT THE RANCH*
PASTORAL, ISN'T IT?
g()
Grover
f()
Alfred
ALMOST FINISHED?
YAH
THAT TICKLES
NOT BAD SHAPE FOR ITS 20,000 EPSILON OVERHAUL
AWK! HERE COMES MELVIN'S MARAUDERS!
lim g(x) x→2
LRW
Good Vibes
TOOLS
QUICK! CIRCLE THE LIMIT MACHINES!
NO, NO, YOU IDIOT!
lim g(x) x→2
Good Vibes

* SEE PAGE 160

INFIDELS! THEY DON'T BELIEVE!
EPSILON WAY TO TIPPERARY
Good Vibes
lim f()
x→3
g()
2
LRW
f()
AlfRed
BOO
HISS
GRONK
YOU INVENTED ALL THAT STUFF JUST TO TORTURE STUDENTS!
YOU HURT?
NAW, THIS IS JUST A COMIC STRIP
Good Vibes
UNBELIEVERS!
FRAUD!
AN'A LEFT AN'A RIGHT AN'A LEFT AN SWING YER PARTNER
Good Vibes
lim g()
x→2
LRW
SPLORT
δε ME, WILL YA? HAH!
f()
ALFRed

KAPOKKETA
KAPOKKETA
QUEEP
QUEEP
MACHINA EX DEUS
N
E
S
W
VEGA
EXOTIC PLUTO
BINARYS HAVE MORE FUN
SIRIUS
UH HO
Good Vibes

WHATS GOING ON?
I THOUGHT THERE WAS INTELLIGENT LIFE HERE!
FLEEP?
CLUB M.E.D.
NO MILK TODAY
ULP
ER...
limF()
X→3
WOW
LRW
WELL, SET MY PHASERS TO STUN.
AN INTELLECHOOL DIS-CUSSION
f()
ALFRED

OK, YOU TWO– SO WHAT'S THE FIGHT ABOUT?
¿QUEEP?
HE'S A TROUBLEMAKER!
WHY THE ANIMALISTIC BEHAVIOUR?
I...ER...UH... I'M RD, YOU SEE... I'M A DUCK? DOES THAT COUNT?
NO
?
WATCH YOUR STEP
WATCH YOUR MIND
WIND YOUR WATCH
WATCH YOUR WATCH
YOUR WATCH WATCHES
f()
ALFRED

HE JUST GUESSES LIMITS! THEN HE SAYS THAT IF HE HAS THIS HERE LIMIT MACHINE IT PROVES THAT HIS GUESS IS RIGHT!
OH, WE KNOW ALL ABOUT LIMITS-'CAUSE WE HAVE THIS HERE HYPERTHANGAMILATOR NAMED CHOMP THAT'S GOT NANOMICROVISION
lim F() x→3
LRW
LEMME SHOW YOU-C'MERE, WIDGET!
WIDGET WILL GET lim x→a ψ[x]
FWEEP
ψ[·]
a
CHOMP!
ψ[·]
WOW THAT'S COOL
DON'T YOU HAVE ONE OF THESE?
NOOO.....

AND WHAT DOES THIS PECULIAR LIMIT MACHINE DO?
GOOD 'OL $\lim_{x \to a} f(x)$ HERE PROCESSES ANY $\epsilon > 0$ TO GET SOME $\delta(\epsilon) > 0$ *Guaranteed* TO SATISFY IF $0 < |x-a| < \delta(\epsilon)$, THEN $|f(x) - L| < \epsilon$.
ALL MEANINGLESS, OF COURSE
$\lim_{x \to a} f(\;)$
OH, THOSE. WE USED TO FOOL AROUND WITH THEM– THEY WERE GOOD FOR CHECKING ERROR TOLERANCES.
HAMMY, SEE IF YOU CAN FIND ONE OF OUR OLD LIMIT MACHINES IN THE BELFRY
DOES YOUR LIMIT MACHINE WORK WITH THE SAME NUMBER **L** PRODUCED BY THE HYPER-THANGAMILATOR?
WHY, YES– IT OUGHT TO, SINCE IF YOU HAVE A LIMIT, YOU SHOULD HAVE ONE OF THESE LIMIT MACHINES.
HERE!
DO YOU BELONG TO THE FAITHFUL SIDEKICK'S UNION, TOO?

* SEE PAGE 139 AND SECTION II.9

BUT THAT NUMBER L IS THE SAME AS THE NUMBER PRODUCED BY THE HYPERTHANGAMILATOR!
HMMM... VERY STRANGE PROOF TECHNIQUE... CRUDE, BUT IT'S RIGHT. AMAZING—WITH THIS YOUR GUESSES ARE RIGHT EVEN THOUGH YOU DON'T HAVE NANOMICROVISION!
DON'T TELL ME YOU'RE BUYING IT, TOO!
WE GLUED ALL OUR CALCULUS TOGETHER WITH THESE LIMITS. THE SAME NUMBERS AS YOURS!
MOST PECULIAR. SO YOUR SCIENCE IS RIGHT EVEN THOUGH YOU DON'T REALLY KNOW WHAT YOU'RE DOING. WAIT 'TILL THE FOLKS BACK ON SIRIUS HEAR ABOUT THIS!
IS THIS THE YELLOW BRICK ROAD?
BAH
LUNCH
lim Ψ[x]
x → a
LRW
Good Vibes
f( )
qed
I'M HUNGRY. WANNA EAT?
SILLY BOY

IF NOTHING ELSE, WE'VE ALL LEARNED OUR LIMITS!
AT LEAST IT'S NICE TO SEE EVERYONE GETTING ALONG.
SO WHAT YOU GONNA DO FOR YOUR NEXT ROLE?
VERY OFF BROADWAY
IT WAS A CALCULATED RISK!
CATCH! MY FAVORITE GAME!
I PROPOSE A TOAST TO THE 'OL LIMIT MACHINE
NOT BAD, EH? THE 3 BASIC FOOD GROUPS- SUGAR, FAT AND STARCH
YOU'RE NOT TAKING ME SIRIUSLY
TOFUETTES
BROCCOLI ICE CREAM
FLANGE
PEACE IN OUR TIME!
Good Vibes
lim ψ[x]
x→a
lim ψ[·]
x→a
Kelp CRACKERS
RUTABEGA ICE CREAM?
IT'S ALL THE RAGE
g()
f()
ALFRed
ψ[·]
CHOCOLATE COVERED ANCHOVIES
COOKIES
JOHNSON

the end
STARRING PROF. E. McSQUARED
WITH ALFRED, GROVER, R.D., PIGGY, GOOD VIBES, THETA, OMEGA, "X", AND MALICIOUS MELVIN. ADMIRAL ETHALYNE DUCKLEEN, HAMMY, SIGMA FRAUD, STANISLAW LIM, HORATIO, MAJOR JAEGER, EPSILON, ZETA AND WIDGET. CASTING BY PLASTIER OF PARIS.
DIRECTED BY H. SWANN & J. JOHNSON
PRODUCED BY J. JOHNSON & H. SWANN
WRITTEN BY H. SWANN & J. JOHNSON
SCRIPT BASED ON A CONCEPT BY I. NEWTON, G.W. LEIBNIZ & K. WEIERSTRASS
ARRANGEMENT BY H. SWANN
GRAPHICS BY J. JOHNSON, TEAPOT GRAPHICS
EDITED BY PAT BREWER, LORING WOODMAN, WILLIAM KAUFMANN AND THOUSANDS OF STUDENTS
NARRATED BY E. McSQUARED
PUBLISHED BY JANSON PUBLICATIONS, INC.
PARTIALLY WRITTEN ON LOCATION AT CAP'T. PSYCHE'S ELECTRIC DUDE RANCH, BOX 511, JACKSON, WYOMING 83001
AN ENDLESS PRODUCTION
SHUDDUP
SLURP

ANSWERS
Good Vibes
TRASH
NO PARKING

I.2.1 Suppose $A = \{\text{[cube]}, \text{[can]}, 6, \theta\}$, $B = \{\theta, 2, \Omega\}$, $C = \{3, 6, \Omega, \text{[can]}\}$ and $D = \{\text{[can]}, 3, \text{[cube]}\}$.

Figure out

(a) $A \cap C = \{?\} = \{6, \text{[can]}\}$

(b) $B \cup C = \{?\} = \{\theta, 2, \Omega, 3, 6, \text{[can]}\}$

(c) $B \cap D = \{?\} = \emptyset = \{\ \}$

(d) $C \cap \emptyset = \{?\} = \emptyset = \{\ \}$

(e) $C \cup \emptyset = \{?\} = C = \{3, 6, \Omega, \text{[can]}\}$

I.2.2 $A \cup (B \cap C) =$
$\{\text{[cube]}, \text{[can]}, 6, \theta\} \cup \{\Omega\} =$
$\{\text{[cube]}, \text{[can]}, 6, \theta, \Omega\}$
This result should be different from $(A \cup B) \cap C$ - you have to be careful about moving parentheses around.

I.2.3 $(A \cap B) \cap C =$
$\{\theta\} \cap \{3, 6, \Omega, \text{[can]}\} = \emptyset$

I.2.4 $(A \cap C) \cup (B \cap C) =$
$\{6, \text{[can]}\} \cup \{\Omega\} = \{6, \text{[can]}, \Omega\}$

I.2.5 $(A \cup B) \cap (A \cup C) =$
$\{\text{[cube]}, \text{[can]}, 6, \theta, 2, \Omega\} \cap$
$\{\text{[cube]}, \text{[can]}, 6, \theta, 3, \Omega\}$
$= \{\text{[cube]}, \text{[can]}, 6, \theta, \Omega\}$

Suppose $A = (-3, 2)$, $B = [-1, 3]$ and $C = (-2, -1]$. Indicate A, B and C on the number line and figure out the following - write the answers using both the squiggly bracket ($\{$and$\}$) set notation and the short notation with [,] , (and) .

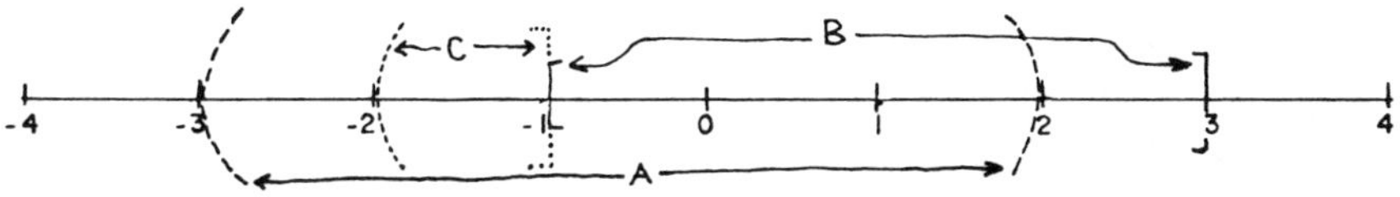

I.3.1 $A \cap C = C = \{x \mid x \in \mathbb{R}; -2 < x \leq -1\}$

I.3.2 $A \cup C = A = \{x \mid x \in \mathbb{R}; -3 < x < 2\}$

I.3.3 $B \cap C = [-1, -1] = \{-1\}$

I.3.4 $B \cup C = (-2, 3] = \{x \mid x \in \mathbb{R}; -2 < x \leq 3\}$

I.3.5 $(B \cup C) \cap A = (-2, 2) = \{x \mid x \in \mathbb{R}; -2 < x < 2\}$

I.3.6 $(B \cap C) \cap A = [-1, -1] = \{-1\}$

I.3.7 $A \cup (B \cap C) = A$

I.3.8 $(A \cap B) \cup C = (-2, 2) = \{x \mid x \in \mathbb{R}; -2 < x < 2\}$

I.3.9 If x's satisfy
$-2 < x \leq 1$,
what inequality will the set of 2x's satisfy?
$-4 < 2x \leq 2$

What inequality will the set of $(\frac{1}{2}x)$'s satisfy?
$-1 < \frac{1}{2}x \leq \frac{1}{2}$

What inequality will the set of $(-3x)$'s satisfy? $6 > -3x \geq -3$
OR, $-3 \leq -3x < 6$

I.3.10 If x's satisfy
$-1 \geq x > -\frac{3}{2}$,
what inequality will the set of 2x's satisfy? $-2 \geq 2x > -3$
OR, $-3 < 2x \leq -2$

What inequality will the set of $(\frac{1}{2}x)$'s satisfy? $-\frac{1}{2} \geq \frac{1}{2}x > -\frac{3}{4}$
OR, $-\frac{3}{4} < \frac{1}{2}x \leq -\frac{1}{2}$

What inequality will the set of $(-3x)$'s satisfy?
$3 \leq -3x < \frac{9}{2}$

I.3.11 Add -4 to every number in
$\{x \mid x \in \mathbb{R}; -5 \leq x < 4\}$.
What interval do you get?
$\{x - 4 \mid x \in \mathbb{R}; -5 \leq x < 4\} = [-9, 0)$

I.3.12 Add $\frac{3}{2}$ to every number in
$\{x \mid x \in \mathbb{R}; 2 \geq x > -1\}$.
What interval do you get?
$\{x + \frac{3}{2} \mid x \in \mathbb{R}; 2 \geq x > -1\} =$
$(\frac{1}{2}, \frac{7}{2}]$

I.3.13 What number interval does the inequality
$|1 - x| < 2$
describe? CHANGE TO $-2 < 1 - x < 2$.
THEN $-3 < -x < 1$
OR, $3 > x > -1$; INTERVAL $= (-1, 3)$

I.3.14 What number interval does the inequality
$|2x + 3| < 4$
describe? CHANGE TO $-4 < 2x + 3 < 4$
THEN $-7 < 2x < 1$ OR,
$-\frac{7}{2} < x < \frac{1}{2}$: INTERVAL $= (-\frac{7}{2}, \frac{1}{2})$

I.3.15 What number interval does the inequality
$|-3 - \frac{1}{2}x| < 5$
describe? CHANGE: $-5 < -3 - \frac{1}{2}x < 5$:
THEN $-2 < -\frac{1}{2}x < 8$
OR, $4 > x > -16$: INTERVAL $= (-16, 4)$

I.3.16 What number interval does the inequality
$|\frac{2}{3}x + 4| < \frac{1}{2}$
describe? CHANGE: $-\frac{1}{2} < \frac{2}{3}x + 4 < \frac{1}{2}$
$-\frac{9}{2} < \frac{2}{3}x < -\frac{7}{2}$;
$(\frac{3}{2})(-\frac{9}{2}) < x < (\frac{3}{2})(-\frac{7}{2})$: INTERVAL $(-\frac{27}{4}, -\frac{21}{4})$

Here is Grover in charge of another function g( ).

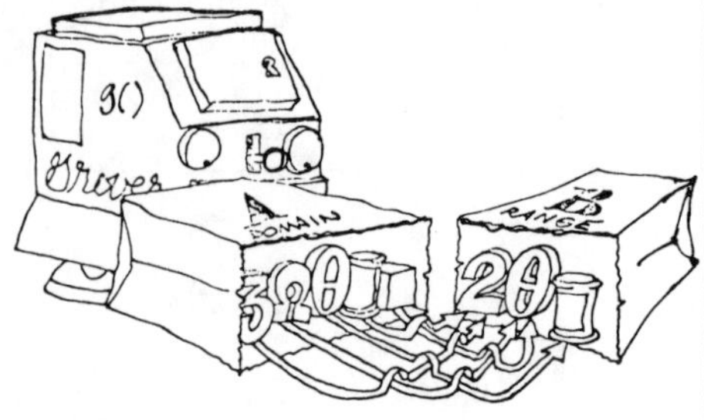

See if you can fill in for all "?."

g(▯) = ? 2   g(3) = ? 2

g(▢) = ? θ   g(θ ?) = ▯

g(⬡ ?) = θ .

Can you find another answer to

g(Ω ?) = θ ?

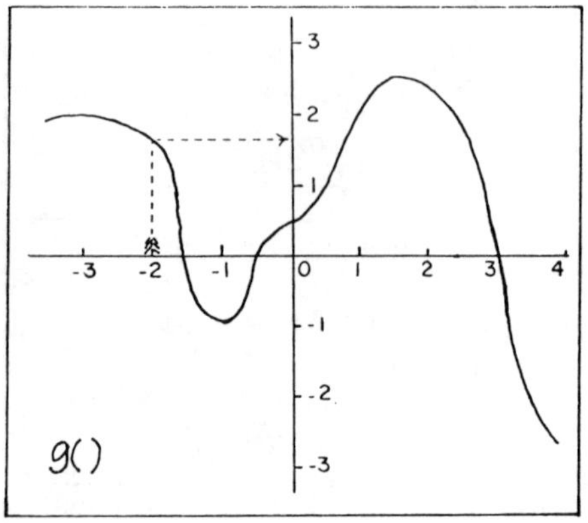

It turns out that

g(-2) is about 5/3 here.

Try a few?

g(-1) = -1 ?   g(0) = ½ ?

g(½) = 1 ?   g(3) = 0 ?

g(3½?) = -2   g({3, -½, -3½}?) = 0

g(1 ?) = 2   g(2½??) = 2

g(-3 ???) = 2

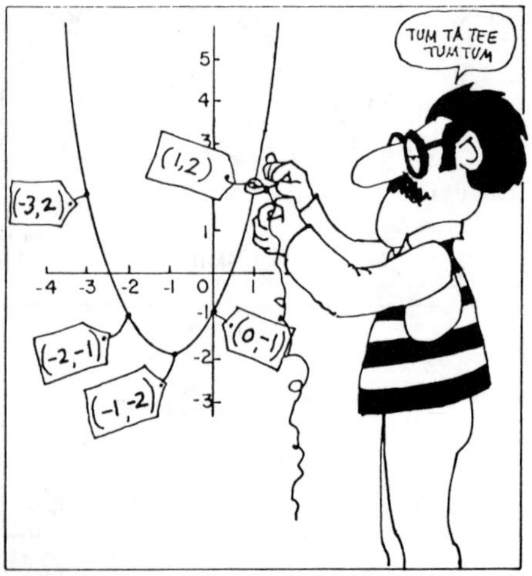

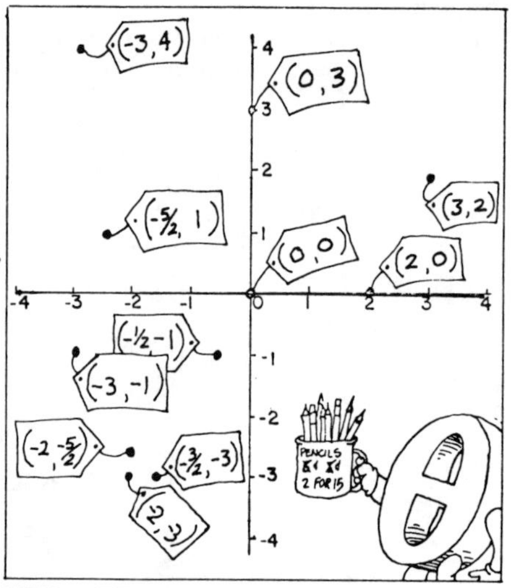

I.4.1

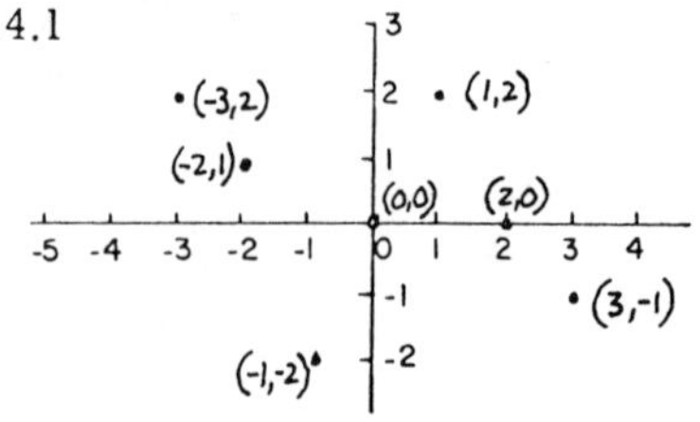

(a) Label the dots.

(b) If the dots represent a function, what is

f(3) = −1 ?

f(-3) = 2 ?

f(1) = 2 ?

f(2) = 0 ?

f(0) = 0 ?

(c) Put in the appropriate arrows to show what corresponds to what in this function.

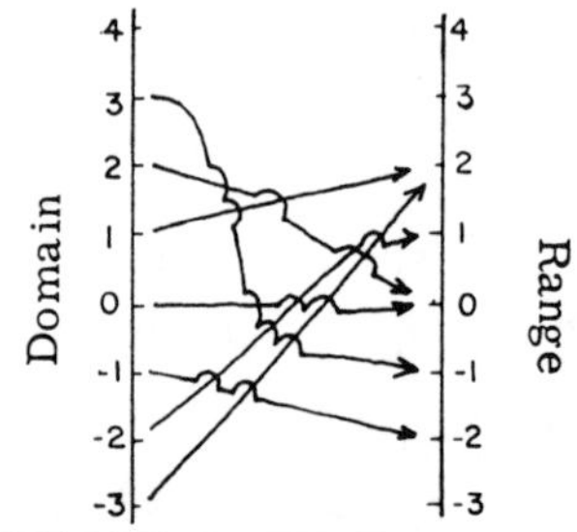

I.4.2 If f( ) is this correspondence;

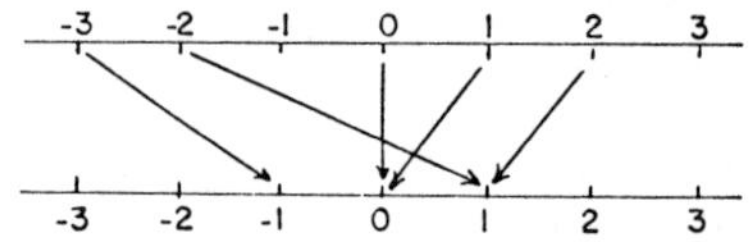

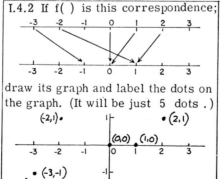

draw its graph and label the dots on the graph. (It will be just 5 dots .)

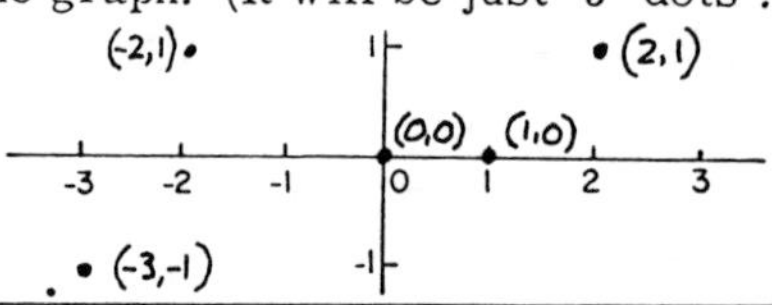

I.4.3 If the domain of f( ) is

$\{-1, 0, \tfrac{1}{2}, 2, 3\}$

and f(-1) = -1, f(0) = 0 , f(½) = 1 , f(2) = -2 and f(3) = -2 ,

(a) Put in the appropriate arrows to show what corresponds to what in this function.

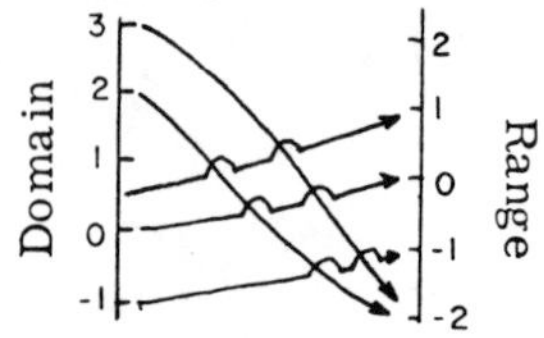

(b) Draw the graph of f( ). (It will just be five dots.)

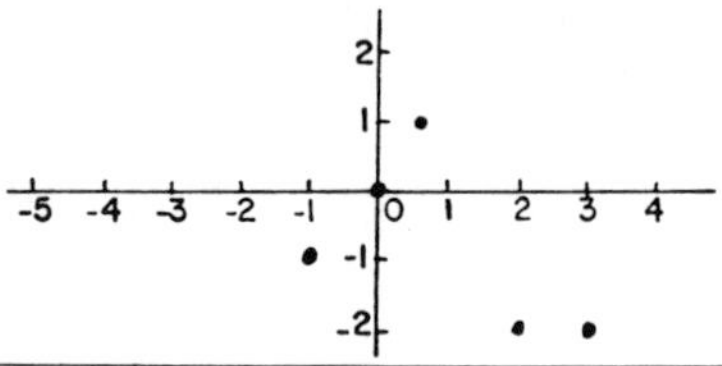

I.4.4 If this is the graph of f( ) ,

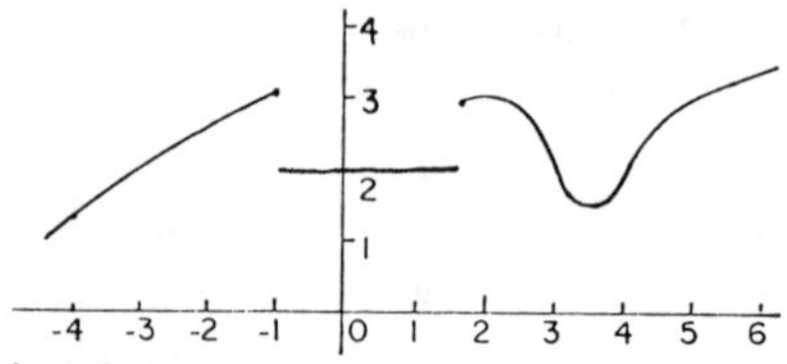

what is

(a) f(2) = 3 ?

(b) f(-2) = 2½ ?

(c) If x is any number in (0, 1) = $\{x \mid x \in \mathbb{R} ;\ 0 < x < 1\}$,

what is f(x) = 2 ?

(d) f(?) = 3 . ANS: -1 OR 2 OR 5

(e) f(?) = 5/2 . ANS: -2 OR 2¾ OR 4½

(f) f(?) = 2 . ANS: ANY X IN (-1, 1⅔) OR X = -3, +3 OR 4.

I.4.5

Formula - symbol:

f( ) = 2·( ) .

Show some of the correspondence-arrows?

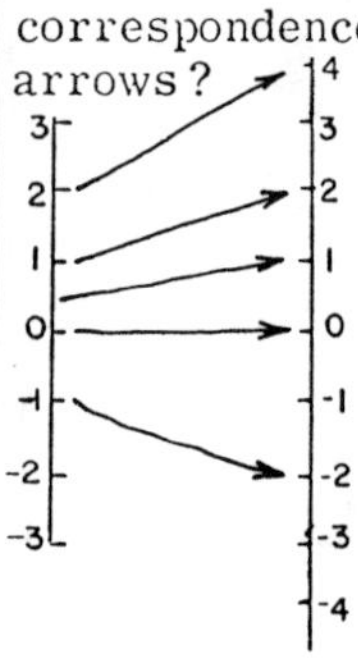

Graph?

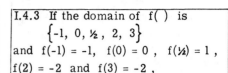

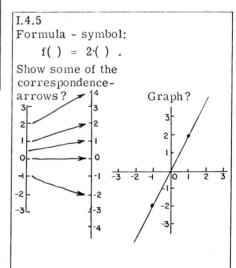

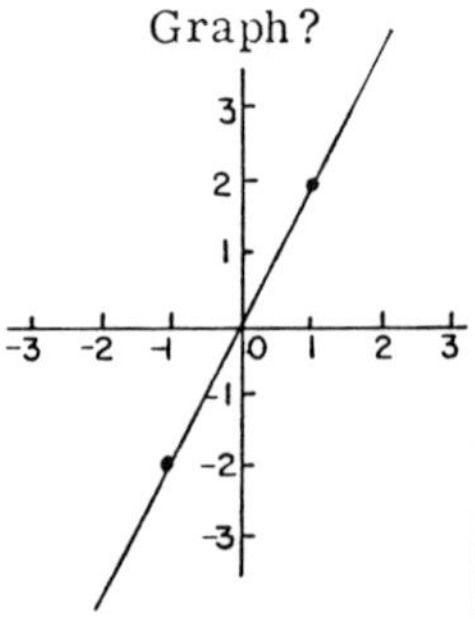

I.4.6

Assuming that g( ) has a straight line as its graph, here are some of the correspondence-arrows of g( ).

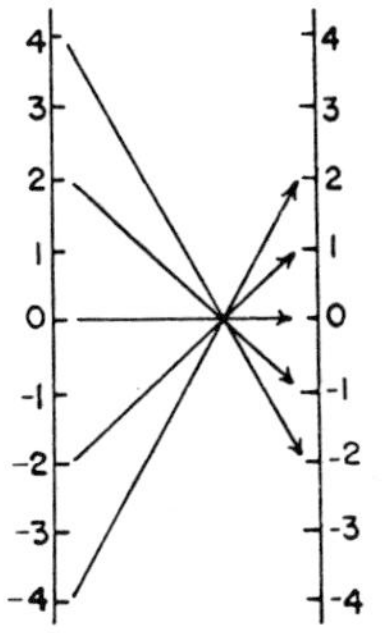

Graph?

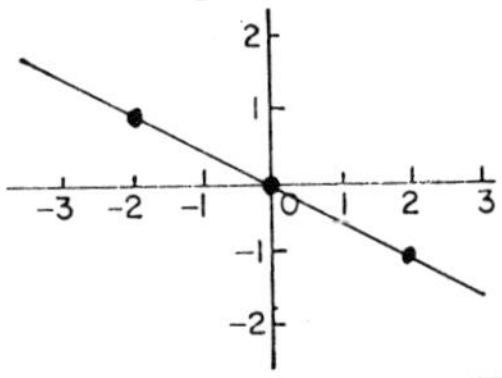

Formula - symbol?

g( ) = $-\frac{1}{2}$( )

?

I.4.7

Graph:

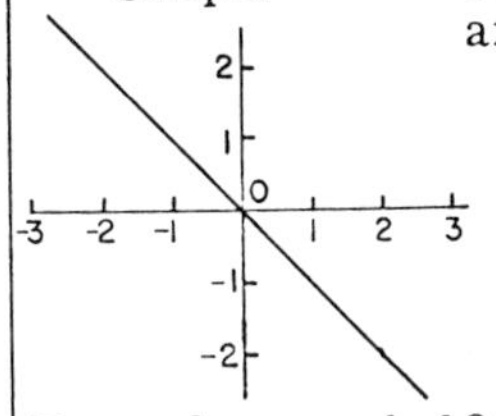

Show some of the correspondence-arrows?

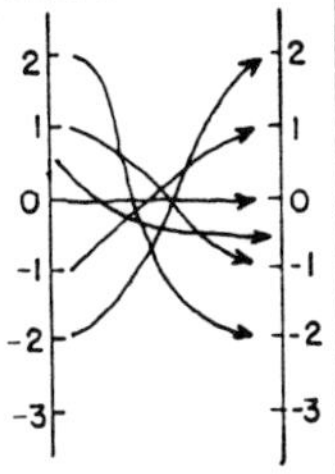

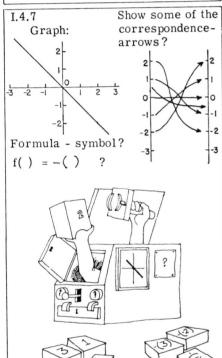

Formula - symbol?

f( ) = −( ) ?

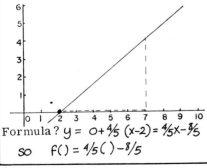

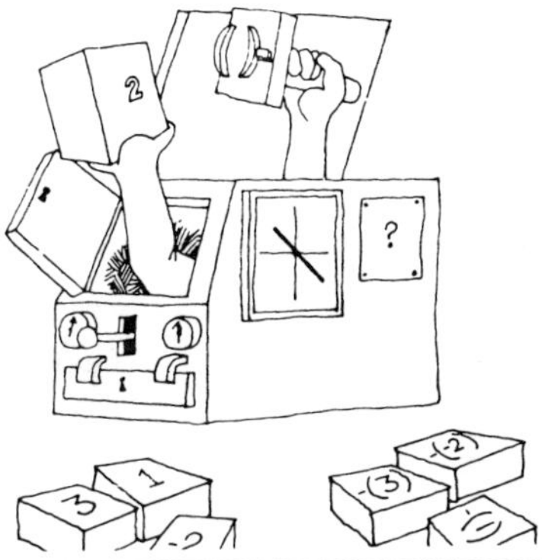

I.5.2 Draw the line through point (2, 0) with slope $\frac{4}{5}$ and find the formula of the function that has the line as its graph.

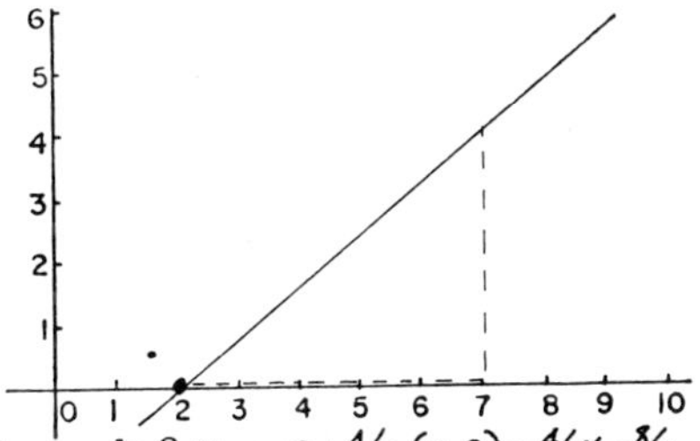

Formula? $y = 0 + \frac{4}{5}(x-2) = \frac{4}{5}x - \frac{8}{5}$

so $f(\ ) = \frac{4}{5}(\ ) - \frac{8}{5}$

I.5.3 Draw the line through point (−2, −3) with slope $1\frac{1}{5}$ and find the formula of the function that has the line as its graph.

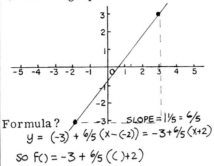

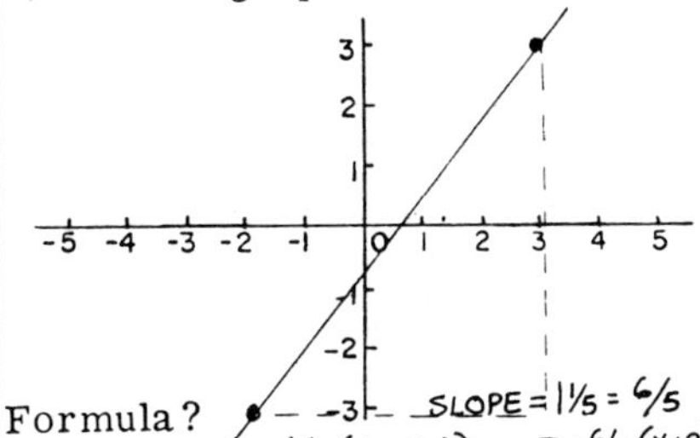

SLOPE $= 1\frac{1}{5} = \frac{6}{5}$

Formula? $y = (-3) + \frac{6}{5}(x-(-2)) = -3 + \frac{6}{5}(x+2)$

so $f(\ ) = -3 + \frac{6}{5}((\ )+2)$

I.5.4 Draw the line through point (1, 3) with slope −2 and find the formula of the function that has the line as its graph.

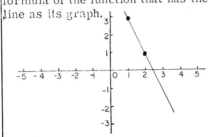

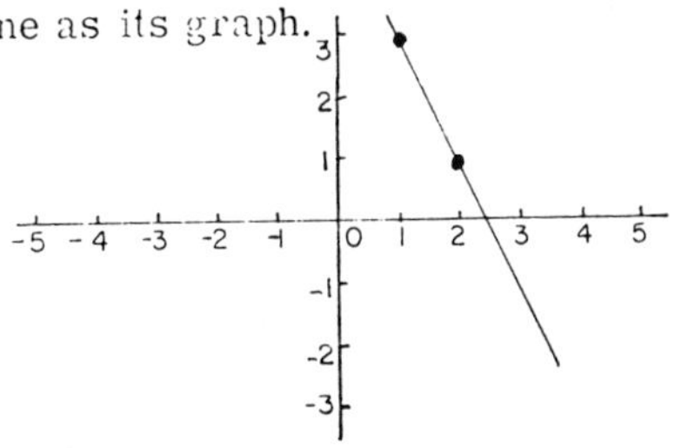

To draw this line, since

$$\text{slope} = -2 = \frac{-2}{1} = \frac{\text{rise}}{\text{run}},$$

just use triangle  to move over to point (2, 1) that also should be on the line.

NEGATIVE SLOPES ALWAYS mean that the line SLANTS UP TO THE LEFT!

Formula? $y = 3 + (-2)(x-1) = 3 - 2x + 2$
$= 5 - 2x$

so $f(\ ) = 5 - 2(\ )$

I.5.5 Draw the line through point $(-1, -\frac{3}{2})$ with slope $-\frac{2}{3}$ and find the formula of the function that has the line as its graph.

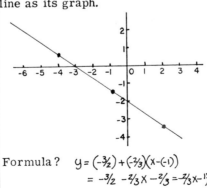

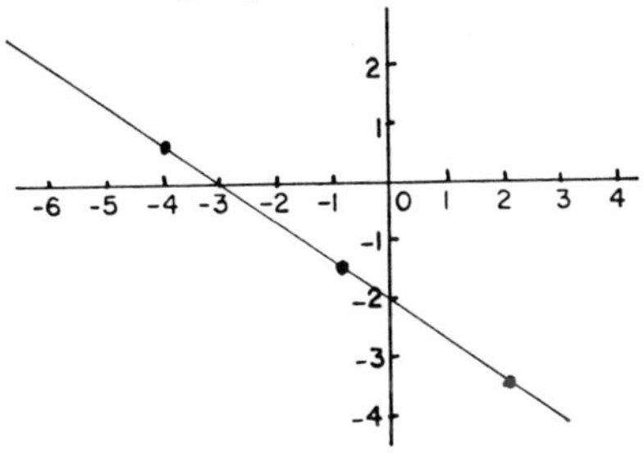

Formula? $y = (-\frac{3}{2}) + (-\frac{2}{3})(x-(-1))$
$= -\frac{3}{2} - \frac{2}{3}x - \frac{2}{3} = -\frac{2}{3}x - \frac{13}{6}$

so $f(\ ) = -\frac{2}{3}(\ ) - \frac{13}{6}$

I.5.6 Draw the line that has formula

$y = -\frac{1}{3}x + 4$.

What is the slope of the line? $-\frac{1}{3}$

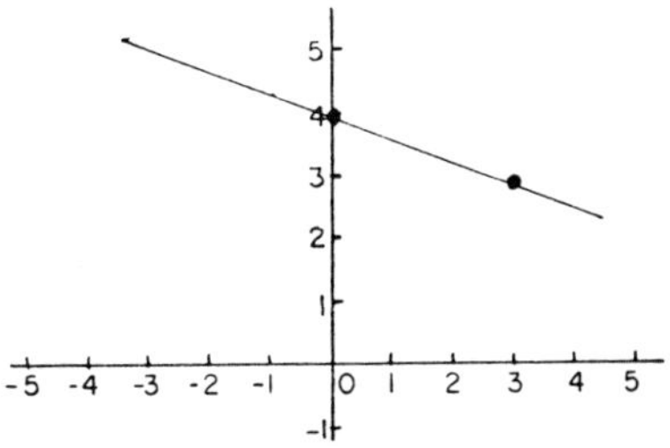

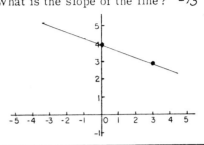

I.5.7 Draw the line described by the formula

$y = -\frac{1}{2}x + 2$.

What is its slope? SLOPE $= -\frac{1}{2}$

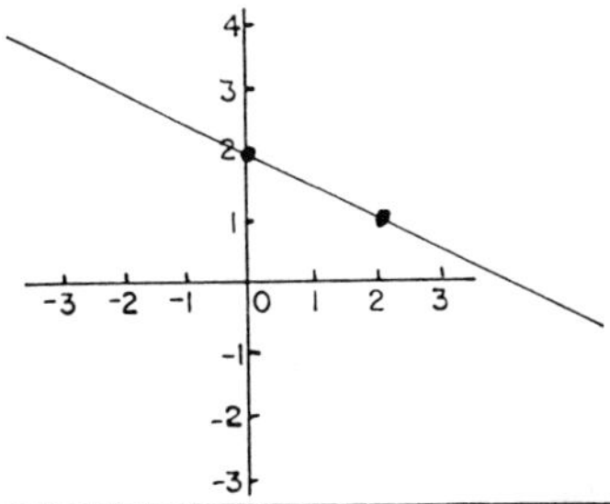

I.5.8 Draw the line described by the formula

$y = \frac{3}{2} - 2x$.

What is its slope?

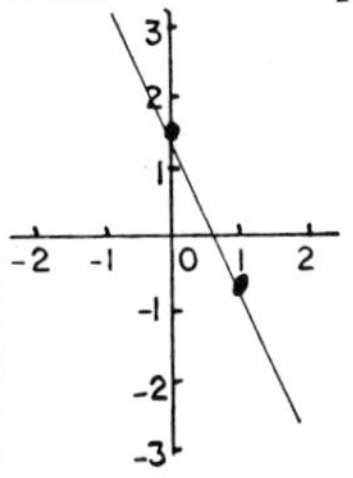

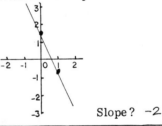

Slope? −2

I.5.9 (Hard!) Draw the line containing points (x, y) that satisfy the formula

$2y = 3x - 2$.

What is its slope? $\frac{3}{2}$

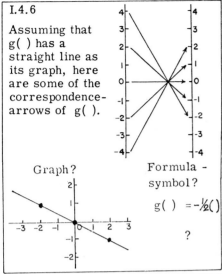

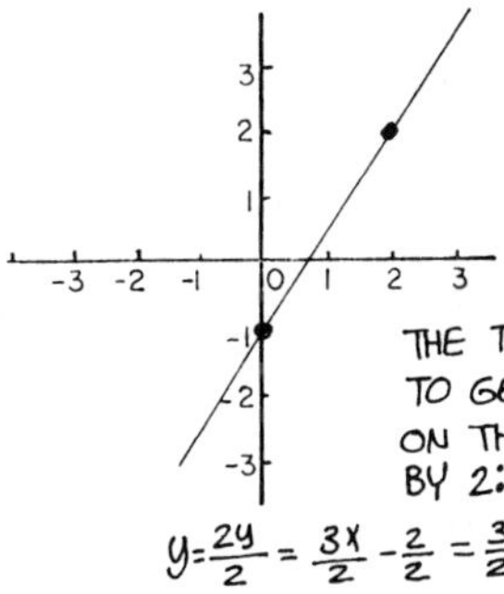

THE TRICK HERE IS TO GET Y ALL ALONE ON THE LEFT: DIVIDE BY 2:

$y = \frac{2y}{2} = \frac{3x}{2} - \frac{2}{2} = \frac{3}{2}x - 1$

SO, $f(\ ) = \frac{3}{2}(\ ) - 1$

I.5.10 (Harder!) Draw the line containing points (x, y) that satisfy the formula

$$2x = 3y - 2$$

What is its slope?

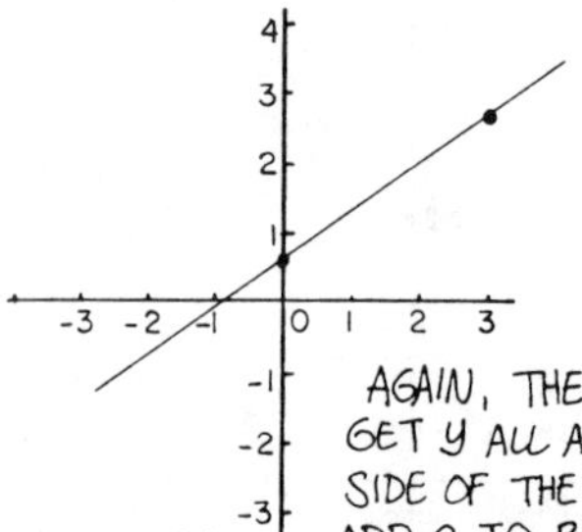

Slope? $\frac{2}{3}$

AGAIN, THE TRICK IS TO GET y ALL ALONE ON ONE SIDE OF THE EQUATION. ADD 2 TO BOTH SIDES: $2+2x = 3y-2+2 = 3y$ THEN $\frac{2}{3} + \frac{2}{3}x = y$; THE REST IS EASY

I.5.11 Find a formula for the line through points (-1, 4) and (1, 1).

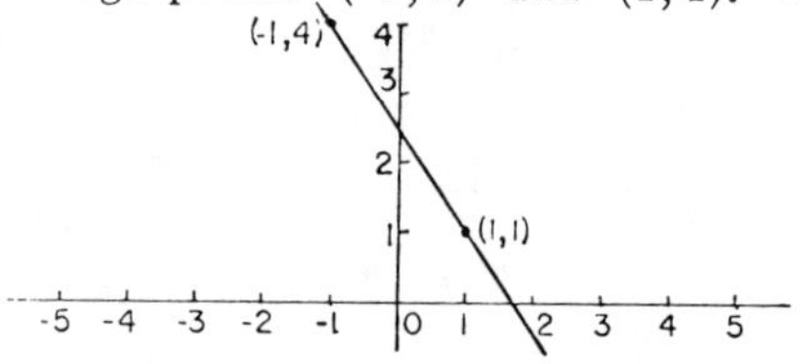

(-1,4)
(1,1)

$$\text{slope} = \frac{(1, 1) - (-1, 4)}{(1, 1) - (-1, 4)} = \frac{1-4}{1-(-1)} = \frac{-3}{2} = -\frac{3}{2}$$

Formula? $y = 4 + (-\frac{3}{2})(x-(-1)) = 4 - \frac{3}{2}x - \frac{3}{2} = \frac{5}{2} - \frac{3}{2}x$

OR $y = 1 + (-\frac{3}{2})(x-1) = 1 - \frac{3}{2}x + \frac{3}{2} = \frac{5}{2} - \frac{3}{2}x$

EITHER WAY, $y = \frac{5}{2} - \frac{3}{2}x$

I.5.12 Find a formula for the line through points (0, 1) and (2, 2).

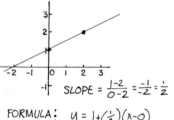

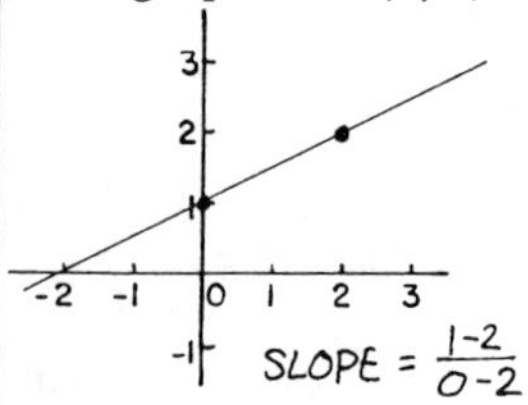

SLOPE $= \frac{1-2}{0-2} = \frac{-1}{-2} = \frac{1}{2}$

FORMULA: $y = 1 + (\frac{1}{2})(x-0)$

$= 1 + \frac{1}{2}x$

I.5.13 Find a formula for the line through points (-1, -2) and (1, 0).

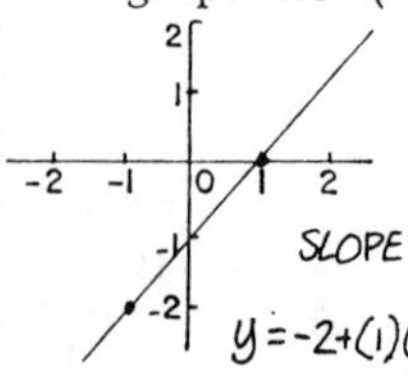

SLOPE $= \frac{-2-0}{-1-1} = \frac{-2}{-2} = 1$

$y = -2 + (1)(x-(-1)) = -2 + x + 1 = x - 1$

OR $y = 0 + (1)(x-1) = x - 1$

AGAIN, EITHER (-1,-2) OR (1,0) CAN BE USED AS "ORIGINAL POINT" THE END RESULTS AGREE

I.5.14 Find a formula for the line through points $(-3, \frac{2}{3})$ and $(-1, -\frac{1}{3})$.

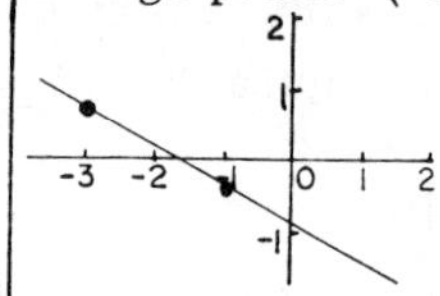

SLOPE $= \frac{\frac{2}{3} - (-\frac{1}{3})}{(-3)-(-1)} = \frac{1}{-2}$

$y = (\frac{2}{3}) + (-\frac{1}{2})(x-(-3))$

$= \frac{2}{3} - \frac{1}{2}x - \frac{3}{2} = -\frac{1}{2}x - \frac{5}{6}$

I.5.15 Is the point (4, -1) on the line that goes through points $(-1, \frac{3}{2})$ and $(1, \frac{1}{2})$?

LOOKS PRETTY CLOSE. TO BE SURE, FIND THE EQUATION FOR THE LINE AND SEE IF (4,-1) SATISFIES IT.

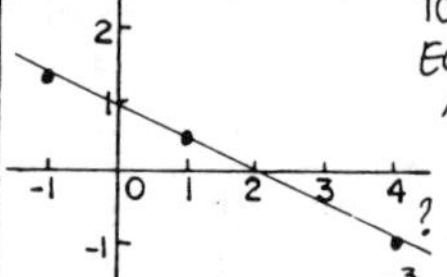

SLOPE $= \frac{\frac{3}{2} - \frac{1}{2}}{-1 - 1} = \frac{1}{-2} = -\frac{1}{2}$

$y = \frac{1}{2} + (-\frac{1}{2})(x-1) = \frac{1}{2} - \frac{1}{2}x + \frac{1}{2} = 1 - \frac{1}{2}x$

NOW SEE IF (4,-1) SATISFIES THIS:

$-1 \stackrel{?}{=} 1 - \frac{1}{2} \cdot 4 = 1 - 2 = -1$; YES, IT CHECKS!

I.6.1 Sketch the graph of the function

$f(\ ) = \frac{1}{2}(\ )^2 + 2(\ ) + 1 \quad (y = \frac{1}{2}x^2 + 2x + 1)$

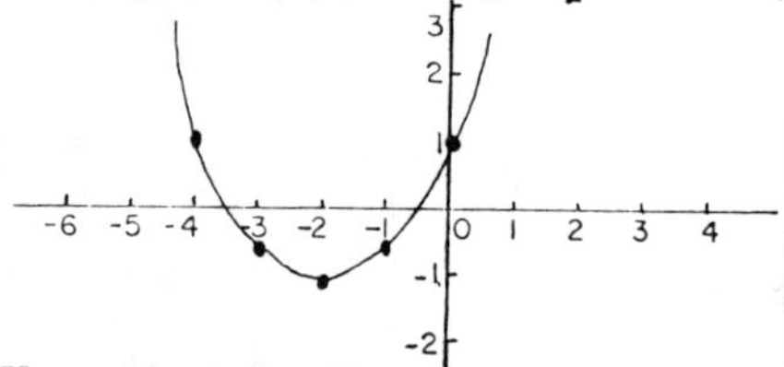

Up or down? UP

Fat or thin? FAT

Nose at? X-VALUE $= \frac{-2}{2(\frac{1}{2})} = -2$

Some points? $(-2,-1)$ = NOSE;

$(-1,-\frac{1}{2})$; $(-3,-\frac{1}{2})$, $(0,1)$; $(-4,1)$

I.6.2 Sketch the graph of the function

$f(\ ) = -2(\ )^2 + 4(\ ) + 1 \quad (y = -2x^2 + 4x + 1)$.

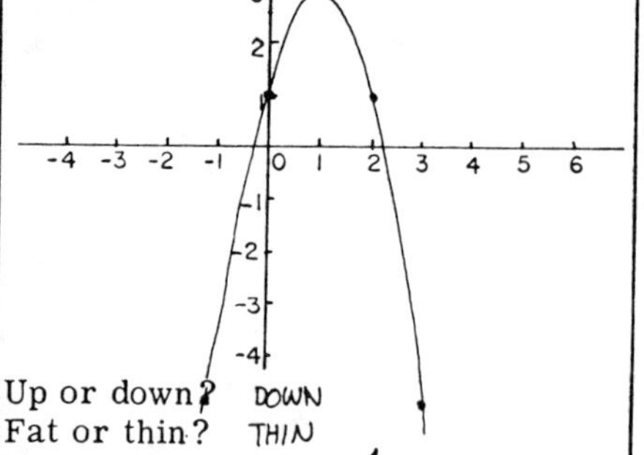

Up or down? DOWN

Fat or thin? THIN

Nose at? X-VALUE $= \frac{-4}{2(-2)} = 1$

Some points? NOSE: (1,3)

(0,1); (2,1); (-1,-5); (3,-5)

I.6.3 Sketch the graph of the function

$f(\ ) = -\frac{1}{3}(\ )^2 - 2(\ ) - 1 \quad (y = -\frac{1}{3}x^2 - 2x - 1)$

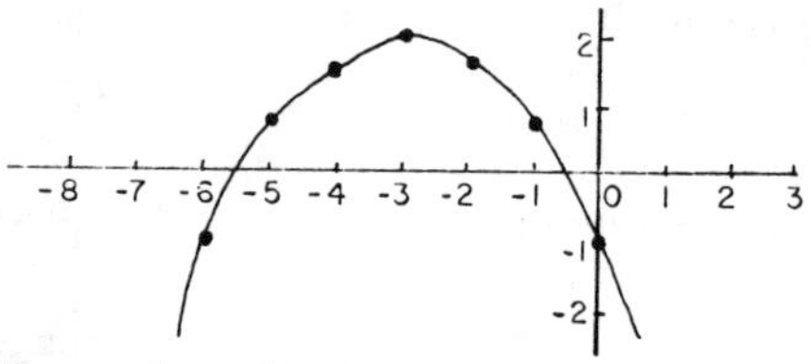

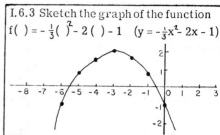

Up or down? DOWN

Fat or thin? FAT

Nose at? X-VALUE $= \frac{-(-2)}{2(-\frac{1}{3})} = \frac{2}{-\frac{2}{3}} = -3$

Some points? NOSE: (-3,2)

$(-2,\frac{5}{3})$; $(-4,\frac{5}{3})$; $(-1,\frac{2}{3})$; $(-5,\frac{2}{3})$

(0,-1); (-6,-1)

I.6.4 Sketch the graph of the function

$f(\ ) = -12(\ ) + 4(\ )^2 + 5 \quad (y = -12x + 4x^2 + 5)$

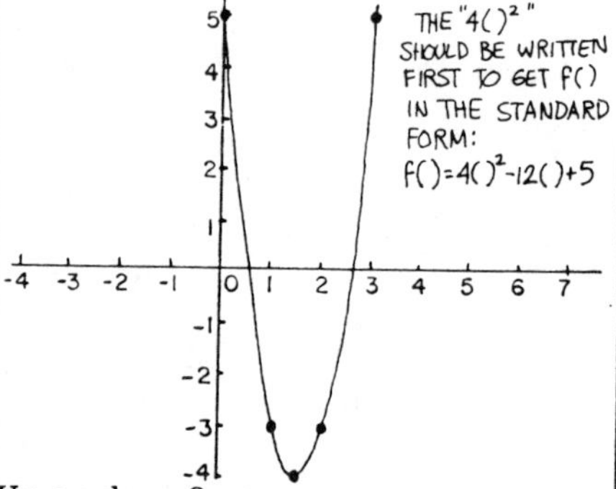

THE "$4(\ )^2$" SHOULD BE WRITTEN FIRST TO GET $f(\ )$ IN THE STANDARD FORM:
$f(\ ) = 4(\ )^2 - 12(\ ) + 5$

Up or down? UP

Fat or thin? THIN

Nose at? X-VALUE $= \frac{-(-12)}{2 \cdot 4} = \frac{3}{2}$

Some points? NOSE: $(\frac{3}{2},-4)$

(1,-3); (2,-3); (0,5); (3,5)

I.6.5 Sketch the graph of the function

$f(\ ) = -(\ )^3 - 6(\ )^2 - 9(\ ) - 3$.

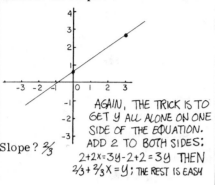

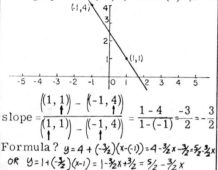

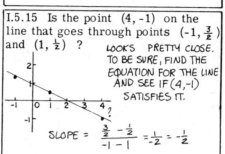

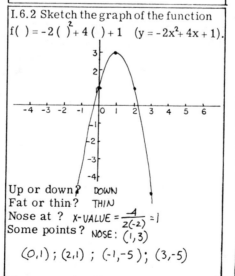

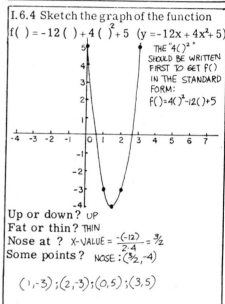

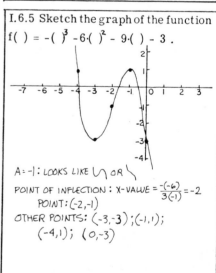

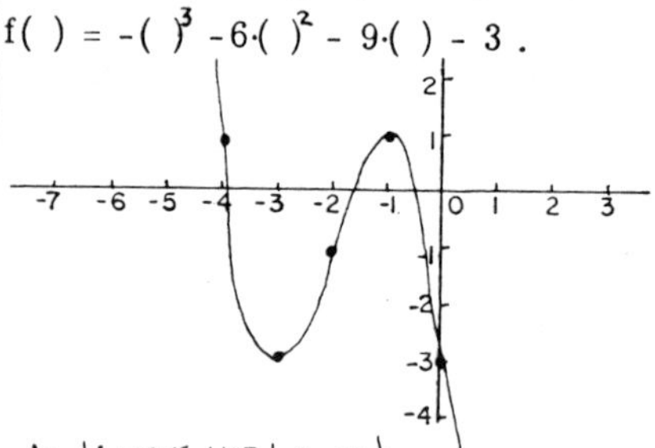

A = -1: LOOKS LIKE ∩∪ OR ∖

POINT OF INFLECTION: X-VALUE $= \frac{-(-6)}{3(-1)} = -2$

POINT: (-2,-1)

OTHER POINTS: (-3,-3); (-1,1); (-4,1); (0,-3)

I.6.6 Sketch the graph of the function
$f(\ ) = \frac{1}{2}(\ )^3 + \frac{3}{2}(\ )^2 + 2\cdot(\ ) + 2$.

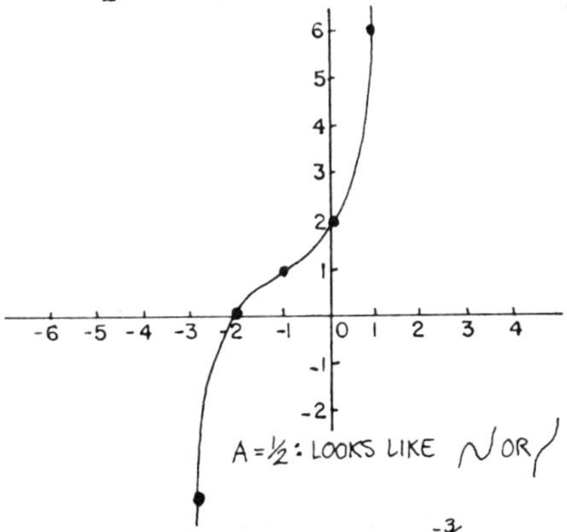

A = ½ : LOOKS LIKE ∿ OR /

POINT OF INFLECTION: X-VALUE $\frac{-\frac{3}{2}}{3(\frac{1}{2})} = -1$

POINT: (-1,1)

OTHER POINTS: (0,2);(-2,0);(1,6);(-3,-4)

I.6.7 Sketch the graph of the function
$f(\ ) = \frac{1}{2}(\ )^3 - 3\cdot(\ )^2 + \frac{9}{2}(\ ) + 1$.

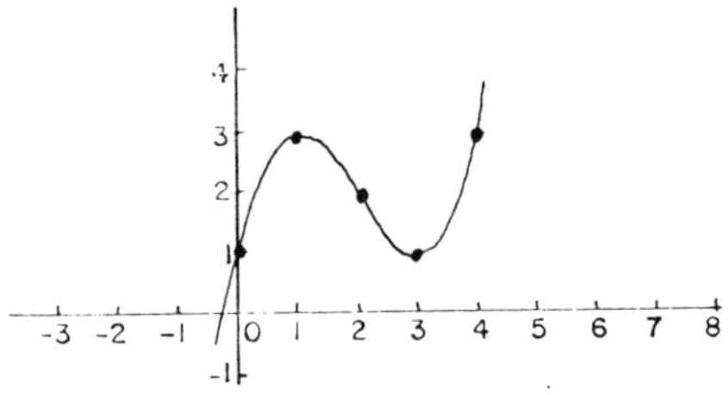

A = ½ : LOOKS LIKE ∿ OR /

POINT OF INFLECTION: X-VALUE: $\frac{-(-3)}{3(\frac{1}{2})} = 2$

POINT: (2,2)

OTHER POINTS: (1,3);(3,1);(0,1);(4,3)

I.6.8 Sketch the graph of the function
$f(\ ) = -(\ )^3 + 3\cdot(\ ) + 2$.

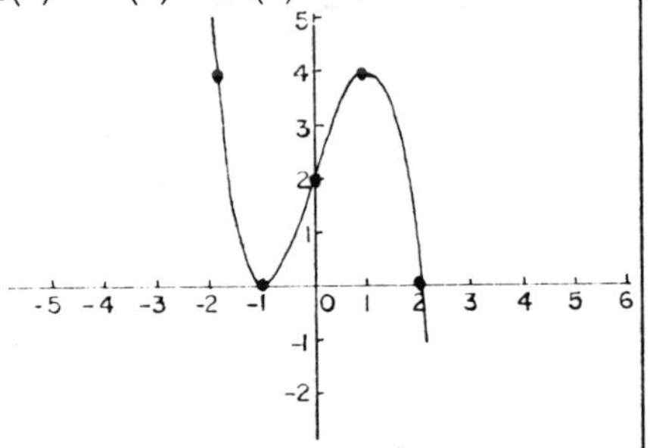

A = -1 : LOOKS LIKE ∿ OR ∿ .

POINT OF INFLECTION: "B" = 0 HERE!

SO X-VALUE = $\frac{-(0)}{3\cdot(-1)} = 0$:

POINT: (0,2)

OTHER POINTS: (1,4);(-1,0);(2,0);(-2,4)

II.2.1 When McSquared put ⅛ into 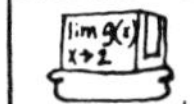 , out came ¼ . Check the guarantee, i.e. argue from IF to THEN using algebra.

IF $2 - \frac14 < x < 2 + \frac14$, (MULTIPLY BY ½)

$\frac12\cdot(2) - \frac12\cdot(\frac14) < \frac12 x < \frac12(2) + \frac12\cdot(\frac14)$

$1 - \frac18 < \frac12 x < 1 + \frac18$

(ADD 2) $3 - \frac18 < \frac12 x + 2 < 3 + \frac18$

CHECKS WITH →

THEN $3 - \frac18 < \frac12 x + 2 < 3 + \frac18$.

II.2.2 When  suggested 1/1000 as an error tolerance challenge number, R. D. said that $\lim_{x\to2} g(x)$ would just put out a guaranteed 1/500 . Check to see that this is really guaranteed.

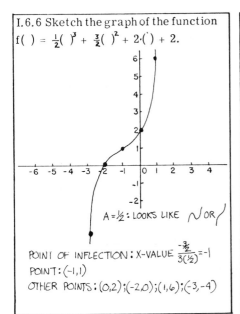

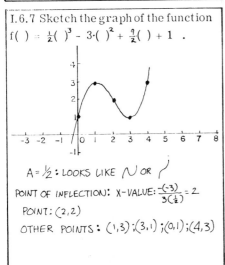

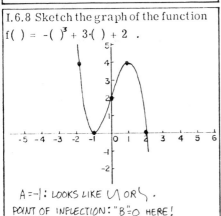

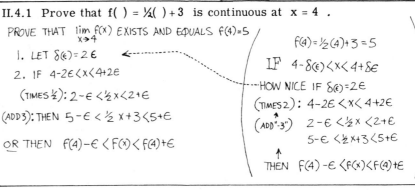

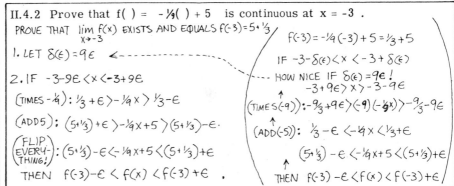

IF $2 - \frac{1}{500} < x < 2 + \frac{1}{500}$

(TIMES ½) $1 - \frac{1}{1000} < \frac12 x < 1 + \frac{1}{1000}$

THEN

(ADD 2) : $3 - \frac{1}{1000} < \frac12 x + 2 < 3 + \frac{1}{1000}$

OR,

THEN $g(2) - \frac{1}{1000} < g(x) < g(2) + \frac{1}{1000}$

II.3.1 If f( ) = 3( )-1, $\lim_{x\to2} f(x)$ puts out $\delta(\epsilon)$ with formula $\delta(\epsilon) = \frac13\epsilon$ Check the guarantee; i.e., letting "a"=2, reason from

IF $2 - \delta(\epsilon) < x < 2 + \delta(\epsilon)$ ($\delta(\epsilon) = \frac13\epsilon$)

SO, $2 - \frac13\epsilon < x < 2 + \frac13\epsilon$

(TIMES 3) $6 - \epsilon < 3x < 6 + \epsilon$

(ADD -1) $5 - \epsilon < 3x - 1 < 5 + \epsilon$

OR (SINCE f(2) = 3(2)-1 = 5),

~~to~~ THEN $f(2) - \epsilon < f(x) < f(2) + \epsilon$.

II.3.2 If f( ) = -3( )+5, $\lim_{x\to1} f(x)$ puts out $\delta(\epsilon) = \frac13\epsilon$. Check the GUARANTEE. (Don't forget that multiplying inequalities by a negative number flips the inequality signs!) ($\delta(\epsilon) = \frac13\epsilon$) SO,

IF $1 - \frac13\epsilon < x < 1 + \frac13\epsilon$

(TIMES -3) $-3 + \epsilon > -3x > -3 - \epsilon$

(ADD 5) $2 + \epsilon > -3x + 5 > 2 - \epsilon$

OR THEN $2 - \epsilon < -3x + 5 < 2 + \epsilon$

OR $f(1) - \epsilon < f(x) < f(1) + \epsilon$

II.3.3 Using the same function as II.3.2 , suppose that $\lim_{x\to1} f(x)$ puts out $\delta(\epsilon) = \frac16\epsilon$ instead. Show that this can still be guaranteed. (Hint: use the fact that $\frac12\epsilon < \epsilon$ since $\epsilon > 0$.)

IF $1 - \frac16\epsilon < x < 1 + \frac16\epsilon$ (TIMES -3)

$-3 + \frac12\epsilon > -3x > -3 - \frac12\epsilon$ (ADD 5)

$2 + \frac12\epsilon > -3x + 5 > 2 - \frac12\epsilon$

OR $2 - \frac12\epsilon < -3x + 5 < 2 + \frac12\epsilon$

(BUT $\frac12\epsilon < \epsilon$ SO, $(-1)\frac12\epsilon > (-1)\epsilon$ OR $-\frac12\epsilon > -\epsilon$, TOO)

THEN $2 - \epsilon < 2 - \frac12\epsilon < -3x + 5 < 2 + \frac12\epsilon < 2 + \epsilon$

OR

THEN $2 - \epsilon < -3(x) + 5 < 2 + \epsilon$!

II.4.1 Prove that $f(\ ) = \frac12(\ ) + 3$ is continuous at x = 4 .

PROVE THAT $\lim_{x\to4} f(x)$ EXISTS AND EQUALS f(4)=5

1. LET $\delta(\epsilon) = 2\epsilon$
2. IF $4 - 2\epsilon < x < 4 + 2\epsilon$

(TIMES ½): $2 - \epsilon < \frac12 x < 2 + \epsilon$

(ADD 3): THEN $5 - \epsilon < \frac12 x + 3 < 5 + \epsilon$

OR THEN $f(4) - \epsilon < f(x) < f(4) + \epsilon$

$f(4) = \frac12(4) + 3 = 5$

IF $4 - \delta(\epsilon) < x < 4 + \delta(\epsilon)$

HOW NICE IF $\delta(\epsilon) = 2\epsilon$

(TIMES 2): $4 - 2\epsilon < x < 4 + 2\epsilon$

(ADD "-3") $2 - \epsilon < \frac12 x < 2 + \epsilon$

$5 - \epsilon < \frac12 x + 3 < 5 + \epsilon$

THEN $f(4) - \epsilon < f(x) < f(4) + \epsilon$

II.4.2 Prove that $f(\ ) = -\frac19(\ ) + 5$ is continuous at x = -3 .

PROVE THAT $\lim_{x\to-3} f(x)$ EXISTS AND EQUALS $f(-3) = 5 + \frac13$

1. LET $\delta(\epsilon) = 9\epsilon$
2. IF $-3 - 9\epsilon < x < -3 + 9\epsilon$

(TIMES $-\frac19$): $\frac13 + \epsilon > -\frac19 x > \frac13 - \epsilon$

(ADD 5): $(5 + \frac13) + \epsilon > -\frac19 x + 5 > (5 + \frac13) - \epsilon$.

(FLIP EVERY-THING!): $(5 + \frac13) - \epsilon < -\frac19 x + 5 < (5 + \frac13) + \epsilon$

THEN $f(-3) - \epsilon < f(x) < f(-3) + \epsilon$.

$f(-3) = -\frac19(-3) + 5 = \frac13 + 5$

IF $-3 - \delta(\epsilon) < x < -3 + \delta(\epsilon)$

HOW NICE IF $\delta(\epsilon) = 9\epsilon$!

$-3 + 9\epsilon > x > -3 - 9\epsilon$

(TIMES (-9)): $-\frac93 + 9\epsilon > (-9)(-\frac19 x) > -\frac93 - 9\epsilon$

(ADD (-5)): $\frac13 - \epsilon < -\frac19 x < \frac13 + \epsilon$

$(5 + \frac13) - \epsilon < -\frac19 x + 5 < (5 + \frac13) + \epsilon$

THEN $f(-3) - \epsilon < f(x) < f(-3) + \epsilon$

II.4.3 Prove that f( ) = 27 - 1/10( ) is continuous at x = 5 .

PROVE THAT $\lim_{x\to 5} f(x)$ EXISTS AND EQUALS $f(5)=27-\frac{1}{2}$

1. LET $\delta(\epsilon)=10\epsilon$ ←

2. IF $5-10\epsilon<x<5+10\epsilon$

(TIMES $-\frac{1}{10}$): $-\frac{1}{2}+\epsilon>-\frac{1}{10}x>-\frac{1}{2}-\epsilon$

(ADD 27): $(27-\frac{1}{2})+\epsilon>27-\frac{1}{10}x>(27-\frac{1}{2})-\epsilon$

OR, $(27-\frac{1}{2})-\epsilon<27-\frac{1}{10}x<(27-\frac{1}{2})+\epsilon$

THEN $f(5)-\epsilon<f(x)<f(5)+\epsilon$

$f(5)=27-\frac{1}{10}(5)=27-\frac{1}{2}$

IF $5-\delta(\epsilon)<x<5+\delta(\epsilon)$

HOW NICE IF $\delta(\epsilon)=10\epsilon$!

$5-10\epsilon<x<5+10\epsilon$

↑

(TIMES -10): $5+10\epsilon>x>5-10\epsilon$

↑

(ADD -27): $-\frac{1}{2}-\epsilon<-\frac{1}{10}x<-\frac{1}{2}+\epsilon$

$(27-\frac{1}{2})-\epsilon<27-\frac{1}{10}x<(27-\frac{1}{2})+\epsilon$

↑

THEN $f(5)-\epsilon<f(x)<f(5)+\epsilon$

II.4.4 Prove that f( ) = 3 - 18( ) is continuous at x = -6.

PROVE THAT $\lim_{x\to -6} f(x)$ EXISTS AND EQUALS $f(-6)=111$.

1. LET $\delta(\epsilon)=\frac{1}{18}\epsilon$. ←

2. IF $-6-\frac{1}{18}\epsilon<x<-6+\frac{1}{18}\epsilon$

(TIMES -18): $108+\epsilon>-18x>108-\epsilon$

(ADD 3) $111+\epsilon>3-18x>111-\epsilon$

OR $111-\epsilon<3-18x<111+\epsilon$

THEN $f(-6)-\epsilon<f(x)<f(-6)+\epsilon$

IF $-6-\delta(\epsilon)<x<-6+\delta(\epsilon)$

HOW NICE IF $\delta(\epsilon)=\frac{1}{18}\epsilon$

$-6-\frac{1}{18}\epsilon<x<-6+\frac{1}{18}\epsilon$

↑

(TIMES $\frac{1}{-18}$): $-6+\frac{1}{18}\epsilon>x>-6-\frac{1}{18}\epsilon$

↑

(ADD -3): $108-\epsilon<-18x<108+\epsilon$

↑

$111-\epsilon<3-18x<111+\epsilon$

↑

THEN $f(-6)-\epsilon<f(x)<f(-6)+\epsilon$

II.4.5 Prove that f( ) = $-\frac{2}{5}$( ) + $\frac{7}{5}$ is continuous at x = $-\frac{13}{2}$.

PROVE THAT $\lim_{x\to -\frac{13}{2}} f(x)$ EXISTS AND EQUALS $f(\frac{-13}{2})=\frac{-2}{5}(\frac{-13}{2})+\frac{7}{5}=4$

1. LET $\delta(\epsilon)=\frac{5}{2}\epsilon$. ←

2. IF $-\frac{13}{2}-\frac{5}{2}\epsilon<x<\frac{-13}{2}+\frac{5}{2}\epsilon$

(TIMES $-\frac{2}{5}$): $\frac{13}{5}+\epsilon>-\frac{2}{5}x>\frac{13}{5}-\epsilon$

(ADD $\frac{7}{5}$): $4+\epsilon>\frac{7}{5}-\frac{2}{5}x>4-\epsilon$

OR, $4-\epsilon<\frac{7}{5}-\frac{2}{5}x<4+\epsilon$

THEN $f(-\frac{13}{2})-\epsilon<\frac{7}{5}-\frac{2}{5}x<f(\frac{-13}{2})+\epsilon$

IF $-\frac{13}{2}-\delta(\epsilon)<x<-\frac{13}{2}+\delta(\epsilon)$

HOW NICE IF $\delta(\epsilon)=\frac{5}{2}\epsilon$

$-\frac{13}{2}-\frac{5}{2}\epsilon<x<\frac{-13}{2}+\frac{5}{2}\epsilon$.

↑

(TIMES $-\frac{5}{2}$): $-\frac{13}{2}+\frac{5}{2}\epsilon>x>-\frac{13}{2}-\frac{5}{2}\epsilon$

↑

(ADD $-\frac{7}{5}$): $\frac{13}{5}-\epsilon<-\frac{2}{5}x<\frac{13}{5}+\epsilon$

$4-\epsilon<\frac{7}{5}-\frac{2}{5}x<4+\epsilon$

↑

THEN $f(\frac{-13}{2})-\epsilon<f(x)<f(\frac{-13}{2})+\epsilon$

II.4.6 Prove that f( ) = 55( ) + 44 is continuous at x = 11 .

PROVE THAT $\lim_{x\to 11} f(x)$ EXISTS AND EQUALS $f(11)=55(11)+44=649$.

1. LET $\delta(\epsilon)=\frac{1}{55}\epsilon$.

2. IF $11-\frac{1}{55}\epsilon<x<11+\frac{1}{55}\epsilon$

(TIMES 55): $605-\epsilon<55x<605+\epsilon$

(ADD 44): $649-\epsilon<55x+44<649+\epsilon$

THEN $f(11)-\epsilon<f(x)<f(11)+\epsilon$

II.4.7 Prove that f( ) = 2( ) + 7 is continuous at x = a for ANY real number "a."

THIS WILL WORK JUST LIKE THE OTHERS EVEN THOUGH WE DON'T KNOW WHAT "a" IS.

PROVE THAT $\lim_{x\to a} f(x)$ EXISTS AND EQUALS $f(a) = 2a+7$

1. LET $\delta(\epsilon) = \frac{1}{2}\epsilon$. ←

2. IF $a - \frac{1}{2}\epsilon < x < a + \frac{1}{2}\epsilon$

(TIMES 2): $2a - \epsilon < 2x < 2a + \epsilon$

(ADD 7) $(2a+7) - \epsilon < 2x+7 < (2a+7) + \epsilon$

OR, THEN $f(a) - \epsilon < f(x) < f(a) + \epsilon$

ALGEBRA SAYS THAT THIS WILL HOLD FOR ANY "a"!

(IF $a - \delta(\epsilon) < x < a + \delta(\epsilon)$

HOW NICE IF $\delta(\epsilon) = \frac{1}{2}\epsilon$

(TIMES $\frac{1}{2}$): $a - \frac{1}{2}\epsilon < x < a + \frac{1}{2}\epsilon$

(add −7) $2a - \epsilon < 2x < 2a + \epsilon$

$(2a+7) - \epsilon < 2x+7 < (2a+7) + \epsilon$

THEN $f(a) - \epsilon < f(x) < f(a) + \epsilon$)

II.4.8 Prove that f( ) = ( ) (same as y = f(x) = x) is continuous at x = a for any real number "a."

PROVE THAT $\lim_{x\to a} f(x)$ EXISTS AND EQUALS $f(a) = a$.

1. LET $\delta(\epsilon) = \epsilon$ ←

2. IF $a - \epsilon < x < a + \epsilon$

THEN $a - \epsilon < x < a + \epsilon$

SO, $f(a) - \epsilon < f(x) < f(a) + \epsilon$.

(A MOST PECULIAR PROOF! DEPENDS ON A TRUTH LIKE "IF FIDO HAS FLEAS THEN FIDO HAS FLEAS" – CERTAINLY TRUE!)

(IF $a - \delta(\epsilon) < x < a + \delta(\epsilon)$

HOW NICE IF $\delta(\epsilon) = \epsilon$!

HOW NICE IF $a - \epsilon < x < a + \epsilon$

THEN $f(a) - \epsilon < f(x) < f(a) + \epsilon$)

II.5.1 Prove that $3x^2-4x-1$ is continuous at x = 2.

SHOW THAT $\lim_{x\to 2} (3x^2-4x-1)$ EXISTS AND EQUALS $3\cdot(2)^2 - 4\cdot(2) - 1 = 3$

PROOF: 1. FOR ANY $\epsilon > 0$, LET $\delta(\epsilon) = \min(1, \epsilon/11)$

2. IF $2 - \delta(\epsilon) < x < 2 + \delta(\epsilon)$,

SO $-\delta(\epsilon) < x - 2 < \delta(\epsilon)$, OR

THEN $|x-2| < \delta(\epsilon)$.

SINCE $\delta(\epsilon) \le 1$, $|x-2| < \delta(\epsilon)$ MEANS $|x-2| < 1$ OR $-1 < x-2 < 1$

OR $1 < x < 3$ OR $3 < 3x < 9$

OR $-11 < 5 < 3x+2 < 11$, SO $|3x+2| < 11$. SINCE $\delta(\epsilon) = \min(1, \epsilon/11)$,

THEN $|x-2| < \epsilon/11$ ALSO, SO

$|x-2|\cdot|3x+2| \le |x-2|\cdot 11 < \epsilon/11 \cdot 11 = \epsilon$,

OR, $|x-2|\cdot|3x+2| < \epsilon$

OR $|3x^2-4x-4| < \epsilon$

OR $-\epsilon < 3x^2-4x-4 < \epsilon$

THEN, $3 - \epsilon < 3x^2-4x-1 < 3 + \epsilon$

II.5.2 Prove that $3-x-2x^2$ is continuous at x = −1.

SHOW THAT $\lim_{x\to -1} (3-x-2x^2)$ EXISTS AND EQUALS $3-(-1)-2(-1)^2 = 2$.

PROOF: 1. FOR ANY $\epsilon > 0$, LET $\delta(\epsilon) = \min(1, \epsilon/5)$

2. IF $-1 - \delta(\epsilon) < x < -1 + \delta(\epsilon)$,

SO $-\delta(\epsilon) < x+1 < \delta(\epsilon)$, OR

THEN $|x+1| < \delta(\epsilon)$.

SINCE $\delta(\epsilon) \le 1$, $|x+1| < \delta(\epsilon)$ MEANS $|x+1| < 1$ OR $-1 < x+1 < 1$

OR $-2 < x < 0$ OR (TIMES "−2")

$4 > -2x > 0$ OR $5 > |-2x> 1 > -5$

SO $|1-2x| < 5$. SINCE $\delta(\epsilon) = \min(1, \epsilon/5)$,

$|x+1| < \delta(\epsilon)$ MEANS $|x+1| < \epsilon/5$ ALSO, SO

$|x+1|\cdot|1-2x| \le |x+1|\cdot 5 < \epsilon/5 \cdot 5 = \epsilon$,

OR, $|x+1|\cdot|1-2x| < \epsilon$

OR, $|1+x|\cdot|1-2x| < \epsilon$

OR, $|1-x-2x^2| < \epsilon$

OR, $-\epsilon < 1-x-2x^2 < \epsilon$

THEN $2 - \epsilon < 3-x-2x^2 < 2 + \epsilon$

II.5.3 Prove that $2x^2+x-1$ is continuous at $x = \frac{1}{2}$.

SHOW THAT $\lim_{x\to 1/2} (2x^2+x-1)$ EXISTS AND EQUALS $2(\frac{1}{2})^2 + \frac{1}{2} - 1 = 0$

PROOF: 1. FOR ANY $\epsilon > 0$, LET $\delta(\epsilon) = \min(1, \epsilon/5)$.

2. IF $\frac{1}{2} - \delta(\epsilon) < x < \frac{1}{2} + \delta(\epsilon)$,

SO $-\delta(\epsilon) < x - \frac{1}{2} < \delta(\epsilon)$,

OR, THEN $|x - \frac{1}{2}| < \delta(\epsilon)$.

SINCE $\delta(\epsilon) \le 1$, $|x-\frac{1}{2}| < \delta(\epsilon)$ MEANS $|x-\frac{1}{2}| < 1$ OR $-1 < x-\frac{1}{2} < 1$

OR $-\frac{1}{2} < x < \frac{3}{2}$ OR, $-1 < 2x < 3$

OR $-5 < 1 < 2x+2 < 5$... SO $|2x+2| < 5$. SINCE $\delta(\epsilon) = \min(1, \epsilon/5)$,

$|x-\frac{1}{2}| < \delta(\epsilon)$ MEANS $|x-\frac{1}{2}| < \epsilon/5$ ALSO, SO...

$|2x+2|\,|x-\frac{1}{2}| \le 5\cdot|x-\frac{1}{2}| < 5\cdot\epsilon/5 = \epsilon$,

OR, $|2x+2|\cdot|x-\frac{1}{2}| < \epsilon$

OR, $|2(x+1)(x-\frac{1}{2})| < \epsilon$

OR, $|(x+1)(2x-1)| < \epsilon$

OR, $|2x^2+x-1| < \epsilon$

THEN $0 - \epsilon < 2x^2+x-1 < 0 + \epsilon$

II.5.4 Prove that $f(x) = x^2$ is continuous at x = a for any real number 'a'.

SHOW THAT $\lim_{x\to a} x^2$ EXISTS AND EQUALS a^2.

PROOF: 1. FOR ANY $\epsilon > 0$, LET $\delta(\epsilon) = \min(1, \epsilon/(2|a|+1))$

2. IF $a - \delta(\epsilon) < x < a + \delta(\epsilon)$

SO, $-\delta(\epsilon) < x - a < \delta(\epsilon)$, OR

THEN $|x-a| < \delta(\epsilon)$

SINCE $\delta(\epsilon) \le 1$, $|x-a| < \delta(\epsilon)$ MEANS $|x-a| < 1$ OR, $-1 < x-a < 1$

OR, $a-1 < x < a+1$ OR,

$2a-1 < x+a < 2a+1$ OR,

$-(2|a|+1) \le 2a-1 < x+a < 2a+1 \le 2|a|+1$,

SO $|x+a| < 2|a|+1$. (WE NEED ABSOLUTE VALUES TO MAKE SURE "$2|a|+1$" IS POSITIVE.)

SINCE $\delta(\epsilon) = \min(1, \epsilon/(2|a|+1))$,

$|x-a| < \delta(\epsilon)$ MEANS $|x-a| < \epsilon/(2|a|+1)$ ALSO, SO

$|x-a|\cdot|x+a| \le |x-a|\cdot(2|a|+1) < \frac{\epsilon}{(2|a|+1)}\cdot(2|a|+1) = \epsilon$,

OR, $|x-a|\cdot|x+a| < \epsilon$

OR, $|x^2-a^2| < \epsilon$

OR, $-\epsilon < x^2-a^2 < \epsilon$

OR, $a^2 - \epsilon < x^2 < a^2 + \epsilon$.

II.5.6 Prove that $f(x) = x^3 - 2x^2 + 3x - 4$ is continuous at $x = 2$.

SHOW THAT $\lim_{x \to 2} (x^3 - 2x^2 + 3x - 4)$ EXISTS AND EQUALS $(2)^3 - 2(2)^2 + 3(2) - 4 = 2$.

PROOF: 1. FOR ANY $\epsilon > 0$, LET $\delta(\epsilon) = \text{MIN}(1, \epsilon/12)$.

2. IF $2 - \delta(\epsilon) < x < 2 + \delta(\epsilon)$ SO $-\delta(\epsilon) < x - 2 < \delta(\epsilon)$, OR THEN $|x-2| < \delta(\epsilon)$. SINCE $\delta(\epsilon) \leq 1$, $|x-2| < \delta(\epsilon)$ MEANS $|x-2| < 1$, OR $-1 < x-2 < 1$, OR $1 < x < 3$ OR $|x| < 3$, SO $x^2 = |x|^2 < 9$. THEN $-12 < 3 \leq x^2 + 3 < 9 + 3 = 12$ OR $|x^2 + 3| < 12$. SINCE $\delta(\epsilon) = \text{MIN}(1, \epsilon/12)$, $|x-2| < \delta(\epsilon)$ MEANS $|x-2| < \epsilon/12$ ALSO, SO $|x-2| \cdot |x^2+3| \leq |x-2| \cdot 12 < \epsilon/12 \cdot 12 = \epsilon$

OR, $|x-2| \cdot |x^2+3| < \epsilon$

OR, $|x^3 - 2x^2 + 3x - 6| < \epsilon$

OR, $-\epsilon < x^3 - 2x^2 + 3x - 6 < \epsilon$, OR

THEN $2 - \epsilon < x^3 - 2x^2 + 3x - 4 < 2 + \epsilon$.

II.6.1 If

$$f(x) = \begin{cases} 2 & \text{if } x > 1 \\ 3 & \text{if } x = 1 \\ 4 & \text{if } x < 1 \end{cases}$$

prove that f( ) is not continuous at $x = 1$.

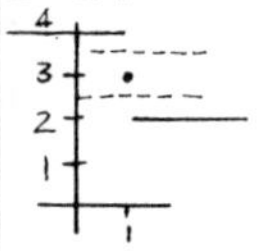

LET $\epsilon = \frac{1}{2}$. IF THERE IS ANY $\delta(\frac{1}{2}) > 0$, THEN $1 + \frac{1}{2}\delta(\frac{1}{2}) > 1$, SO $f(1 + \frac{1}{2}\delta(\frac{1}{2})) = 2$. THEN $1 + \frac{1}{2}\delta(\frac{1}{2})$ SATISFIES $1 - \delta(\frac{1}{2}) < 1 + \frac{1}{2}\delta(\frac{1}{2}) < 1 + \delta(\frac{1}{2})$, YET

$$\underset{f(1)}{3} - \tfrac{1}{2} \underset{\text{FALSE}}{<} \underbrace{f(1+\tfrac{1}{2}\delta(\tfrac{1}{2}))}_{2} < 3 + \tfrac{1}{2}$$

IS FALSE, SO NO $\delta(\frac{1}{2})$ CAN EXIST.

II.6.2 If

$$f(x) = \begin{cases} 3 & \text{if } x > 1 \\ 2 & \text{if } x \leq 1 \end{cases}$$

prove that f( ) is not continuous at $x = 1$.

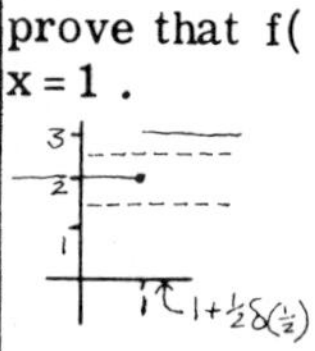

LET $\epsilon = \frac{1}{2}$. IF THERE IS ANY $\delta(\frac{1}{2}) > 0$, THEN $1 + \frac{1}{2}\delta(\frac{1}{2}) > 1$, SO $f(1 + \frac{1}{2}\delta(\frac{1}{2})) = 3$. THEN $1 + \frac{1}{2}\delta(\frac{1}{2})$ SATISFIES $1 - \delta(\frac{1}{2}) < 1 + \frac{1}{2}\delta(\frac{1}{2}) < 1 + \delta(\frac{1}{2})$, YET

$$\underset{f(1)}{2} - \tfrac{1}{2} < \underbrace{f(1+\tfrac{1}{2}\delta(\tfrac{1}{2}))}_{3} \underset{\text{FALSE}}{<} 2 + \tfrac{1}{2}$$

IS FALSE. SO NO $\delta(\frac{1}{2})$ CAN EXIST, AND THUS f( ) IS NOT CONTINUOUS AT $x = 1$.

II.6.3 If

$$f(x) = \begin{cases} 3 & \text{if } x \geq 1 \\ 2 & \text{if } x < 1 \end{cases}$$

prove that f( ) is not continuous at $x = 1$.

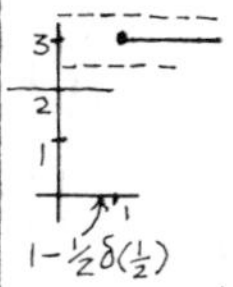

LET $\epsilon = \frac{1}{2}$. IF THERE IS ANY $\delta(\frac{1}{2}) > 0$, THEN $1 - \frac{1}{2}\delta(\frac{1}{2}) < 1$, SO $f(1 - \frac{1}{2}\delta(\frac{1}{2})) = 2$. THEN $1 - \frac{1}{2}\delta(\frac{1}{2})$ SATISFIES $1 - \delta(\frac{1}{2}) < 1 - \frac{1}{2}\delta(\frac{1}{2}) < 1 + \delta(\frac{1}{2})$,

$$\text{YET } \underset{f(1)}{3} - \tfrac{1}{2} \underset{\text{FALSE}}{<} \underbrace{f(1-\tfrac{1}{2}\delta(\tfrac{1}{2}))}_{2} < 3 + \tfrac{1}{2}$$

IS FALSE. SO NO $\delta(\frac{1}{2})$ CAN EXIST, AND THUS f( ) IS NOT CONTINUOUS AT 1.

II.6.4 If

$$f(x) = \begin{cases} 1.02 & \text{if } x > 2 \\ 1.00 & \text{if } x \leq 2 \end{cases}$$

prove that f( ) is not continuous at $x = 2$.

MAGNIFY:

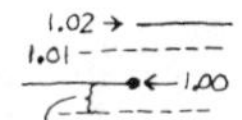

LET $\epsilon = .01$. IF THERE IS ANY $\delta(.01) > 0$, THEN $2 + \frac{1}{2}\delta(.01) > 2$, SO $f(2 + \frac{1}{2}\delta(.01)) = 1.02$. THEN $2 + \frac{1}{2}\delta(.01)$ SATISFIES $2 - \delta(.01) < 2 + \frac{1}{2}\delta(.01) < 2 + \delta(.01)$,

$$\text{YET } \underset{f(2)}{1} - .01 < \underbrace{f(2+\tfrac{1}{2}\delta(.01))}_{1.02} \underset{\text{FALSE}}{<} 1 + .01$$

IS FALSE. SO NO $\delta(.01)$ CAN EXIST.

II.7.1 Prove that $f(x) = 2x^2$ is continuous at every c ; $0 < c < 4$.

PROOF: 1. FOR ANY $\epsilon > 0$, LET $\delta(\epsilon) = \epsilon/16$.

2. IF $c - \delta(\epsilon) < x < c + \delta(\epsilon)$ THEN $|x - c| < \delta(\epsilon)$. NOW, SINCE $0 < c < 4$, AND WE ONLY WANT TO CONSIDER X'S SATISFYING $0 < x < 4$, WE CAN MULTIPLY BOTH INEQUALITIES BY 2: $0 < 2c < 8$, $0 < 2x < 8$;

THEN ADD: $0 < 2x + 2c < 16$

OR $|2x + 2c| < 16$. SINCE $\delta(\epsilon) = \epsilon/16$, $|x - c| < \delta(\epsilon) = \epsilon/16$ GIVES $|x-c| \cdot |2x + 2c| \leq |x - c| \cdot 16 < \frac{\epsilon}{16} \cdot 16 = \epsilon$

OR, $|x-c| \cdot |2x+2c| < \epsilon$

OR, $|2x^2 - 2c^2| < \epsilon$

OR, $-\epsilon < 2x^2 - 2c^2 < \epsilon$

OR, $2c^2 - \epsilon < 2x^2 < 2c^2 + \epsilon$.

II.7.2 Prove that $f(x) = x^2 + 3x - 7$ is continuous at every c ; $1 < c < 5$.

PROOF: 1. FOR ANY $\epsilon > 0$, LET $\delta(\epsilon) = \epsilon/13$.

2. IF $c - \delta(\epsilon) < x < c + \delta(\epsilon)$ THEN, $|x - c| < \delta(\epsilon)$. NOW, SINCE $1 < c < 5$ AND WE ONLY WANT TO CONSIDER X'S SATISFYING $1 < x < 5$ AS WELL, WE CAN ADD BOTH INEQUALITIES TO GET...

II.7.2 (CONTINUED)

$\ldots 2 < x + c < 10$, SO $-13 < 5 < x + c + 3 < 13$, OR $|x + c + 3| < 13$. SINCE $\delta(\epsilon) = \epsilon/13$, $|x - c| < \delta(\epsilon) = \epsilon/13$ GIVES $|x-c| \cdot |x+c+3| \leq |x-c| \cdot 13 < \epsilon/13 \cdot 13 = \epsilon$

OR, $|x-c| \cdot |x+c+3| < \epsilon$

OR, $|x^2 - c^2 + 3(x-c)| < \epsilon$

OR, $|x^2 + 3x - 7 - (c^2 + 3c - 7)| < \epsilon$

OR, $-\epsilon < x^2 + 3x - 7 - (c^2 + 3c - 7) < \epsilon$, OR

$(c^2 + 3c - 7) - \epsilon < x^2 + 3x - 7 < (c^2 + 3c - 7) + \epsilon$

II.8.1 Suppose f( ) and g( ) are functions such that

1. $f(x) = g(x)$ if $x \neq a$ and
2. f( ) and g( ) are continuous at $x = a$.

Prove that $g(a) > f(a)$ is impossible.

PROOF: IF $g(a) > f(a)$, LET $\epsilon = \frac{1}{2}(g(a) - f(a))$. SINCE f( ) IS CONTINUOUS AT "a", THERE IS A $\delta^f(\frac{1}{2}(g(a) - f(a))) > 0$ SUCH THAT, FOR X'S SATISFYING

(i) $a - \delta^f(\frac{1}{2}(g(a)-f(a))) < x < a + \delta^f(\frac{1}{2}(g(a)-f(a)))$,

$f(a) - \frac{1}{2}(g(a) - f(a)) < f(x) < f(a) + \frac{1}{2}(g(a) - f(a))$

SO, $f(x) < \frac{1}{2}(f(a) + g(a))$.

SINCE g( ) IS CONTINUOUS AT "a", THERE IS A $\delta^g(\frac{1}{2}(g(a)-f(a)))$ SUCH THAT, FOR X'S SATISFYING

(ii) $a - \delta^g(\frac{1}{2}(g(a)-f(a))) < x < a + \delta^g(\frac{1}{2}(g(a)-f(a)))$,

$g(a) - \frac{1}{2}(g(a) - f(a)) < g(x) < g(a) + \frac{1}{2}(g(a) - f(a))$

SO $\frac{1}{2}(f(a) + g(a)) < g(x)$. NOW,

$\min(\delta^f(\frac{1}{2}(g(a)-f(a))), \delta^g(\frac{1}{2}(g(a)-f(a)))) > 0$,

SO $c = a + \frac{1}{2}\min(\delta^f(\frac{1}{2}(g(a)-f(a))), \delta^g(\frac{1}{2}(g(a)-f(a)))) > a$

THEREFORE $f(c) = g(c)$, AND FURTHERMORE, c IS IN THE GUARANTEED INTERVALS GIVEN BY (i) AND (ii). AS A RESULT, $f(c) < \frac{1}{2}(f(a) + g(a))$ AND $\frac{1}{2}(f(a) + g(a)) < g(c)$. PUT THESE TWO TOGETHER:

$f(c) < \frac{1}{2}(f(a) + g(a)) < g(c) = f(c)$, A CONTRADICTION.

II.9.1
Prove that $\lim_{x\to 3} \frac{x^2-x-6}{x-3}$ exists.

PROOF: LET $L=5$. FOR ANY $\epsilon>0$, LET $\delta(\epsilon)=\epsilon$.
IF $3-\epsilon < x < 3+\epsilon$ AND $x\neq 3$
THEN $5-\epsilon < x+2 < 5+\epsilon$.
SINCE $x\neq 3$, $\frac{x-3}{x-3}=1$, SO
THEN $5-\epsilon < \frac{(x+2)(x-3)}{x-3} < 5+\epsilon$
OR $5-\epsilon < \frac{x^2-x-6}{x-3} < 5+\epsilon$.

II.9.2
Prove that $\lim_{x\to -4} \frac{-2x^2-5x+12}{x+4}$ exists.

PROOF: LET $L=11$. FOR ANY $\epsilon>0$, LET $\delta(\epsilon)=\epsilon/2$.
IF $-4-\frac{\epsilon}{2} < x < -4+\epsilon/2$ AND $x\neq -4$
THEN $8+\epsilon > -2x > 8-\epsilon$
SO $11+\epsilon > -2x+3 > 11-\epsilon$.
SINCE $x\neq -4$, $\frac{x+4}{x+4}=1$, SO
THEN $11+\epsilon > \frac{(-2x+3)(x+4)}{x+4} > 11-\epsilon$
OR $11+\epsilon > \frac{-2x^2-5x+12}{x+4} > 11-\epsilon$.

II.9.3
Prove that $\lim_{x\to 2} \frac{x^2-4x+4}{x-2}$ exists.

PROOF: LET $L=0$. FOR ANY $\epsilon>0$, LET $\delta(\epsilon)=\epsilon$.
IF $2-\epsilon < x < 2+\epsilon$ AND $x\neq 2$,
THEN $0-\epsilon < x-2 < 0+\epsilon$.
SINCE $x\neq 2$, $\frac{x-2}{x-2}=1$, SO
THEN $-\epsilon < \frac{(x-2)(x-2)}{(x-2)} < \epsilon$
OR $-\epsilon < \frac{x^2-4x+4}{x-2} < \epsilon$.

II.9.4
Prove that $\lim_{x\to 0} \frac{0}{x}$ exists.

PROOF: LET $L=0$. FOR ANY $\epsilon>0$, LET $\delta(\epsilon)=$ (ANYTHING POSITIVE)
IF $0-\delta(\epsilon) < x < 0+\delta(\epsilon)$, AND $x\neq 0$,
IT DOESN'T MATTER WHAT THIS SAYS, SINCE THIS IS ALWAYS TRUE:
NOW $0-\epsilon < 0 < 0+\epsilon$, AND SINCE $x\neq 0$,
SO $0-\epsilon < \frac{0\cdot x}{x} < 0+\epsilon$. BUT $0\cdot x=0$,
SO $0-\epsilon < \frac{0}{x} < 0+\epsilon$.

II.9.5
Prove that $\lim_{h\to 0} \frac{2h^3+3h^2+h}{h}$ exists.

PROOF: LET $L=1$. FOR ANY $\epsilon>0$, LET $\delta(\epsilon)=\min(1,\epsilon/5)$
IF $0-\delta(\epsilon) < h < 0+\delta(\epsilon)$ AND $h\neq 0$,
THEN $|h| < \delta(\epsilon) \leq 1$ GIVES $-1<h<1$
OR $-2<2h<2$ OR $-5<1<2h+3<5$ OR $|2h+3|<5$. SINCE $\delta(\epsilon)=\min(1,\epsilon/5)$,
$|h|<\delta(\epsilon)$ GIVES $|h|<\epsilon/5$ ALSO. SO
$|h|\cdot|2h+3| \leq |h|\cdot 5 < \epsilon/5\cdot 5=\epsilon$
OR $|h|\cdot|2h+3|<\epsilon$ OR $|2h^2+3h|<\epsilon$
OR $-\epsilon<2h^2+3h<\epsilon$ OR $1-\epsilon<2h^2+3h+1<1+\epsilon$
SINCE $h\neq 0$, $\frac{h}{h}=1$, SO $1-\epsilon<\frac{2h^3+3h^2+h}{h}<1+\epsilon$.

II.9.6
Prove that $\lim_{x\to 2} \frac{x^3-4x^2+7x-6}{x-2}$ exists.

PROOF: LET $L=3$. FOR ANY $\epsilon>0$, LET $\delta(\epsilon)=\min(1,\epsilon/3)$
IF $2-\delta(\epsilon) < x < 2+\delta(\epsilon)$ AND $x\neq 2$,
THEN $-\delta(\epsilon) < x-2 < \delta(\epsilon)$ OR $|x-2|<\delta(\epsilon)$.
SINCE $\delta(\epsilon)\leq 1$, $|x-2|\leq 1$ OR $-1<x-2<1$
OR $1<x<3$ OR $|x|<3$.
SINCE $\delta(\epsilon)=\min(1,\epsilon/3)$, $|x-2|<\delta(\epsilon)$
MEANS $|x-2|<\epsilon/3$ ALSO. SO
$|x|\cdot|x-2| \leq 3\cdot|x-2| < 3\cdot\epsilon/3=\epsilon$,
OR, $|x^2-2x|<\epsilon$
OR, $-\epsilon<x^2-2x<\epsilon$
OR, $3-\epsilon<x^2-2x+3<3+\epsilon$. SINCE
$x\neq 2$, $\frac{x-2}{x-2}=1$, SO
$3-\epsilon < \frac{(x^2-2x+3)(x-2)}{x-2} < 3+\epsilon$
OR THEN $3-\epsilon < \frac{x^3-4x^2+7x-6}{x-2} < 3+\epsilon$.

III.1.1 Prove that
$f(x) = 3x^2 - 6x + 2$
is differentiable at $x = -3$. Find the derivative $Df|_{-3}$.

STEP 1: $\Delta(h) = \frac{f(-3+h)-f(-3)}{h} =$
$\frac{[3(-3+h)^2-6(-3+h)+2]-[3(-3)^2-6(-3)+2]}{h}$
$= \frac{[3(9-6h+h^2)+18-6h+2]-[27+18+2]}{h}$
$= \frac{3h^2-24h}{h} = \begin{cases} 3h-24 & \text{IF } h\neq 0. \\ \text{UNDEFINED} & \text{IF } h=0. \end{cases}$
STEP 2: $Df|_{-3} \overset{?}{=} -24$ SO LET $L=-24$
STEP 3: 1. FOR ANY $\epsilon>0$, LET $\delta(\epsilon)=\epsilon/3$.
2. IF $0-\epsilon/3 < h < 0+\epsilon/3$ AND $h\neq 0$
THEN $-\epsilon < 3h < \epsilon$
SO $-24-\epsilon < 3h-24 < -24+\epsilon$
SINCE $h\neq 0$, $\frac{h}{h}=1$, SO
$-24-\epsilon < \frac{3h^2-24h}{h} < -24+\epsilon$
OR $-24-\epsilon < \Delta(h) < -24+\epsilon$.

III.1.2 Prove that
$f(x) = 4-3x-x^2$
is differentiable at $x=2$. Find the derivative $Df|_2$.

STEP 1: $\Delta(h) = \frac{f(2+h)-f(2)}{h} =$
$\frac{[4-3(2+h)-(2+h)^2]-[4-3(2)-(2)^2]}{h}$
$= \frac{[4-6-3h-4-4h-h^2]-[-6]}{h}$
$= \frac{-h^2-7h}{h} = \begin{cases} -h-7 & \text{IF } h\neq 0 \\ \text{UNDEFINED} & \text{IF } h=0 \end{cases}$
STEP 2: $Df|_2 \overset{?}{=} -7$ SO LET $L=-7$
STEP 3: FOR ANY $\epsilon>0$, LET $\delta(\epsilon)=\epsilon$.
IF $0-\epsilon<h<0+\epsilon$ AND $h\neq 0$, $\epsilon>-h>-\epsilon$.
THEN $-7-\epsilon<-h-7<-7+\epsilon$. SINCE $h\neq 0$,
$\frac{h}{h}=1$, SO $-7-\epsilon<\frac{-h^2-7h}{h}<-7+\epsilon$
OR, THEN $-7-\epsilon<\Delta(h)<-7+\epsilon$.

III.1.3 Prove that $f(x) = x$ is differentiable at $x=a$ for any real number a. Verify that $Dx|_a = 1$.

STEP 1: $\Delta(h) = \frac{f(a+h)-f(a)}{h} =$
$\frac{(a+h)-(a)}{h} = \frac{h}{h} = \begin{cases} 1 & \text{IF } h\neq 0. \\ \text{UNDEFINED} & \text{IF } h=0. \end{cases}$
STEP 2: $Dx|_a \overset{?}{=} 1$ SO LET $L=1$.
STEP 3: FOR ANY $\epsilon>0$, LET $\delta(\epsilon)=$ (ANYTHING POSITIVE)
IF $0-\delta(\epsilon) < h < 0+\delta(\epsilon)$, AND $h\neq 0$,
IT DOESN'T MATTER WHAT THIS SAYS, SINCE THIS IS ALWAYS TRUE
$1-\epsilon<1<1+\epsilon$. SINCE $h\neq 0$ AND $\frac{h}{h}=1$, SO
$1-\epsilon<\frac{h}{h}<1+\epsilon$ OR $1-\epsilon<\Delta(h)<1+\epsilon$.

III.1.4 Prove that
$f(x) = 2x^3-3x-5$
is differentiable at $x=2$. Find the derivative $Df|_2$.

STEP 1: $\Delta(h) = \frac{f(2+h)-f(2)}{h} =$
$\frac{[2(2+h)^3-3(2+h)-5]-[2(2)^3-3(2)-5]}{h}$
$= \frac{21h+12h^2+2h^3}{h} = \begin{cases} 21+12h+2h^2 & \text{IF } h\neq 0 \\ \text{UNDEFINED} & \text{IF } h=0 \end{cases}$
STEP 2: $Df|_2 \overset{?}{=} 21$
STEP 3: LET $L=21$. FOR ANY $\epsilon>0$,
LET $\delta(\epsilon)=\min(1,\epsilon/14)$.
IF $0-\delta(\epsilon)<h<0+\delta(\epsilon)$ AND $h\neq 0$,
THEN $|h|<\delta(\epsilon)$. SINCE $\delta(\epsilon)\leq 1$, $|h|<1$
OR $-1<h<1$ OR $-2<2h<2$ OR
$-14<10<12+2h<14$ OR $|12+2h|<14$.
SINCE $\delta(\epsilon)\leq\epsilon/14$, $|h|<\delta(\epsilon)$ MEANS $|h|<\epsilon/14$.
SO $|h|\cdot|12+2h| \leq |h|\cdot 14 < \epsilon/14\cdot 14=\epsilon$
OR, $|h|\cdot|12+2h|<\epsilon$
OR, $|12h+2h^2|<\epsilon$
OR, $-\epsilon<12h+2h^2<\epsilon$
OR, $21-\epsilon<21+12h+2h^2<21+\epsilon$
OR, (SINCE $h\neq 0$ AND SO $\frac{h}{h}=1$)
$21-\epsilon<\frac{21h+12h^2+2h^3}{h}<21+\epsilon$
OR, $21-\epsilon<\Delta(h)<21+\epsilon$.

III.1.5 Prove that $f(x) = x^2$ is differentiable at $x=a$ for any number a. Verify that $Dx^2|_a = 2a$.

STEP 1: $\Delta(h) = \frac{f(a+h)-f(a)}{h} =$
$\frac{(a+h)^2-(a)^2}{h} = \frac{a^2+2ah+h^2-a^2}{h} = \frac{2ah+h^2}{h}$
$= \begin{cases} 2a+h & \text{IF } h\neq 0 \\ \text{UNDEFINED} & \text{IF } h=0. \end{cases}$
STEP 2: $Dx^2|_a \overset{?}{=} 2a$ SO LET $L=2a$.
STEP 3: FOR ANY $\epsilon>0$, LET $\delta(\epsilon)=\epsilon$.
IF $0-\epsilon<h<0+\epsilon$ AND $h\neq 0$, THEN
$2a-\epsilon<2a+h<2a+\epsilon$.
SINCE $h\neq 0$, $\frac{h}{h}=1$, SO
$2a-\epsilon<\frac{2ah+h^2}{h}<2a+\epsilon$
OR $2a-\epsilon<\Delta(h)<2a+\epsilon$ ... ON ALTERNATE OCTOBERS ONLY, ALFRED SAID.

III.1.6 Prove that $f(x)=x^3$ is differentiable at x=a for any number a. Verify that $Dx^3|_a = 3a^2$.

STEP 1: $\Delta(h) = \frac{f(a+h)-f(a)}{h} = \frac{(a+h)^3-(a)^3}{h} = \frac{3a^2h+3ah^2+h^3}{h} = \begin{cases} 3a^2+3ah+h^2 & \text{IF } h\neq 0 \\ \text{UNDEFINED} & \text{IF } h=0. \end{cases}$

STEP 2: $Dx^3|_a \stackrel{?}{=} 3a^2$ SO LET $L=3a^2$.

STEP 3: FOR ANY $\epsilon>0$, LET $\delta(\epsilon)=\min\left(1,\frac{\epsilon}{3|a|+1}\right)$

IF $0-\delta(\epsilon)<h<0+\delta(\epsilon)$ AND $h\neq 0$,

THEN $|h|<\delta(\epsilon)$. SINCE $\delta(\epsilon)\leq 1$, $|h|<1$

OR $-1<h<1$ OR,

$-(3|a|+1)\leq 3a-1<3a+h<3a+1\leq 3|a|+1$.

SINCE $\delta(\epsilon)\leq\frac{\epsilon}{3|a|+1}$, $|h|<\delta(\epsilon)$ MEANS

$|h|<\frac{\epsilon}{3|a|+1}$ ALSO, SO...

$|h|\cdot|3a+h|\leq|h|\cdot(3|a|+1)<\frac{\epsilon}{3|a|+1}\cdot(3|a|+1)=\epsilon$.

THEN $|h|\cdot|3a+h|<\epsilon$

OR, $|3ah+h^2|<\epsilon$

OR, $-\epsilon<3ah+h^2<\epsilon$

OR, $3a^2-\epsilon<3a^2+3ah+h^2<3a^2+\epsilon$

OR, $3a^2-\epsilon<\frac{3a^2h+3ah^2+h^3}{h}<3a^2+\epsilon$

OR, $3a^2-\epsilon<\Delta(h)<3a^2+\epsilon$.

III.1.7

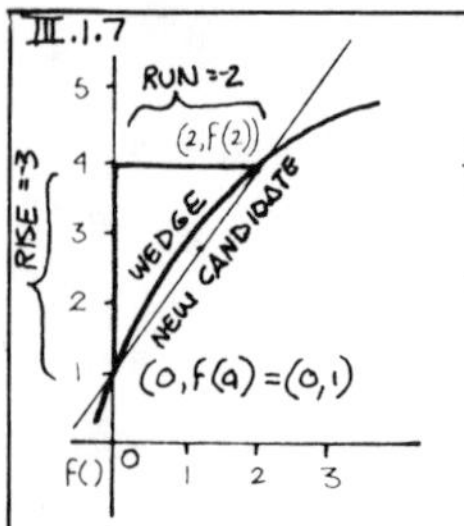

SLOPE OF WEDGE = $\frac{\text{RISE}}{\text{RUN}} = \frac{-3}{-2} = \frac{3}{2}$

$\Delta(-2) = 1-\frac{1}{4}(-2) = 1+\frac{1}{2} = \frac{3}{2}$.

they agree!

III.2.1 Compute $D(3x^2-2x-17)|_a$ = $D(3x^2)|_a + D(-2x)|_a + D(-17)|_a$ = $3\cdot(Dx^2|_a)-2\cdot(Dx|_a)+0 = 3(2a)-2\cdot 1 = 6a-2$.

Find a formula for the line tangent to the graph of $f(x)=3x^2-2x-17$ at point $(3,f(3))$. SLOPE = $Df|_3 = 6\cdot 3-2 = 16$.

$f(3) = 3\cdot(3)^2-2\cdot 3-17 = 4$. SO, FROM PAGE 52, THE FORMULA IS $y = 4+16(x-3) = 16x-44$

III.2.2 Compute the derivatives of the following functions at any point "a."

(i) $f(x)=x^3+3(x-1)^2+2x-1 =$ $x^3+3(x^2-2x+1)+2x-1 = x^3+3x^2-4x+2$.

SO, $Df|_a = Dx^3|_a+3\cdot(Dx^2|_a)-4(Dx|_a)+D(2)|_a$ $= 3a^2+6a-4$.

(ii) $f(x) = \frac{x^2+2x}{5} = \frac{1}{5}x^2+\frac{2}{5}x$.

SO, $Df|_a = \frac{1}{5}\cdot(Dx^2|_a)+\frac{2}{5}(Dx|_a) = \frac{2a}{5}+\frac{2}{5}\cdot 1$ $= \frac{2a+2}{5}$.

III.2.3 If $f(x) = 2x^3+3x^2-36x+5$, find the points where the tangent line to the graph of f( ) is flat (has slope zero).

$Df|_a = 6a^2+6a-36$.

NOW FIND THE VALUES OF "a" SUCH THAT

$0 = Df|_a = 6(a^2+a-6) = 6(a+3)(a-2)$

SOLUTION: $a=2, a=-3$.

III.2.4 Find a formula for the line tangent to the graph of

$f(x)=-3x^2+12x-5$

at point $(2,f(2))$.

SLOPE = $Df|_2 = -3(2.2)+12=0$

$F(2) = -3(2)^2+12\cdot 2-5 = 7$

FORMULA: $y = 7+0\cdot(x-2) = 7$

III.2.5 If $f(x) = x^3+3x^2-6x-9$, find the points where the line tangent to the graph of f( ) will have slope equal to 3.

$Df|_a = 3a^2+6a-6$

NOW, FIND THE VALUES OF "a" SUCH THAT

$Df|_a = 3a^2+6a-6 = 3$ = SLOPE

OR, $3a^2+6a-6-3=0$

OR, $3a^2+6a-9=0$

BUT $3a^2+6a-9 = 3(a+3)(a-1)$

AND THIS WILL BE ZERO WHEN $a=1$ OR $a=-3$.

III.3.1 Compute the following

$D(x^5-4x^3+2x+97)|_2 =$

$5\cdot(2)^4-4\cdot 3\cdot(2)^2+2\cdot 1+0 = 34$

$D(x^4-3x^9+2x)|_{-1} =$

$4(-1)^3-3\cdot 9\cdot(-1)^8+2\cdot 1 = -29$

$D(\frac{2}{3}x^3-x^{14}+3(x+1)^2)|_1 =$

$\frac{2}{3}\cdot 3(1)^2-14(1)^{13}+3(D(x+1)^2|_1)$

$= 2-14+3(D(x^2+2x+1)|_1)$

$= -12+3(2\cdot(1)+2+0) = 0$

$D(22x^{47}-3x^{17})|_a = 22\cdot 47a^{46}-3\cdot 17a^{16}$

III.3.2 Find any relative maxima or minima of

$f(x) = -x^3-6x^2-9x-3$.

Which is which? SINCE THE COEFFICIENT OF x^3 IS (-1), THIS CUBIC WILL LOOK LIKE 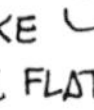 OR 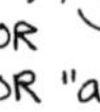. A TANGENT LINE WILL BE FLAT FOR "a's" WHERE

$0 = Df|_a = -3a^2-12a-9 = -3(a+1)(a+3)$

FROM THE SHAPE OF THE CUBIC, WE WILL HAVE A RELATIVE MINIMUM WHEN $a=-3$ AND A RELATIVE MAXIMUM WHEN $a=-1$.

III.3.3 Find any relative maxima or minima of

$f(x) = x^3-3x+5$.

Which is which? SINCE THE COEFFICIENT OF x^3 IS (+1), THIS CUBIC WILL LOOK LIKE 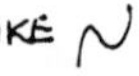 OR . A TANGENT LINE WILL BE FLAT FOR "a's" WHERE

$0 = Df|_a = 3a^2-3 = 3(a+1)(a-1)$.

FROM THE SHAPE OF THE CUBIC, WE WILL HAVE A REL. MAX. WHEN $a=-1$ AND A REL. MIN. WHEN $a=1$.

III.3.4 The fact that the tangent line to the graph of a function has slope zero at a point doesn't necessarily mean that the point is a relative maximum or minimum. To see what else can happen, let $f(x) = x^3-6x^2+12x-6$. Find where the derivative is zero and compare your answer with the location of the point of inflection. Carefully sketch the graph.

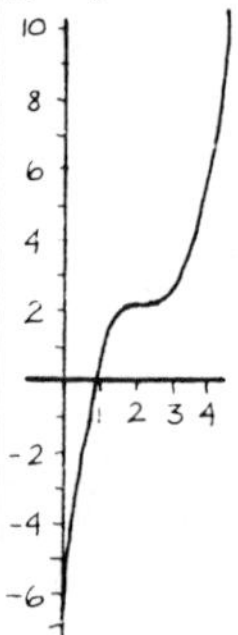

$Df|_a = 3a^2-12a+12$

$= 3(a^2-4a+4) = 3(a-2)^2$.

So, $Df|_a = 0$ ONLY IF $a=2$. THE POINT OF INFLECTION IS ABOVE $x = -\frac{(-6)}{3\cdot 1} = 2$.

SO, THE POINT OF INFLECTION IS ALSO THE ONLY POINT WHERE THE TANGENT LINE IS FLAT.

III.3.6 The Classy Carton Corp. has just bought a load of square cardboard 12 inches on a side. Find the dimensions of the open-topped box of greatest volume that they can make out of the cardboard by cutting square pieces from the corners.

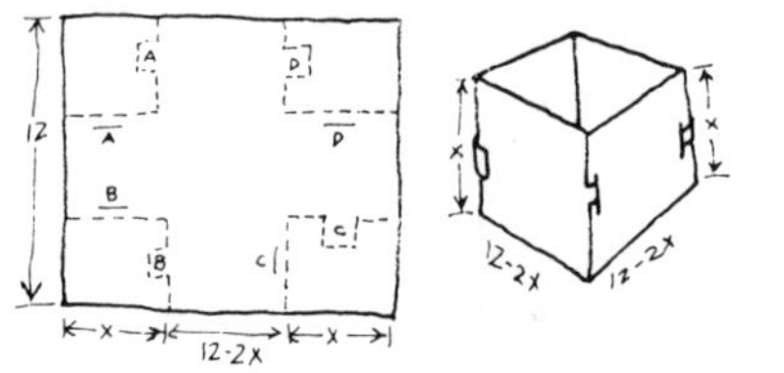

LET x = LENGTH OF A SIDE OF ONE OF THE PIECES TO BE CUT OUT = HEIGHT OF BOX.

V = VOLUME = (AREA OF BASE)(HEIGHT)

$= (12-2x)^2x = 144x-48x^2+4x^3$

THIS IS A CUBIC THAT LOOKS LIKE→ 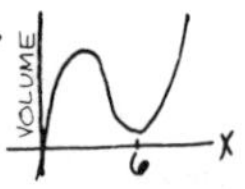

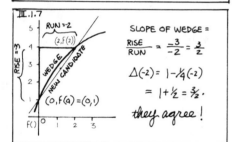

SINCE THE COEFFICIENT OF x^3 IS POSITIVE (4 IN THIS CASE).

NOW FIND THE RELATIVE MAXIMUM:

$DV|_a = 12a^2-96a+144 = 12\cdot(a-2)(a-6)$.

FROM THE SHAPE OF THE CUBIC, THE REL. MAX. IS ABOVE $a=2$ (WHERE $DV|_a=0$). SO THE HEIGHT FOR THE MAX. VOLUME MUST BE 2, (SINCE WE MUST HAVE $0<x<6$), AND THE SIDE OF THE BASE IS $12-2\cdot 2=8$.

III.4.1 If the motion of a point is described by the formula

$p(t) = t^3 - 6t^2 + 9t$ inches; $0 \le t \le 5$ (min.)

1. How fast was the point moving at the end of the fourth minute?

$\dot{p}(t) = Dp|_t = 3t^2 - 12t + 9$

SO $\dot{p}(4) = 3\cdot 4^2 - 12\cdot 4 + 9 = 9$ in./min.

2. When did the point reverse direction? WHEN $\dot{p}(t)$ EQUALED 0.

SINCE $\dot{p}(t) = 3(t^2 - 4t + 3) = 3(t-3)(t-1)$,

$\dot{p}(t) = 0$ IF AND ONLY IF

$t = 1$ min OR $t = 3$ min.

3. What was the maximum distance from "start" that the point reached during the five minutes? SINCE THE GRAPH OF $p(t)$ LOOKS LIKE → THE ANSWER WILL BE THE LARGER OF $p(1)$ AND $p(5)$: $p(1) = 4$

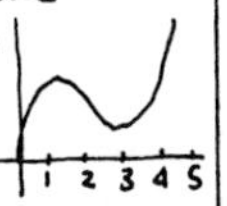

$p(5) = 20$ IN. ANS.

4. When was the point traveling most rapidly backward?

WE HAVE TO FIND THE MOST NEGATIVE POINT OF $\dot{p}(t) = 3t^2 - 12t + 9$.

SINCE $\dot{p}(t)$ LOOKS LIKE →

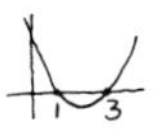

THIS POINT WILL BE WHERE

$0 = D\dot{p}|_t = 6t - 12$, IE., WHEN $t = 2$ MIN.

III.4.2 A coin is dropped in "Ye Olde Wishinge Welle." Three seconds later a "plink" is heard as it hits the bottom. How deep is the well? How fast was the coin going when it hit bottom?

USE THE FORMULA:

SET $S = 0$ AND $v = 0$;

THEN $p(t) = 0 + 0\cdot t - 16t^2 = -16t^2$.

WHEN $t = 3$, $p(3) = -16\cdot(3)^2 =$ **−144 FEET ANS.**

THE VELOCITY $\dot{p}(t) = -32t$. WHEN $t = 3$, $\dot{p}(3) = -32\cdot(3) =$ **−96 FT/SEC.**

III.4.3 An avid birdwatcher is on the top of a cliff 48 feet above the water. She tosses a chunk of bread up to a seagull. The gull misses and the bread falls back down to the water. If the initial velocity of the bread is 32 ft./sec. ,

1. How many seconds before the bread hits the water?
2. How high does the bread get before it falls in the water?
3. When does the bread pass its starting point on the way back down? Compare the velocity at this time with the starting velocity.

USE THE FORMULA:

SET $S = 0$ AND $v = 32$ FT/SEC.

SO $p(t) = 0 + 32t - 16t^2$.

1. WITH STARTING HEIGHT $S = 0$, THE WATER LEVEL IS AT −48 FEET, SO WE SOLVE $p(t) = -48$ FOR t:

$32t - 16t^2 = -48$ OR $48 + 32t - 16t^2 = 0$

OR, $16(3-t)(1+t) = 0$. SOLUTIONS TO THIS ARE $t = 3$ AND $t = -1$; t MUST BE POSITIVE, SO THE ANSWER IS **$t = 3$ SEC.**

2. FIND THE MAXIMUM OF $p(t)$, WHICH LOOKS LIKE

SET $\dot{p}(t) = 0$ AND SOLVE FOR t:

$\dot{p}(t) = 32 - 32t = 0$ WHEN **$t = 1$ SEC.**

3. SINCE WE SET THE STARTING HEIGHT $S = 0$, PASSING THE STARTING POINT ON THE WAY BACK DOWN CORRESPONDS TO HAVING $p(t) = 0$. SOLVE FOR t:

$0 = p(t) = 32t - 16t^2 = 16t(2-t)$ HAPPENS WHEN $t = 0$ (STARTING TIME) AND **$t = 2$ SEC.**

THE VELOCITY AT 2 SEC. IS $\dot{p}(2) = 32 - 32\cdot(2) = -32$ FT/SEC., EQUAL IN ABSOLUTE VALUE AND OPPOSITE IN SIGN TO THE STARTING VELOCITY.

III.4.4 El Zippo, the Human Cannonball, is to be projected straight up to go through a flaming hoop at the top of his flight. He then falls back down to end his flight by diving into a tub of water. If the center of the hoop is 25 feet above the cannon's mouth, what must the cannon's muzzle velocity be so that Zippo will go through the hoop? If the tub is 49 feet below the hoop, what will Zippo's velocity be when he hits the water?

SET STARTING HEIGHT "S" = 0. WE MUST FIND THE APPROPRIATE VALUE FOR THE INITIAL VELOCITY "v" SO THAT ZIPPO'S TRIP WILL BE A SUCCESS. THE TOP OF ZIPPO'S FLIGHT OCCURS WHEN

$\dot{p}(t) = 0 = v - 32t$, I.E., WHEN $t = \frac{v}{32}$.

TO MAKE SURE THAT ZIPPO GOES THROUGH THE CENTER OF THE HOOP, WE MUST HAVE

$25 = p(\frac{v}{32}) = 0 + v(\frac{v}{32}) - 16(\frac{v}{32})^2$

$= \frac{v^2}{64}$, OR $v^2 = 25\cdot 64$

OR **$v = 40'$/SEC. ANS.**

NOW WE HAVE THE FORMULA FOR ZIPPO'S POSITION:

$p(t) = 0 + 40t - 16t^2$. TO FIND ZIPPO'S VELOCITY AT THE END OF HIS FLIGHT, WE FIRST NOTE THAT ZIPPO WILL HIT THE TUB WHEN $p(t) = -24$ (SINCE THE TUB IS 24' BELOW THE STARTING POINT) SOLVE THIS FOR t:

$40t - 16t^2 = -24$ OR,

$0 = 24 + 40t - 16t^2 = 8(3-t)(1+2t)$.

SO, $t = 3$ OR $t = -\frac{1}{2}$: t MUST BE POSITIVE, SO GOOD 'OL ZIPPO'S ARRIVAL TIME AT THE TUB IS 3 SEC. HIS VELOCITY THEN WILL BE

$\dot{p}(3) = 40 - 16\cdot 2(3) = -56'$/SEC.

Selected Solutions and Answers for the Extra Exercises and Excursions.

Some Answers for the Extra Exercises for Chapter I:

EI.3.2:

(a) $0 \leq x^2 < 9$.

(b) $0 \leq x^2 < 9$.

(c) $0 \leq x^2 \leq 9$.

EI.3.3.

x's satisfying $-3 < x < -2$ will be negative; -3 and -2 are negative, so if we multiply the inequality by any one of them, this will flip the inequality signs. We get $9 = (-3)\cdot(-3) > -3x$;

$$-3x > x^2 > -2x \text{ and}$$
$$-2x > (-2)\cdot(-2) = 4. \text{ So}$$

$$9 > -3x > x^2 > -2x > 4$$

or $4 < x^2 < 9$.

EI.3.4.

$$8 \leq x^2 < 27.$$

EI.3.5.

Using the answer to EI.3.3, we know that if $-3 < x < -2$, then

$4 < x^2 < 9$. With x negative, we multiply this by x to get

$4x > x^3 > 9x$. 4 and 9 are positive, so $4x < 4\cdot(-2) = -8$ and $-27 = 9\cdot(-3) < 9x$. So

$$-27 < 9x < x^3 < 4x < -8,$$

or $-27 < x^3 < -8$.

EI.3.7.

(a) $22 < 7x+2x^2 < 39$.

(b) $-4 < 7x-2x^2 < 13$.

(c) $|7x-2x^2| < 13$.

EI.3.8.

(a) $1 < x < 2$.

(d) $|4x+x^2| < 12$.

EI.3.11.

$|a-b| = |(-1)(b-a)| = |-1|\cdot|b-a| = 1\cdot|b-a| = |b-a|$.

EI.4.4.

(a) The domain is $[-5,-1]\cup[0,4]\cup\{5\}$.
The range is $[-2,0]\cup[1,3]$.

EI.4.5. The formula is $f(x) = x-1$.

EI.4.6. The formula is $f(x) = 1 - 2x$.

EI.5.1. Yes.

EI.6.2. The parabola is

$$f(x) = x^2+x-1.$$

Some Hints for some of the Extra Exercises for Chapter II:

EII.5.5. Hint:

$$x^4-1 = (x-1)(x^3+x^2+x+1).$$

EII.7.2. Hint:

$$x^3-c^3 = (x-c)(x^2+cx+c^2).$$

Some Answers for the Extra Exercises for Chapter III:

EIII.1.1. $Df|_{-2} = -1$.

EIII.1.2. $Df|_{-1} = 21$.

EIII.1.2. $Df|_{2} = 0$.

EIII.1.4. $Df|_{-2} = 0$.

EIII.2.1. The tangent line has formula $y = 14x + 17$.

EIII.2.2. The line crosses the x-axis at point $(5/4, 0)$.

EIII.2.3. The slope of the line tangent at any $(x_1, f(x_1))$ will be the same as the line tangent at $(-x_1, f(-x_1))$.

EIII.2.11. $f'(-1) = -274$.

EIII.3.2. There is a relative maximum for f(x) at (-1,12) and a relative minimum at (2,-15).

EIII.3.4. The dimensions should be 65′×65′.

EIII.3.5. x=10; y=10.

EIII.3.9. 3″×3″×3″.

EIII.3.10. 80′×160′.

EIII.3.11. 800′×1600′.

EIII.3.12. 160′×240′.

EIII.3.13. 600′×400′.

EIII.3.14. The tank should be 7′×7′×7′.

EIII.3.15. The ticket price should be $11.00.

EIII.3.16. Setting the ticket price to $12.00 should exactly fill the arena.

EIII.3.17. Raise the price to $24.50.

EIII.3.18. No. Keep the price at $20.00.

EIII.4.1.
(1) 12′/sec. forwards.
(2) The point reversed direction at t=2 sec. and t=5 sec.
(3) The maximum distance the point reached from 'start' was 52′ at time 2 sec. when p(2)=0.
(4) At t=3.5 sec. when the velocity was 13.5′/sec.

EIII.4.2. Wait 15 seconds; the velocity will be -480′/sec.

EIII.4.3. It will take 4 seconds to catch the mouse; the cat will travel 48′.

EIII.4.4. The maximum speed of the train was 60 m.p.h.; the train travelled 80 miles.

EIII.4.5. For $0 \le t \le 2$, the yo-yo reverses direction only at t=1.
The string must be 3.75′ long.

SCRATCH WORK

SCRATCH WORK

SCRATCH WORK